DEVELOPMENTS IN PRIMATOLOGY: PROGRESS AND PROSPECTS

Series Editor:
Russell H. Tuttle
University of Chicago, Chicago, Illinois

This peer-reviewed book series melds the facts of organic diversity with the continuity of the evolutionary process. The volumes in this series exemplify the diversity of theoretical perspectives and methodological approaches currently employed by primatologists and physical anthropologists. Specific coverage includes: primate behavior in natural habitats and captive settings primate ecology and conservation; functional morphology and developmental biology of primates primate systematics genetic and phenotypic differences among living primates and paleoprimatology.

COMPARATIVE VERTEBRATE COGNITION
Edited by Lesley J. Rogers and Gisela Kaplan

ANTHROPOID ORIGINS: NEW VISIONS
Edited by Callum F. Ross and Richard F. Kay

MODERN MORPHOMETRICS IN PHYSICAL ANTHROPOLOGY
Edited by Dennis E. Slice

BEHAVIORAL FLEXIBILITY IN PRIMATES: CAUSES AND CONSEQUENCES
By Clara B. Jones

NURSERY REARING OF NONHUMAN PRIMATES IN THE 21ST CENTURY
Edited by Gene P. Sackett, Gerald C. Ruppenthal and Kate Elias

NEW PERSPECTIVES IN THE STUDY OF MESOAMERICAN PRIMATES: DISTRIBUTION, ECOLOGY, BEHAVIOR, AND CONSERVATION
Edited by Paul Garber, Alejandro Estrada, Mary Pavelka and LeAndra Luecke

HUMAN ORIGINS AND ENVIRONMENTAL BACKGROUNDS
Edited by Hidemi Ishida, Martin Pickford, Naomichi Ogihara and Masato Nakatsukasa

PRIMATE BIOGEOGRAPHY
Edited by Shawn M. Lehman and John Fleagle

REPRODUCTION AND FITNESS IN BABOONS: BEHAVIORAL, ECOLOGICAL, AND LIFE HISTORY PERSPECTIVES
Edited by Larissa Swedell and Steven R. Leigh

PRIMATE ORIGINS: ADAPTATIONS AND EVOLUTION
Edited by Matthew J. Ravosa and Marian Dagosto

RINGTAILED LEMUR BIOLOGY: *LEMUR CATTA* IN MADAGASCAR
Edited by Alison Jolly, Robert W. Sussman, Naoki Koyama and Hantanirina Rasamimanana

RINGTAILED LEMUR BIOLOGY

Lemur Catta in Madagascar

Edited by

Alison Jolly
University of Sussex
Brighton, United Kingdom

Robert W. Sussman
Washington University
St. Louis, Missouri, USA

Naoki Koyama
Kyoto University
Kyoto, Japan

Hantanirina Rasamimanana
University of Antananarivo
Antananarivo, Madagascar

Alison Jolly
Department of Biology and
 Environmental Sciences
University of Sussex
Lewes BN7 1HX, Brighton
UK
ajolly@sussex.ac.uk

Robert W. Sussman
Department of Anthropology
Washington University
St. Louis, Missouri 63130-4899
USA
rwsussma@artsci.wustl.edu

Naoki Koyama
Center for African Area Studies
Kyoto University
Kyoto 606-8501
JAPAN
nk05401@mx.cable-net.ne.jp

Hantanirina Rasamimanana
École Normale Supérieure
University of Antananarivo
BP 881
Antananarivo 101
MADAGASCAR
hantani1@yahoo.fr

Library of Congress Control Number: 2006921354

ISBN 10: 0-387-32669-3
ISBN 13: 978-0387-32669-6

Printed on acid-free paper.

© 2006 Springer Science+Business Media, LLC
All rights reserved. This work may not be translated or copied in whole or in part without the written permission of the publisher (Springer Science+Business Media, LLC, 233 Spring Street, New York, NY 10013, USA), except for brief excerpts in connection with reviews or scholarly analysis. Use in connection with any form of information storage and retrieval, electronic adaptation, computer software, or by similar or dissimilar methodology now known or hereafter developed is forbidden.
The use in this publication of trade names, trademarks, service marks, and similar terms, even if they are not identified as such, is not to be taken as an expression of opinion as to whether or not they are subject to proprietary rights.
While the advice and information in this book are believed to be true and accurate at the date of going to press, neither the authors nor the editors nor the publisher can accept any legal responsibility for any errors or omissions that may be made. The publisher makes no warranty, express or implied, with respect to the material contained herein.

Printed in the United States of America.   (SPi/IBT)

9 8 7 6 5 4 3 2 1

springer.com

Preface

This book is a truly international collaboration, with editors based on four continents and first authors from Canada, France, Japan, Madagascar, the United Kingdom, and the United States. Clearly, there is something attractive about studying *Lemur catta*, the ringtailed lemur. Perhaps it is the lemurs themselves!

Why study ringtailed lemurs? Because lemurs are a separate radiation of primates from the monkeys, apes, and humans. Because ringtails live in the largest social groups of any known lemur and therefore offer the closest comparison with other social primates, including ourselves. And also because ringtails have become the flagship species of Madagascar. Some 70 species or subspecies of lemurs inhabit Madagascar. Each kind has its own fascinating story, but ringtails are the ones that everyone thinks they know. That black-and-white tail adorns tourist brochures and school notebooks and banknotes. All the same, after 40 years of field study, we don't know nearly enough. In this book, we make a first estimate on how many might be alive today, but we do not know how many ringtailed lemurs and how many southern forests will survive.

This book explores part of what we do know. Its four sections are (I) Distribution: Ringtailed Lemurs in Madagascar, (II) Ringtails and Their Forests: Feeding and Ranging Behavior, (III) Social Behavior Within and Between Troops, and finally (IV) Health and Disease. Of course, all these issues are interrelated. We would like to point out some cross-cutting themes that emerge from the chapters if read as a whole.

The first is that the southern area of Madagascar, where the ringtails live, spans a huge variety of habitats and a huge variation of year-to-year fluctuations in climate. The range of ringtails lies mainly in very dry spiny forests and marginal scrub, but they also flourish in rich, tamarind-dominated gallery forests, in the fortress canyons of the Isalo, and even above the tree-line on the Andringitra Massif (Goodman et al. and Sussman et al. on range). Beza Mahafaly Reserve and Berenty Reserve, the only two long-term study sites, Both lie in gallery forest and thus represent the homes of lemur plutocrats rather than more widespread and more challenging habitats (Sussman et al. on Beza, Jolly et al. on Berenty). Even within such favored areas though, the effect of recurrent drought can be catastrophic. It shapes everything – from population size to life history to microevolution (Jolly et al. on territoriality as bet hedging; Cuozzo and Sauther on tooth microevolution). Furthermore, the small patches of gallery forest have always been discontinuous, and they have shifted in space over a timescale of decades or centuries even before people began to clear them (Blumenfeld-Jones et al. on tamarind recruitment). Thus, any overview of the species needs the widest possible spread in space and time, because no one year, and not even any one forest, can be called "typical."

In fact, ringtailed lemurs as a species are adapted to difference and challenge. This is one reason why they are so well known. Zoos love to exhibit them, not only as striking animals but also because they are semiterrestrial, diurnal, at home in discontinuous habitat, and individually as tough as old boots.

The second theme is the importance of tamarind trees, *Tamarindus indica*. This keystone resource provides food year round, whether green pods in the dry, cold winter season when females are gestating, ripe fruit as the young are born and lactation begins, or fruit and leaves when the young are weaned. Years when the fruit fails are disastrous for the lemurs. Long-term patterns of tamarind growth and regeneration determine lemur use of the richest of their forest habitats If the ringtails' access to the tamarinds is limited by competing brown lemurs, this is also likely to have a major impact on their populations and distribution. (Simmen et al., plant food species; Blumenfeld-Jones, tamarind recruitment; Koyama et al., tamarinds and home range; Mertl-Millhollen et al., tamarind quality; Simmen et al., taste thresholds; Cuozzo and Sauther, tooth microevolution; Pinkus et al., brown lemur competition).

The third theme is the extreme role played by females in troop coherence, troop rivalry, and resource defense. Female dominance over males does not reflect differential energy expenditure (Rasamimanana et al., energetic strategy). However, it is clear that intertroop female resource defense is crucial in gallery forest (Pride et al., group size and defense). Within a troop, females vie directly for status (Pereira, agonistic power). Status in turn is fundamental to continued troop membership. Subordinates exiled by targeted aggression may roam as all-female nomadic groups for many months before they succeed in establising defended ranges (Ichino et al., social changes). Even then the ranges are usually inferior to those held by former dominants: inequality may be perpetuated over generations (Jolly et al., territory as bet hedging). Immigrant males as postulants for troop membership seem secondary to the troop structure, though crucial for the mixing of genes (Gould et al., male migration). But males are still an unsolved riddle: why do such combative, polygamous animals not evolve dominance over females?

A fourth theme is the role of health and disease, particularly in human-altered habitats. At Berenty Reserve, the lemurs have adopted introduced species as foods, which have benefited them, allowing massive population growth (Soma et al., introduced trees; Simmen et al., food plants). However, Leucaena, a leguminous tree originating from Central America that has been planted worldwide by foresters for fuel wood and livestock forage, actually poisons some Berenty lemurs (Crawford et al., bald lemur syndrome; Soma et al., introduced trees) Even in the more natural forest of Beza Mahafaly, lemurs put on weight when they have access to campsite foods, though they pay a price in tooth decay (Sauther et al., health and disease). Tooth wear reflects interactions between the lemurs, their society, the impact of drought, and their keystone tamarind resources (Cuozzo and Sauther, tooth microevolution). The interaction between health and habitat change will be an ever-increasing concern of the emerging field of conservation medicine.

Finally, there is the overarching question of ringtailed lemur conservation. Sussman et al. estimate that between 1985 and 2000, 9.5% of forest habitat suitable for ringtailed lemurs has been lost. In this period, the population may have fallen from a possible 930,000 ringtailed lemurs down to 750,000, a 20% reduction, even under the unlikely assumption that all suitable habitat is occupied by lemurs. Add to this the pressures from natural changes (Blumenfeld-Jones et al., Koyama et al., Cuozzo et al., Jolly et al.) and the pressures brought by even well-meaning human intervention, let alone hunting and forest fragmentation (Sauther et al.; Crawford et al.; Pinkus et al.), many forests that look suitable to an orbiting satellite will not in fact hold any lemurs. Still, the species is both widespread and adaptable, classified as vulnerable, not endangered.

The future of ringtailed lemurs is not a story of their own biology. It is a story of how people value their habitat and their survival. Long may they remain a flagship species for Madagascar!

The Editors

Contents

Contributors xiii

PART I: DISTRIBUTION:
        RINGTAILED LEMURS IN MADAGASCAR 1

1. The Distribution and Biogeography of the
   Ringtailed Lemur (*Lemur catta*) in Madagascar 3
   Steven M. Goodman, Soava V. Rakotoarisoa, and Lucienne Wilmé

2. A Preliminary Estimate of *Lemur catta*
   Population Density Using Satellite Imagery.................. 16
   *Robert W. Sussman, Sean Sweeney, Glen M. Green,
   Ingrid Porton, O.L. Andrianasolondraibe,
   and Joelisoa Ratsirarson*

3. Berenty Reserve: A Research Site in
   Southern Madagascar.. 32
   *Alison Jolly, Naoki Koyama, Hantanirina Rasamimanana,
   Helen Crowley, and George Williams*

4. Beza Mahafaly Special Reserve: A Research
   Site in Southwestern Madagascar 43
   Robert W. Sussman and Joelisoa Ratsirarson

PART II: RINGTAILS AND THEIR FORESTS:
        FEEDING AND RANGING 53

5. Plant Species Fed on by *Lemur catta* in Gallery Forests of the
   Southern Domain of Madagascar............................. 55
   *Bruno Simmen, Michelle L. Sauther, Takayo Soma,
   Hantanirina Rasamimanana, Robert. W. Sussman, Alison Jolly,
   Laurent Tarnaud, and Annette Hladik*

6. **Tamarind Recruitment and Long-Term Stability in the Gallery Forest at Berenty, Madagascar** 69
   Kathryn Blumenfeld-Jones, Tahirihasina M. Randriamboavonjy, George Williams, Anne S. Mertl-Millhollen, Susan Pinkus, and Hantanirina Rasamimanana

7. **Home Ranges of Ringtailed Lemur Troops and the Density of Large Trees at Berenty Reserve, Madagascar** .. 86
   Naoki Koyama, Takayo Soma, Shinichiro Ichino, and Y. Takahata

8. **The Influence of Tamarind Tree Quality and Quantity on *Lemur catta* Behavior** 102
   Anne S. Mertl-Millhollen, Hajarimanitra Rambeloarivony, Wendy Miles, Veronica A. Kaiser, Lisa Gray, Loretta T. Dorn, George Williams, and Hantanirina Rasamimanana

9. **Feeding Competition Between Introduced *Eulemur fulvus* and Native *Lemur catta* During the Birth Season at Berenty Reserve, Southern Madagascar** 119
   Susan Pinkus, James N.M. Smith, and Alison Jolly

10. **Tradition and Novelty: *Lemur catta* Feeding Strategy on Introduced Tree Species at Berenty Reserve** 141
    Takayo Soma

11. **Diet Quality and Taste Perception of Plant Secondary Metabolites by *Lemur catta*** 160
    B. Simmen, S. Peronny, M. Jeanson, A. Hladik, and A. Marez

PART III: SOCIAL BEHAVIOR WITHIN AND BETWEEN TROOPS 185

12. **Territory as Bet-hedging: *Lemur catta* in a Rich Forest and an Erratic Climate** 187
    Alison Jolly, Hantanirina Rasamimanana, Marisa Braun, Tracy Dubovick, Christopher Mills, and George Williams

13. **Resource Defense in *Lemur catta*: The Importance of Group Size** 208
    R. Ethan Pride, Dina Felantsoa, Tahiry Randriamboavonjy, and Randriambelona

14. Social Changes in a Wild Population of
    Ringtailed Lemurs (*Lemur catta*) at Berenty, Madagascar 233
    Shinichiro Ichino and Naoki Koyama

15. Obsession with Agonistic Power 245
    Michael. E. Pereira

16. Male and Female Ringtailed Lemurs' Energetic Strategy
    Does Not Explain Female Dominance 271
    *Hantanirina Rasamimanana, Vonjy N. Andrianome, Hajarimanitra
    Rambeloarivony, and Patrick Pasquet*

17. Male Sociality and Integration During the Dispersal Process
    in *Lemur catta*: A Case Study 296
    Lisa Gould

PART IV: HEALTH AND DISEASE 311

18. Patterns of Health, Disease, and Behavior
    Among Wild Ringtailed Lemurs, *Lemur catta*:
    Effects of Habitat and Sex 313
    *Michelle L. Sauther, Krista D. Fish, Frank P. Cuozzo,
    David S. Miller, Mandala Hunter-Ishikawa, and
    Heather Culbertson*

19. Bald Lemur Syndrome and the Miracle Tree:
    Alopecia Associated with *Leucaena leucocephala*
    at Berenty Reserve, Madagascar 332
    *Graham C. Crawford, Louis-Expert Andriafaneva,
    Kathryn Blumenfeld-Jones, Gary Calaba, Linda Clarke,
    Lisa Gray, Shinichiro Ichino, Alison Jolly, Naoki Koyama,
    Anne Mertl-Millhollen, Susan Ostpak, R. Ethan Pride,
    Hantanirina Rasamimanana, Bruno Simmen, Takayo Soma,
    Laurent Tarnaud, Alison Tew, and George Williams*

20. Temporal Change in Tooth Size Among Ringtailed Lemurs
    (*Lemur catta*) at the Beza Mahafaly Special Reserve,
    Madagascar: Effects of an Environmental Fluctuation 343
    Frank P. Cuozzo and Michelle L. Sauther

Index 367

Contributors

Andriafaneva, L.-E. c/o Berenty Reserve, BP 54, Tolagnaro (Fort Dauphin), Madagascar
Andrianasolondraibé, O.L., Département de Paléontologie, Anthropologie et Biologie, Université d'Antananarivo (101), Madagascar, harivony@wanadoo.mg
Andrianome, V.N., École Normale Supérieur, University of Antananarivo, BP 881, Antananarivo, Madagascar, andrianomenirina@yahoo.fr
Blumenfeld-Jones, K. Dept. of Anthropology, Arizona State University, Tempe, Arizona, corbeau@linuxmail.org, dbj@asu.edu
Braun, M. Graduate School of Medicine, George Washington University, Washington DC 20052 USA, mbraun@gwu.edu
Calaba, G. 4141 Laddington Ct., Portland, OR 97232, USA, calaba@ipns.com
Clarke, L. San Francisco Zoological Society, San Francisco CA 94132-1098, USA
Crawford, G.C. San Francisco Zoological Society, San Francisco CA 94132-1098, USA, hospital@sfzoo.org
Crowley, H. Wildlife Conservation Society, BP 8500, Antananarivo, Madagascar, hcrowley@wcs.org
Culbertson, H. College of Veterinary Medicine, Cornell University, Ithaca, NY 14850, USA, hjculbertson@yahoo.com
Cuozzo, F. Department of Anthropology, University of North Dakota, Grand Forks, ND 58202, USA, frank.cuozzo@und.nodak.edu
Dorn, L.T. Department of Chemistry, Fort Hays State University, Hays, KS, USA, ldorn@fhsu.edu
Dubovick, T. School of Veterinary Medicine, University of Pennsylvania, Philadelphia PA 19104, USA, dubovic@vet.upenn.edu
Felantsoa, D. Dept. of Biology, École Normale Superieur, University of Antananarivo, BP 881 Madagascar, felantsoadina@yahoo.fr
Fish, K. University of Colorado Boulder, Department of Anthropology, Box 233, Boulder, CO 80309-0233, USA, Krista.Fish@colorado.edu

Goodman, S.M. Field Museum of Natural History, Chicago, Illinois, 60605, USA, sgoodman@fieldmuseum.org, WWF, B.P. 738, Antananarivo (101), Madagascar, sgoodman@wwf.mg,

Gould, L. Department of Anthropology, University of Victoria, Victoria BC Canada, lgould@uvic.ca

Gray, L. Joint Information Systems Committee, University of Bristol, Bristol, UK, l.gray@bristol.ac.uk

Green, G.M. Center for the Study of Institutions, Population and Environmental Change (CIPEC) Indiana University, Bloomington, IN 47408, USA, glgreen@indiana.edu

Hladik, A. Eco-Anthropologie et Ethnobotanique, Muséum National d'Histoire Naturelle/CNRS, 4 avenue du Petit-Château, 91800 Brunoy, France, hladik@mnhn.fr

Hunter-Ishikawa, M. College of Veterinary Medicine and Biomedical Sciences, Colorado State University, Fort Collins CO 80523-1601, USA, mandalahunter@wildmail.com

Ichino, S. Laboratory of Human Evolution Studies, Graduate School of Science, Kyoto University Kitashirakawaoiwake-cho, Sakyo-ku, Kyoto 606-8502 Japan, ichino@jinrui.zool.kyoto-u.ac.jp

Jeanson, M. Eco-Anthropologie et Ethnobotanique, Muséum National d'Histoire Naturelle/CNRS, 91800 Brunoy, France, marcjeanson@yahoo.fr

Jolly, A. Department of Biology and Environmental Science, University of Sussex, Brighton BN1 9QG, UK, ajolly@sussex.ac.uk

Kaiser, V.A. Department of Biology and Environmental Science, University of Sussex, Brighton BN1 9QG, UK, Veronicakaiser@hotmail.com

Koyama, N. Center for African Area Studies, Kyoto University, Kyoto 606-8501, Japan, nk05401@mx.cable-net.ne.jp

Marez, A. IUT, Département Génie Biologique, Université Paris XII, Avenue du Général de Gaulle, 94010 Créteil, France, amarez@univ-paris12.fr

Mertl-Millhollen, A.S. University of Oregon, Dept. of Anthropology, Eugene, OR 97403 USA, hplam_1998@yahoo.com

Miles, W. Oxford University Center for the Environment, Oxford University, Oxford OX1 2JD, UK, woodsandrain@gmail.com

Miller, D.S. Animal Population Health Institute, College of Veterinary Medicine and Biological Sciences, Colorado State University, Fort Collins CO 80523, USA, dmiller1@colostate.edu

Mills, C. 645 El Dorado Ave # 308, Oakland CA 94611, USA, chrisnmills@yahoo.com

Ostpak, S. San Francisco Zoological Society, San Francisco CA 94132-1098, USA, hospital@sfzoo.org

Pasquet, P. CNRS, Muséum de l'Histoire Naturelle, 75231 Paris, Cedex 05, France, ppasquet@mnhn.fr

Pereira, M.E. The Latin School of Chicago, 59 West North Boulevard, Chicago, IL 60610-1492, mpereira@latinschool.org

Peronny, S. Eco-Anthropologie et Ethnobotanique, Muséum National d'Histoire Naturelle/CNRS, 4 avenue du Petit-Château, 91800 Brunoy, France, fywol@wanadoo.fr

Pinkus, S. Department of Ecology, University of British Columbia, Vancouver, Canada, spinkus@interchange.ubc.ca

Porton, I. St. Louis Zoo, 1 Government Drive, St. Louis, Missouri, 63110-1395, USA, rufflemur@aol.com

Pride, R.E. CAS Biology, Boston University, 5 Cummington St. Boston MA 02215, USA, pride@bu.edu

Rakotaoarisoa, S.V. Programme Alimentaire Mondial (PAM), BP 1348, Antananarivo (101), Madagascar, Soava.Rakotoarisoa@wfp.org

Rambeloarivony, H. Dept. of Biology, Ècole Normale Superieur, University of Antananarivo, BP 881, Madagascar, haja_kely@hotmail.fr

Randriambelona, R. Parc Ivoloina, Madagascar Fauna Group, BP 442, Toamasina (501) Madagascar, tim@savethelemur.org

Randriamboavonjy, T.M. Dept. of Biology, Ècole Normale Superieur, University of Antananarivo, BP 881 Madagascar, tahiry@vodka orange.com

Rasamimanana, H. École Normale Supérieur, Université d'Antananarivo, BP 881, Antananarivo 101, Madagascar, hantani1@yahoo.fr

Ratsirarson, J. Départment des Eaux et Forêts de l'Ecole Supérieur des Sciences Agronomiques, Université d'Antananarivo (101) Madagascar, jratsirarson@simicro.mg

Sauther, M.L. University of Colorado Boulder, Department of Anthropology, Hale Building, Box 233, Boulder, CO 80309-0233, USA, michelle. sauther@colorado.edu

Simmen, B. Eco-Anthropologie et Ethnobotanique, Muséum National d'Histoire Naturelle/CNRS, 4 avenue du Petit-Château, 91800 Brunoy, France, simmen@ccr.jussieu.fr

Smith, J.N.M. Department of Ecology, University of British Columbia, Vancouver, Canada

Soma, T. Center for African Area Studies, Kyoto University, 46 Shimoadachicho, Yoshida, Sakyo-ku, Kyoto 606-8501, Japan, soma@ jambo. africa.kyoto-u.ac.jp

Sussman, R.W. Department of Anthropology, Washington University, Campus Box 114, One Brookings Drive, St. Louis, Missouri 63130-4899 USA, rwsussma@artsci.wustl.edu

Sweeney, S. Center for the Study of Institutions, Population and Environmental Change (CIPEC) Indiana University, 408 N. Indiana Ave., Bloomington, IN 47408, USA, spsweene@indiana.edu

Takahata, Y. Kwansei Gakuin University, Sanda, 669-1337, Japan, z96014 @ksc.kwansei.ac.jp

Tarnaud, L. Eco-Anthropologie et Ethnobotanique, Muséum d'Histoire Naturelle/CNRS, 4 avenue du Petit-Château, 91800 Brunoy, France, Laurent.Tarnaud@free.fr

Tew, A. Department of Anthropology, Oxford Brookes University, Oxford, UK, alilemur@hotmail.com

Williams, G. 44 Alan Road, Santa Barbara, CA 93109, USA, gww@silcom.com

Wilmé, L. Missouri Botanical Garden, BP 3391, Antananarivo (101) Madagascar, lucienne.wilme@simicro.mg

Part I

Distribution: Ringtailed Lemurs in Madagascar

1
The Distribution and Biogeography of the Ringtailed Lemur (*Lemur catta*) in Madagascar

STEVEN M. GOODMAN, SOAVA V. RAKOTOARISOA, AND LUCIENNE WILMÉ

1.1. Introduction

Although *Lemur catta* has been the subject of detailed behavioral and ecological field studies at a few localities in the southern portion of Madagascar and is certainly one of the best known of the island's primates, little has been published about its distribution and the range of habitats it uses. The major exception to this point is a recent assessment of the geographical extent of this species overlaid on anthropogenic habitat degradation (Sussman et al., 2003; see also Sussman et al., Chapter 2, this volume). *L. catta* is often associated with being a denizen of gallery forests of the southern spiny bush. This is natural as the vast majority of information on the life history of this taxon comes from long-term studies at Berenty and the Réserve Spéciale (RS) de Beza Mahafaly and concerns mostly troops living in this forest type. However, as discussed below, this species is the least forest-dwelling of the extant species of lemurs and occurs in a wide range of habitats in the southern third of the island, and the current categorization of certain life-history parameters may be slightly exaggerated given the intensive focus on gallery forest zones.

In this contribution, we address four principal points concerning *L. catta*:

1. Its current distribution.
2. Geographic range associated with limitation of freshwater sources.
3. Aspects of the ecology of the Andringitra high mountain population.
4. A biogeographic scenario to explain its current distribution.

1.2. Sources of Information

During the course of more than a decade, we have been gathering data on the distribution of Malagasy mammals based on our own field excursions, published literature, and specimens held in natural history museum around the world. This database is the principal source of the distributional information presented herein. The paper of Sussman (1977) thoroughly reviewed earlier literature on the geographical range of this species, and these details are not presented here. For information on botanical species, we have used the database of TROPICOS (http://mobot.mobot.org/W3T/Search/vast.html).

1.3. The Distribution of *Lemur catta*

In most general accounts, the range of *Lemur catta* is considered as the spiny bush of the south and southwest and the dry deciduous forest of the lowland central west, and with populations ranging into the distinctly more mesic interior highlands, particularly mountainous areas (e.g., Petter et al., 1977; Tattersall, 1982; Mittermeier et al., 1994; Jolly, 2003). In the southern portion of its range, this lemur tends to be more common in gallery forest. Tattersall (1982, p. 46) noted that the northern limit of this species' distribution is a line connecting Belo sur Mer to Fianarantsoa and then to Tolagnaro.

We are now in a position to add some precision to the known distribution of *L. catta* (Figure 1.1) and to examine some of the parameters potentially associated with these range limits. The northwestern boundary of this species is considerably north of the Mangoky River at Belo sur Mer or Mahababoky (20 °44'S) (Sussman, 1977) and slightly more inland within the Parc National (PN) de Kirindy-Mitea—both sites are in the southern portion of the Menabe region. This is a zone of dry deciduous forest with transitional vegetation elements characteristic of the more southern spiny bush. Central Menabe sites, such as Kirindy (CFPF), which lack the prevalence of spiny bush elements found at Kirindy-Mitea and have been extensively researched—this lemur is unknown to occur in this region (Ganzhorn et al., 1999; Zinner et al., 2001). Little fieldwork has been conducted in the southern portion of the Menabe, between the PN de Kirindy-Mitea and the Morondava River, and it is possible that the limit of this species' range is a bit further north than described here. We propose that this boundary is correlated with some ecological links that are associated with the transition between the southern spiny bush and typical central west dry deciduous forest. An excellent example of this type of correlation in geographical distribution is with members of the endemic plant family Didiereaceae (Figure 1.1). The known northern limit of this family in northwestern Madagascar largely coincides with that of *L. catta*, and we presume that these aspects are closely tied to bioclimatic parameters as described by Cornet (1974), specifically his Subarid Stage, and elaborated on by Schatz (2000).

The southeastern limit of *L. catta*'s distribution ends at the divide between western (here including southern) and eastern watersheds (Figure 1.1). This line (except as noted below in the extreme southeast) is aligned with the division between western dry and eastern humid vegetation formations. From at least the Betroka region southwards, this ecotone is notably abrupt. We are unaware of records of this species at sites within eastern humid forest vegetational formations. In the region of the Ivohibe Massif, dry forest and open anthropogenic savanna occur to the western foot of the mountain, and the shift to more mesic

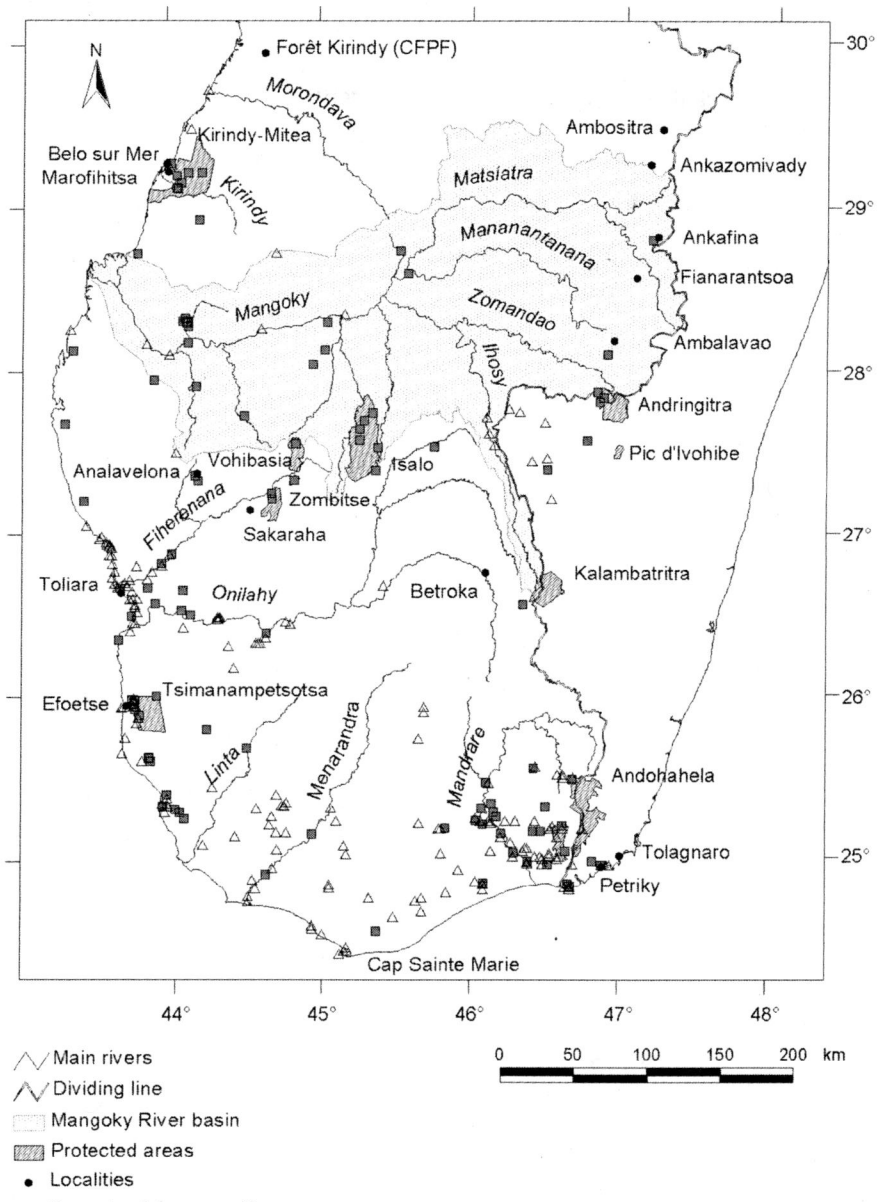

FIGURE 1.1. Map showing the geographical distribution of *Lemur catta* and members of the plant family Didiereaceae in the southern third of Madagascar. Most site names mentioned in the text and large rivers are also shown.

forests abruptly occurs with a slight rise in elevation. Here groups of *L. catta* can be found in the lower, dry formations. The same pattern exists on the Kalambatritra Massif (M. Irwin, pers. comm.), with this species occurring in the lower western gallery forests and zones with more xerophytic vegetation. All records of wild *L. catta* that we are aware of from parcel 1 of the PN d'Andohahela, which is the humid forest zone of the three parcels of this park, are from the extreme western portion, at the ecotone between humid and dry forests. The Nahampoana private reserve, 7 km north of Tolagnaro and in the humid forest sector, has within its animalier *L. catta* that have escaped on occasion over the past few years (J. Ganzhorn, pers. comm.). In the extreme southeastern coastal region, in forest formations largely resting on sand, there is an extension of western vegetation formations, such as Didiereaceae, into zones of eastern watersheds, and several dry forest faunal elements (Goodman et al., 1997a), including *L. catta*, extend their range more easterly. In this portion of Madagascar, the most easterly record we are aware of for this species is at Petriky.

The northeastern limit of *L. catta*'s distribution is notably more complex than in other portions of its range. In the highland regions of the upper Mangoky River and associated tributaries (Figure 1.1), the habitat is a mixture of granite domes, open savanna, and relict natural vegetation formations in valleys, which in some cases contain gallery forest. This is a region of Madagascar where members of the Didiereaceae cross over in highland areas from western to eastern watersheds. This lemur shows the same pattern and can be found in different open habitats of this region from 920 to 2600 m (Goodman and Langrand, 1996), but, once again, it is unrecorded from eastern humid forests. The record from Ankafina, the most northeastern site where it is known to occur, is an animal collected in 1881 (Jenkins, 1987). An individual was found dead in 1993 along Route National No. 7, near the forest of Ankazomivady, south of Ambositra (A. Raselimanana, pers. comm.). This forest has been surveyed, and no evidence of this primate was found (Goodman et al., 1998). This section of Route National No. 7 is the zone where the Direction des Eaux et Forêts has at times released confiscated animals (C. Ravaoarinoromanga, pers. comm.). Further, the possibility exists that this individual was transported in a vehicle from the southern portion of Madagascar and was either released, escaped, or died en route. Given the uncertainties concerning this animal, we do not accept this as a valid locality record.

In the southern portion of *L. catta*'s range, it is often associated with gallery forest. There are considerable areas within this expansive region, away from river margins, where it is unknown to occur, and in many cases faunal research teams have not visited these zones. Thus, the lack of records from these areas is not necessarily indicative of the absence of this species. In the driest and harshest portions of the south, there is evidence of seasonal or erratic movements. For example, in the RS de Cap Sainte Marie, the southern limit of the island, troops have been observed on occasion exploiting available fruits, but they are not permanent residents at this site (Sussman et al., 2003).

In the region between Sakaraha and Toliara is an isolated massif known as Analavelona, rising to slightly more than 1300 m. The upper 300 m of the massif is a distinctly mesic forest with a mixture of eastern and western floristic elements and massive trees and a relatively open understory. The lower portion of the mountain is largely anthropogenic grassland with some remnant of mixed deciduous forest–spiny bush. *L. catta* is relatively common in the upper portion of the massif, which, once again, attests to its ability to adapt to a wide variety of ecological conditions, and in this case those approaching eastern humid forest habitat with regard to the flora and vegetational structure.

1.4. Ecology of High Mountain Population

As described in the previous section, *Lemur catta* is very flexible in its ecological requirements, but one of the more extreme zones in which it can be found is the high mountain area of the Andringitra Massif (Goodman and Langrand, 1996). Here its basic life-history traits are notably different from sites of lowland gallery forest, dry deciduous forest, or spiny bush. At the upper portion of its elevational range on Andringitra, this species occurs in a zone above forest line, which is at about 1950 m and which experiences daily temperature amplitudes of 30–35 °C, with nightly lows reaching −16 °C, and in a vast expanse of exposed vertical rock, with up to 400-m-tall talwegs, surrounded by ericoid savanna. Recent research on *L. catta* conducted on the upper portions of this massif forms the basis of information presented here and important insight into their ecology (Rakotoarisoa, 1999).

The western side of the Andringitra Massif descends abruptly in altitude and holds distinctly drier habitat than the eastern slopes. Within this western zone, *L. catta* have been found within the elevational range from about 900 to 2600 m. Individual troops have home ranges that span considerable elevational gradients. One troop occurred across a zone from 1310 to 2360 m and another troop from 1250 to 2040 m.

The five most important dietary elements of animals living in the high-elevation zone during the cold and dry season (May to September) include (averaged percent based on data from two different troops): *Vaccinium emirnense* (Ericaceae), 50.4%; *Ficus pyrifolia* (Moraceae), 6.6%; *V. secondiflorum*, 6.4%; *Locusta migratoria* (Insecta: Acrididae), 4.6%; and *Asteropeia micraster* (Asteropeiaceae), 4.6%. In contrast, during the warm and wet season (October to April), there is a notable change in diet, and the five most important dietary elements include (averaged percent based on data from two different troops): *Aphloia theiformis* (Aphloiaceae), 27.0%; *Ficus* spp., 14.0%; *Maesa lanceolata* (Myrsinaceae), 13.4%; *Buddleja madagascariensis* (Loganiaceae), 7.0%; and *Solanum auriculatum* (Solanaceae), 5.7%. Fruits make up about 75% of the foods consumed, leaves between 8% and 12%, and twigs, stems, and insects another 6–12%.

The local flora of this zone of Madagascar and the associated diet of *L. catta* are notably different from other regions on the island. These divergences are best demonstrated by a comparison with the plants consumed by one troop at

Andringitra and animals at the RS de Beza Mahafaly (Table 1.1). In this case, there is not a single plant species, genus, and family, with the exception of Fabaceae, shared in common within the diets of these two populations.

L. catta is remarkably dexterous in scaling vertical rock faces, in a very baboon-like fashion. Two captured individuals had notably well-developed and callous footpads. Night sleeping sites are generally in fissures or overhangs that often can only be accessed by scaling vertical surfaces. These sites presumably provide protection from Carnivora predators, such as *Cryptoprocta ferox* that is known above forest line on this massif (Goodman et al., 1997b), and a means to damper the extremely low nightly temperatures. This type of rock-climbing behavior is not limited to the Andringitra animals but has been reported from sites with dry deciduous forests such as Isoky-Vohimena in the PN de Zombitse-Vohibasia complex (Goodman et al., 1997c).

On the basis of some preliminary field observations, it was suggested that the pelage pattern and coloration of the Andringitra high-mountain population differed from that of "typical" *L. catta* (Goodman and Langrand, 1996). The conclusion of Groves (2001) that these differences represented a locally occurring subspecies

TABLE 1.1. Comparison of food elements of *Lemur catta* composing more than 80% of its diet on the upper slopes of the Andringitra Massif (Rakotoarisoa, 1999) and in the gallery forest of the Réserve Spéciale de Beza Mahafaly (Ratsirarson, 1987). Values are presented as percentages.

Species – family	Beza Mahafaly		Andringitra	
	Hot season	Cold season	Hot season	Cold season
Tamarindus indica – Fabaceae	18.1	39.0	—	—
Metaporana spp. – Convolvulaceae	5.2	32.4	—	—
Azima tetracantha – Salvadoraceae	27.1	32.4	—	—
Salvadora angustifolia – Salvadoraceae	23.1	—	—	—
Cedrelopsis grevei – Rutaceae	9.0	—	—	—
Maerua filiformis – Brassicaceae	—	5.8	—	—
Teramnus spp. – Fabaceae	—	4.1	—	—
Vaccinium emirnense – Ericaceae	—	—	2.6	49.3
Ficus pyrifolia – Moraceae	—	—	—	7.4
Vaccinium secondiflorumi – Ericaceae	—	—	—	6.2
Asteropeia micraster – Asteropeiaceae	—	—	—	4.7
Syzygium spp. – Myrtaceae	—	—	—	3.1
Homalium spp. – Salicaceae	—	—	—	3.1
Canthium variistipula – Rubiaceae	—	—	—	2.2
Erica spp. – Ericaceae	—	—	—	1.7
Senecio spp. – Asteraceae	—	—	—	1.6
Ficus spp. – Moraceae	—	—	12.2	1.6
Aphloia theiformis – Aphloiaceae	—	—	30.0	—
Maesa lanceolata – Myrsinaceae	—	—	13.6	—
Buddleja madagascariensis – Loganiaceae	—	—	7.2	—
Solanum auriculatum – Solanaceae	—	—	6.1	—
Vitex spp. – Lamiaceae	—	—	5.5	—
Acacia spp. – Fabaceae	—	—	5.1	—

of *L. catta* is an incorrect extrapolation. Subsequent fieldwork of this species on the Andringitra Massif verified that the tail-ring variation of the locally occurring animals falls within the range of typical members of this species; the thicker coat and lighter coloration may simply be an adaptation associated with the local extreme low temperatures and fading caused by intensive solar radiation (Rakotoarisoa, 1999; Goodman, unpubl. data). External measurements made of two adults trapped above forest-line on the Andringitra Massif fall within the normal range of *L. catta* (Table 1.2). Further, genetic studies of these same individuals indicate that they are typical of populations occurring elsewhere on the island (Yoder et al., 2000).

TABLE 1.2. External measurements of two adult *Lemur catta* from the Andringitra Massif as compared with other populations of this species.

Character	Range of values (minimum–maximum) for typical *Lemur catta* (Tattersall, 1982)	Andringitra: adult male (Rakotoarisoa, 1999)	Andringitra: adult female (Rakotoarisoa, 1999)
Head and body length	385–455 mm	455 mm	450 mm
Tail length	560–624 mm	619 mm	611 mm
Ear length	40–48 mm	42 mm	43 mm
Weight	2295–3488 g	3120 g	2920 g

1.5. Does *Lemur catta* Occur in Areas Where No Freshwater Source Is Available?

Petter et al. (1977, p. 157) remarked that *Lemur catta* exists in the southern portion of the island, in regions where the dry season is not too excessive. The context here is that in certain areas, this species is absent due to water or food limitations. Further, it has been suggested that *L. catta* needs to drink at least occasionally (M. Pidgeon, in Harcourt and Thornback, 1990, p. 105). Thus, one potential limiting factor in the distribution of *L. catta* in the dry portions of Madagascar is the presence of water sources. During the course of the past decade, inventories have been conducted in portions of southwestern Madagascar, in zones of spiny bush, where no freshwater sources occur. We have field data to address the issue of the role this resource plays as a limiting factor in the distribution of this species.

At the western foot of the Mahafaly Plateau, there is a narrow fault zone where subterranean water sources percolate up from the ground (Guyot, 2001). In the section along this fault between the PN de Tsimanampetsotsa and Itampolo, these water sources are slightly saline, after passing through a limestone aquifer. For example, at Mitoho, a site where *L. catta* is common and can be easily observed drinking at the cave entrance, salt-mineral content of the water reaches 1900 mg/L (Demergue, 1974). Other sources in the region reach a salt-mineral content of up 2600 mg/L. Thus, water quality, at least at this level of mineralization, is not a limiting factor in their distribution.

The western portion of the Mahafaly Plateau is a vast expanse of spiny bush, and certain large zones are without water sources, except during the short wet season when temporary rain pools form. For example, the Bemananteza Forest, about 21.5 km northwest of Efoetse and in the eastern portion of the PN de Tsimanampetsotsa, is without a freshwater source. While camped in this area in early March 2002, we obtained water for the field crew from a well 40 km to the north. During this period, there was relatively heavy nightly dew that evaporated before 7:00 a.m. the following morning. A group of travelers from the village of our water supply passed by the camp, laden with an assortment of filled water gourds, and mentioned that to their knowledge there was no source for a considerable distance. While in the Bemananteza Forest for more than a week, no evidence of *L. catta* was found, and local people stated that it does not occur in the immediate area (Goodman et al., 2002).

In another case, we conducted an inventory of forests within the PN de Kirindy-Mitea, 13 km west of Marofihitsa, during the month of November 2002 and before the start of the local rainy season. As in the Bemananteza Forest, there was no known freshwater source in a vast area of slightly degraded deciduous forest surrounding our research site, but the zone experienced notable nightly dew. On several occasions, we encountered troops of *L. catta* at this site and heard them calling. Given the regularity of vocalizing individuals, both in the sense of the time of day and localization in the forest, during the course of the week that we were at this site, it was our impression that they were residents rather than dispersing or migrating individuals. If this assumption is correct, this would be evidence that this species is able to exist in a forested zone without a permanent water source. Presumably they are able to fill their water requirements from early morning dew and from moisture in different foods. Given this extrapolation and the fact that seemingly comparable levels of dew fell each night at the Bemananteza Forest and the site within the PN de Kirindy-Mitea, the absence of this species from the former site may be associated with food resources.

In a study of the home-range patterns of *L. catta* troops in the Berenty area, it was found that certain "land-locked" troops did not have access to river margins, as they would not cross the territories of several neighboring groups (O'Connor, 1987), at least during certain periods of the year. Further, the population in these closed canopy forests is twice that of sites with open canopy and bush and scrub (O'Connor, 1987). Given these points, it is clear that these troops were able to prosper in spite of the fact that they did not have access to freshwater, and they obtain their needed water sustenance from dew or food sources (Randramboavonjy, 2003). As typified by spiny bush dominated by Didiereaceae, water-rich plants in this habitat are distinctly more common in the local flora than in more northerly dry deciduous forest. Hence, the means for *L. catta* to derive their daily water needs simply through their diet may be more plausible in the spiny bush than in dry deciduous forest.

These sorts of data have interesting implications for understanding certain life-history traits of *L. catta*. On the basis of a long-term field study in the RS de Beza

Mahafaly, the local population of *L. catta* was stable during a 15-year period, with the exception of a 2-year drought, when there was a notable decline (Gould et al., 2003). Given the above information from the Kirindy-Mitea Forest and that water sources at Beza Mahafaly are very seasonal, it can be extrapolated that the population decline during the drought was associated with depletion of food sources rather than limited drinking water (Gould et al., 1999; Jolly et al., 2002).

1.6. A Biogeographic Scenario to Explain the Distribution of *Lemur catta*

A considerable amount has been written concerning the role of rivers as dispersal barriers for a wide variety of animals, including lemurs (Martin 1972; Goodman and Ganzhorn, 2004; Ganzhorn et al., 2006). In the context we are addressing here, rivers can act as a physical barrier for random dispersal of individuals and for species expanding their geographic range. The inverse of this is that species use rivers as corridors for dispersal.

Certain aspects of the life history of *L. catta* are interesting with regard to their mode of dispersal. First, it is the most terrestrial of the living lemurs and is known to occur outside of forest. Thus, their dispersal across nonforested areas would not be strictly inhibited as in the case of forest-dependent arboreal species. Second, *L. catta* is known to have higher densities in gallery forest than in other vegetational communities (O'Connor, 1987), particularly relatively undisturbed gallery forest (Raharivololona and Ranaivosoa, 2000). Finally, throughout a portion of its range, particularly the south, the largest river drainage systems are seasonal, and, as such, would not pose a permanent barrier across which this species could not disperse. For example, during the dry season the Mandrare River becomes a series of isolated pools, and it would be a straightforward matter for *L. catta* to simply walk across the river channel. This is the period that males tend to migrate (Jones, 1983). In fact, this species has been observed in the middle of this river channel during the dry season and has been seen crossing the Menarandra River (M. Pidgeon and S. O'Connor, pers. comm.). Given all of these three points, it can be postulated that rivers might act as dispersal corridors, rather than barriers, for this taxon.

An examination of the geographic range of *L. catta* with an overlay of river systems (Figure 1.1) of the southern third of Madagascar illustrates several interesting points. This species is broadly distributed, and its range is not bounded or limited by any river system. Many of the records from the southern portion of its range are from gallery forest habitats, which is in accordance with its proposed habitat preference in the drier extreme southern portion of its range. These aspects are concordant with the hypothesis that rivers do not act as dispersal barriers for this species.

L. catta's occurrence in highland areas of the Fianarantsoa Province (e.g., Andringitra and Ankafina) can be explained by the use of the Mangoky River and associated tributaries as dispersal corridors, a point already mentioned by

Sussman et al. (2003). Across the middle and upper portions of the Mangoky watershed, there are gallery forests and vast granite domes with slightly xerophytic vegetation—habitats used by this species of lemur. However, the food plants it consumes in these highland areas are notably different from the lowland dry forest and spiny bush portions of its range (Table 1.1), which contain a remarkable assortment of primary and secondary compounds (Simmen et al., 1999), further indicating the remarkable adaptability of this species with regard to diet and habitat.

On the basis of the various parameters associated with the distribution of *L. catta*, we propose that this species evolved in dry habitats in southern and southwestern Madagascar and subsequently dispersed to quasi-mesic highland regions using river systems as corridors, particularly the Mangoky watershed. The southern and western limits of its range are bounded by the coastal regions of the island and the northern and eastern limits by bioclimatic factors, which show direct parallels to the family Didiereaceae and may be related to the presence of water-rich plants from which it can derive its daily requirements. To our knowledge, there is no subfossil record of this species outside of its current distribution (Godfrey et al., 1999), which might be an indication that its geographical range has been rather stable in recent geological time.

1.7. Summary

The suggestion that *Lemur catta* is best considered as a highly adaptable "edge" or "weed" species (e.g., Gould et al., 2003) is supported by the information we have presented here. It occurs in a considerable number of habitats in the southern third of Madagascar, including spiny bush, gallery forest, anthropogenic savanna, deciduous forest, rock canyons, and upland inland areas (up to 2600 m) with expanses of granite domains and open ericoid vegetation. This species encounters the most extreme climatic regimes on the island from the hottest and driest (spiny bush) sites to the coldest known locality on the island (Andringitra Massif). Across its range it has a rather varied diet, and in many cases, particularly in the southern gallery forests and spiny bush, consumes plants with high tannin concentrations. In some areas it drinks from water sources, and in other areas the only regularly available moisture is apparently night dew or that from consumed foods. There is evidence that rivers do not bind the distribution of *L. catta*, and it seems to use river basins as dispersal corridors. We suspect that there is considerable movement and exchange between populations of this species. This is supported by some limited genetic work that shows that exchange across the range of this species has been persistent and extensive (Yoder et al., 2000).

Acknowledgments. We are grateful to the following individuals for providing information on the distribution of *Lemur catta* or aid with bibliographic citations: Barry Ferguson, Jörg Ganzhorn, Mitchell Irwin, Alison Jolly, Sheila O'Connor,

Pete Phillipson, Mark Pidgeon, Achille Raselimanana, Michelle Sauther, Bob Sussman, and Dietmar Zinner. Fieldwork associated with some of the data used in this review has been generously supported by Conservation International (CEPF), Ellen Thorne Smith Fund of the Field Museum of Natural History, John D. and Catherine T. MacArthur Foundation, National Geographic Society, Volkswagen Foundation, and WWF-Madagascar. A grant from WWF-USA provided the means for entry of records and refinement of the database. For comments on an earlier version of this paper, we are grateful to Jörg Ganzhorn, Alison Jolly, and Bob Sussman.

References

Cornet, A. (1974). Essai de cartographie bioclimatique à Madagascar. *Notice Explicative de l'ORSTOM* 55:1–28.

Domergue, C. A. (1974). Considérations sur la minéralisation des eaux de l'extrême Sud. *Bulletin de l'Académie Malgache* 52/1–2:119–125.

Ganzhorn, J. U., Fietz, J., Rakotovao, E., Schwab, D., and Zinner, D. (1999). Lemurs and the regeneration of dry deciduous forest in Madagascar. *Conservation Biol.* 13:1–11.

Ganzhorn, J. U., Goodman, S. M., Nash, S., and Thalmann, U. (2006). Lemur biogeography. In: Lehman, S., and Fleagle, J. G. (eds.), *Primate Biogeography*. Plenum/Kluwer Press, New York, pp. 229–254.

Godfrey, L. R., Jungers, W. L., Simons, E. L., Chatrath, P. S., and Rakotosamimanana, B. (1999). Past and present distributions of lemurs in Madagascar. In: Rakotosamimanana, B., Rasamimanana, H., Ganzhorn, J. U., and Goodman, S. M. (eds.), *New Directions in Lemur Studies*. Kluwer Academic/Plenum, New York, pp. 19–53.

Goodman, S. M., and Ganzhorn, J. U. (2004). Biogeography of lemurs in the humid formations of Madagascar: The role of elevational distribution and rivers. *J. Biogeogr.* 31:47–55.

Goodman, S. M., and Langrand, O. (1996). A high mountain population of the ring–tailed lemur *Lemur catta* on the Andringitra Massif, Madagascar. *Oryx* 30:259–268.

Goodman, S. M., Pidgeon, M., Hawkins, A. F. A., and Schulenberg, T. S. (1997a). The birds of southeastern Madagascar. *Fieldiana: Zoology*, new series, 87:1–132.

Goodman, S. M., Langrand, O., and Rasolonandrasana, B. P. N. (1997b). The food habits of *Cryptoprocta ferox* in the high mountain zone of the Andringitra Massif, Madagascar (Carnivora, Viverridae). *Mammalia* 61:185–92.

Goodman, S. M., Langrand, O., and Rasoloarison, R. (1997c). Les lémuriens. In: Langrand, O., and Goodman, S. M. (eds.), Inventaire biologique forêt de Vohibasia et d'Isoky–Vohimena. *Recherches pour le Développement*, série sciences biologiques, Centre d'Information et de Documentation Scientifique et Technique, Antananarivo 12:156–161.

Goodman, S. M., Duplantier, J. M., Rakotomalaza, P. J., Raselimanana, A. P., Rasoloarison, R., Ravokatra, M., Soarimalala, V., and Wilmé, L. (1998). Inventaire biologique de la forêt d'Ankazomivady, Ambositra. *Akon'ny Ala* 24:19–32.

Goodman, S. M., Raherilalao, M. J., Rakotomalala, D., Rakotondravony, D., Raselimanana, A. P., Razakarivony, H. V., and Soarimalala, V. (2002). Inventaire des vertébrés du Parc National de Tsimanampetsotsa (Toliara). *Akon'ny Ala* 28:1–36.

Gould, L., Sussman, R. W., and Sauther, M. L. (1999). Natural disasters and primate populations: The effects of a 2–year drought on a naturally occurring population of ring–tailed lemurs (*Lemur catta*) in southwestern Madagascar. *Int. J. Primatol.* 20:69–84.

Gould, L., Sussman, R. W., and Sauther, M. L. (2003). Demographic and life-history patterns in a population of ring-tailed lemurs (*Lemur catta*) at Beza Mahafaly Reserve, Madagascar: A 15-year perspective. *Am. J. Phys. Anthropol.* 120:182–194.

Groves, C. P. (2001). *Primate Taxonomy*. Smithsonian Institution Press, Washington, D.C.

Guyot, L. (2001). Reconnaissance hydrogéologie pour l'alimentation en eau d'une plaine littorale en milieu semi-aride: Sud Ouest de Madagascar, Thèse de Doctorat, Université de Nantes.

Harcourt, C., and Thornback, J. (1990). *Lemurs of Madagascar and the Comoros. The IUCN Red Data Book*. International Union for Conservation of Nature and Natural Resources, Gland.

Jenkins, P. D. (1987). *Catalogue of Primates in the British Museum (Natural History), Part IV*. British Museum (Natural History), London.

Jolly, A. (2003). *Lemur catta*, ring-tailed lemur. In: Goodman, S. M., and Benstead, J. P. (eds.), *The Natural History of Madagascar*. The University of Chicago Press, Chicago, pp. 1329–1331.

Jolly, A., Dobson, A., Rasamimanana, H., Walker, J., O'Connor, S., Solberg, M., and Perel, V. (2002). Demography of *Lemur catta* at Berenty Reserve, Madagascar: Effects of troop size, habitat, and rainfall. *Int. J. Primatol.* 23:327–353.

Jones, K. C. (1973). Inter-troop transfer in *Lemur catta* at Berenty, Madagascar. *Folia Primatol.* 40:145–160.

Martin, R. D. (1972). Adaptive radiation and behaviour of the Malagasy lemurs. *Philos. Trans. R. Soc. London B* 264:295–352.

Mittermeier, R. A., Tattersall, I., Konstant, W. R., Meyers, D. M., and Mast, R. B. (1994). *Lemurs of Madagascar*. Conservation International, Washington, D.C.

O'Connor, S. M. (1987). *The Effect of Human Impact on Vegetation and the Consequences to Primates in Two Riverine Forests, Southern Madagascar*. Ph.D. thesis, University of Cambridge.

Petter, J-J., Albignac, R., and Rumpler, Y. (1977). *Mammifères Lémuriens (Primates Prosimiens)*. Vol. 44 of *Faune de Madagascar*, ORSTOM/CNRS, Paris.

Raharivololona, B. M, and Ranaivosoa, V. (2000). Suivi écologique de Lémuriens diurnes dan le Parc National d'Andohahela à Fort Dauphin. *Lemur News* 5:8–11.

Rakotoarisoa, S. V. (1999). *Contribution à l'étude de l'adaptation de* Lemur catta *Linnaeus, 1758, aux zones sommitales de la Réserve Naturelle d'Andringitra*, Mémoire D.E.A., Faculté des Sciences, Université d'Antananarivo.

Randramboavonjy, T. M. (2003). Etude des besoins en eau de *Lemur catta* (Linné, 1758) pendant la saison sèche. Mémoire CAPEN, Ecole Normale Supérieur, Antananarivo.

Ratsirarson, J. (1987). Contribution à l'étude comparative de l'éco-éthologie de *Lemur catta* dans deux habitats différents de la Réserve Spéciale de Bezà-Mahafaly. Mémoire de fin d'étude, E.S.S.A. Eaux et Forêts, Université d'Antananarivo.

Schatz, G.E. (2000). Endemism in the Malagasy tree flora. In: Lourenço, W. R., and Goodman, S. M. (eds.), *Diversity and Endemism in Madagascar*. Mémoires de la Société de Biogéographie, Paris, pp. 1–9.

Simmen, B., Hladik, A., Ramasiarisoa, P. L., Iaconelli, S., and Hladik, C. M. (1999). Taste discrimination in lemurs and other primates, and the relationships to distribution of plant allelochemicals in different habitats of Madagascar. In: Rakotosamimanana, B., Rasamimanana, H., Ganzhorn, J. U., and Goodman, S. M. (eds.), *New Directions in Lemur Studies*. Kluwer Academic/Plenum, New York, pp. 201–220.

Sussman, R. W. (1977). Distribution of the Malagasy lemurs. Part 2: *Lemur catta* and *Lemur fulvus* in southern and western Madagascar. *Ann. N. Y. Acad. Sci.* 293:170–184.

Sussman, R. W., Green, G. M., Porton, I., Andrianasolondraibe, O. L., and Ratsirarson, J. (2003). A survey of the habitat of *Lemur catta* in southwestern and southern Madagascar. *Primate Conserv.* 19:32–57.

Tattersall, I. (1982). *The Primates of Madagascar*. Columbia University Press, New York.

Yoder, A. D., Irwin, J. A., Goodman, S. M., and Rakotoarisoa, S. V. (2000). Genetic tests of the taxonomic status of *Lemur catta* from the high mountain zone of the Andringitra Massif, Madagascar. J. *Zool. London* 252:1–9.

Zinner, D., Ostner, J., Dill, A., Razafimanantsoa, L., and Rasoloarison, R. (2001). Results of a reconnaissance expedition in the western dry forests between Morondava and Morombe. *Lemur News* 6:16–18.

2
A Preliminary Estimate of *Lemur catta* Population Density Using Satellite Imagery

ROBERT W. SUSSMAN, SEAN SWEENEY, GLEN M. GREEN, INGRID PORTON, O.L. ANDRIANASOLONDRAIBE, AND JOELISOA RATSIRARSON

2.1. Introduction

Ringtailed lemurs are found in many habitats throughout southwestern and southern Madagascar. As stated by Goodman et al. (this volume), it is the least forest-dwelling of the extant species of lemurs and lives in some of the most xerophytic forests on the island. The dry forests of the south and west are unique and are inhabited by many plants and animals found nowhere else on Earth. Although rain forests have received a great deal of research attention from conservation and development organizations, there has been less focus on dry forests, and there is some indication that these forests are among the most endangered habitats worldwide (Janzen, 1988; Kramer, 1997; Smith, 1997, Cabido and Zak, 1999; Trejo and Dirzo, 2000; Dirzo and Sussman, 2002, Sussman et al., 2003). There is great urgency to document the deforestation, to determine the rate and patterns of habitat loss, and to see how this habitat loss is affecting the unique fauna of southern and southwestern Madagascar. Given that the geographic range of ringtailed lemurs is coincidental with that of these dry forest habitats (Sussman, 1977; Sussman et al., 2003; Goodman et al., this volume), it is important to know the density of *L. catta* populations in various habitat types and how the patterns and processes of deforestation are affecting ringtailed populations currently and how they have done so in the past.

Satellite platforms, most notably the Landsat series, have benefited scientists by enabling them to observe land cover change and the patterns of land use at regional scales. Although use of the Landast platforms provides a rather narrow temporal sampling window, 1970s to present, it captures the several decades where many areas of the world have seen aggressive deforestation episodes, as is the case of dry forests in southwestern Madagascar. Cutting of forests has certainly been detectable in this study area over this breadth of time, as we observe the initial stages of major deforestation events between the early 1970s and 1985. Post-1985, accelerated deforestation has occurred resulting in large areas of contiguous forest being cut to satisfy demands for charcoal and agricultural land, both small- and large-scale (Sussman et al., 2003). Likewise, the greatest reduction in lemur habitat has occurred since 1985. In order to assess the impact that this dynamic has had on specific lemur habitats and population, the imagery acquired in 1985 was selected to represent habitat at T_0, initial habitat for this study. Additionally, image selection was influenced by resolution of the satellite platforms: 2000 (ETM) and 1985 (TM) have cell resolutions resampled to 30 m, whereas 1973 (MSS) has a cell resolution resampled to 60 m, enabling the former two to resolve objects a fraction (one-fourth) the size of the latter. At the conclusion of this analysis, an approximation of deforestation between 1985 and 2000 was calculated for our study area in an initial attempt to assess how this dynamic has affected habitat extent and predicted ringtailed populations. A future stepwise temporal analysis is planned to quantify change in specific habitat extents, conditions, locations, and lemur populations between 1950 and 2005 (using aerial photographs as well as satellite images) in an effort to reconstruct the history of deforestation during the past half-century, to predict land cover trajectories, and to identify areas for conservation efforts.

In this paper, we make a preliminary attempt to determine the population density in 1985 and the relatively current population density of *L. catta* in relationship with gradients of vegetation cover over its entire geographic range. We use a parameter of forest condition, canopy density, derived from satellite imagery, and published information on ringtailed lemur population densities to address this question. We also discuss the methodology used to make our analyses. Goodman et al. (this volume) have pointed to the fact that many of the aspects of *L. catta* life-history parameters may be exaggerated given the intensive focus of past research on gallery forest zones, the richest of ringtailed lemur habitats. We agree with this assessment and discuss how this has impinged on our analysis. In doing so, we describe the data that would be needed to improve our analysis and stress the urgent need for research to be conducted to collect these missing data.

In order to determine accurately the density of ringtailed lemur populations in space and time using remote-sensing technologies, we need to determine the existence and nature of a number of relationships. We must determine whether a relationship can be established between a quantitative measure of vegetated cover and spectral data (satellite DNs) and, if so, if this relationship will allow us to discriminate and map, with confidence, the variety of potential lemur habitats (gallery forest, dry brush and scrub forest, other xerophytic forests). Next, we must determine whether a relationship can be established between lemur densities and satellite spectral data, either directly or via a quantitative measure of vegetated land cover. It is the relationship between satellite data and lemur densities that we investigate in this paper. If these relationships can be established, an estimate of population densities of lemurs in relation to regions with different vegetation cover generally can be proposed using vegetation maps derived from the satellite images.

Given the above, in this paper we develop a methodology using reflectance spectra from Landsat images for calculating a measure of forest canopy density (FCD) and for examining the direct relationship, if any, between FCD (a measure derived from satellite reflectance data) and lemur densities. The relationships that we examine in this paper do not allow us to discriminate between habitats but do enable us to explore the extent of habitat in totality. This information enables us to map lemur habitat in its most generic form throughout southern Madagascar and estimate population using the relationships being explored between FCD and lemur densities. Although it is important to our research, and certainly a component of future research, to be able to identify lemur densities and populations in particular habitats, it is not specifically the aim of this paper. The purpose of this exercise is to explore the advantages and usefulness of incorporating a spatial and temporal mechanism for identifying and mapping, not only location but condition, of primate habitat and how it relates to densities and populations.

Thus, the synoptic view of satellite images provides spatially explicit information on potential lemur habitat, which is then used to focus on the acquisition of higher resolution, more costly field surveys in representative regions of southern forest habitat, thereby providing a robust and extensive monitoring system for *Lemur catta*. Research on the ecology of *Lemur catta* at several sites (Jolly, 1966; Budnitz and Dainis, 1975; O'Connor, 1987; Sussman, 1991; Koyama et al., 2001; Jolly et al., 2002, Gould et al., 2003, Sauther, pers. comm.) has demonstrated that ringtailed lemur density is directly related to habitat quality. However, currently available estimates of the population and distribution of *Lemur catta* are little more than guesswork. In this study, we document a research strategy for a more effective mapping. We recognize that a number of variables, other than forest condition, may affect actual lemur population densities: human activities of hunting or charcoaling, distance to village, distance to road, availability of water, soil composition, and topography. In addition, behavioral factors such as willingness of a group to range, and maximum distance, to multiple forest patches separated by nonforested land cover and the likelihood of reoccupation of a forest previously disturbed could contribute to significant disparities between actual and predicted values.

Thus, the results of this analysis represent a "best case" scenario in which we assume that all potential habitats that are adequate in extent to sustain a lemur population enjoy a lemur presence and that there are no external or behavioral factors adversely affecting lemur density in these areas. Furthermore, we stress that currently available information on ringtailed lemur densities, as mentioned above, come from a very small proportion of the habitats in which they are found.

2.2. Methods

The characterization of forest condition, as it relates to *Lemur catta* habitat, is essential to predicting lemur population in this study area. Employment of a forest canopy density measure allows us not only to identify habitat capable of sustaining a lemur population but also affords us a temporal measure of habitat condition by enabling us to detect change in the percentage of crown closure and therefore a change in area of occupation (Roy et al., 1996). This parameter of forest condition is directly related to lemur density data, and the function representing the relationship is employed to predict population densities for all potential habitats.

2.2.1. Image Pre-processing

Six footprints from the WRS2 reference system cover the study area. Landsat 5 Thematic Mapper (TM) images, acquired January thru February 1985, were selected for use in this analysis. The first step in the pre-processing sequence was to mask all cloud and cloud shadows from each TM scene. The scenes were then geo-referenced to the Laborde projection system using digital topographic base maps as the reference source. Radiometric calibration and atmospheric correction were then performed on each scene in order to relate the digital counts in satellite image data to reflectance at the surface of the earth. The entire data set was mosaiced into a single image as the final step in the initial processing sequence.

2.2.2. Computing Forest Canopy Density

Three indices, advance vegetation index (AVI), bare soil index (BI), and scaled shadow index (SSI), were generated from the TM data and employed as inputs to a forest canopy density model to (1) differentiate habitable land cover (forests) from other and (2) to give us a measure of forest condition (Figure 2.1). The black soil detection component of the processing sequence was omitted from this methodology but is available to assist in differentiating shadow from black soil, particularly burn scars (Rikimaru and Miyatake, 1997). Prior to calculating the indices, the reflectance values of each TM band are normalized over a data range with values 0–255 using the

Linear Transformation: $Y = AX + B$

$$A = (-200) / [(M_i - 2S_i) - (M_i + 2S_i)] = 50 / S_i$$
$$B = -A(M_i - 2S_i) + 20$$

where M is mean, S is standard deviation, and i is Landsat TM band number.

The model component, *advanced vegetation index* (AVI), is used to distinguish subtle differences in canopy density (Jamalabad and Abkar, 2004). After normalization of the TM bands, B_3 is subtracted from B_4 where B_4 is TM band 4 and B_3 is TM band 3. Difference values that are less than or equal to 0 are assigned an AVI value of 0. The following calculation is applied to the remaining pixels with difference values greater than 0.

Advanced Vegetation Index (AVI): $[(B_4 + 1) * (256 - B_3) * (B_4 - B_3)]^{1/3}$

where B_4 is TM band 4 and B_3 is TM band 3. (Note: AVI = 0 if $B_4 < B_3$ after normalization.)

Bare soil index (BI) is a normalized index of the difference of sums used to differentiate vegetated land cover with different background response and due to varying canopy density (Jamalabad and Abkar, 2004).

Bare Soil Index (BI): $[(B_5 + B_3) - (B_4 + B_1)] / [(B_5 + B_3) + (B_4 + B_1)] * 100 + 100$
where B_5 is TM band 5, B_4 is TM band 4, B_3 is TM band 3, and B_1 is TM band 1.

Canopies of forests vary markedly depending on age, early succession to mature, as well as species composition. Differences in canopy structure and

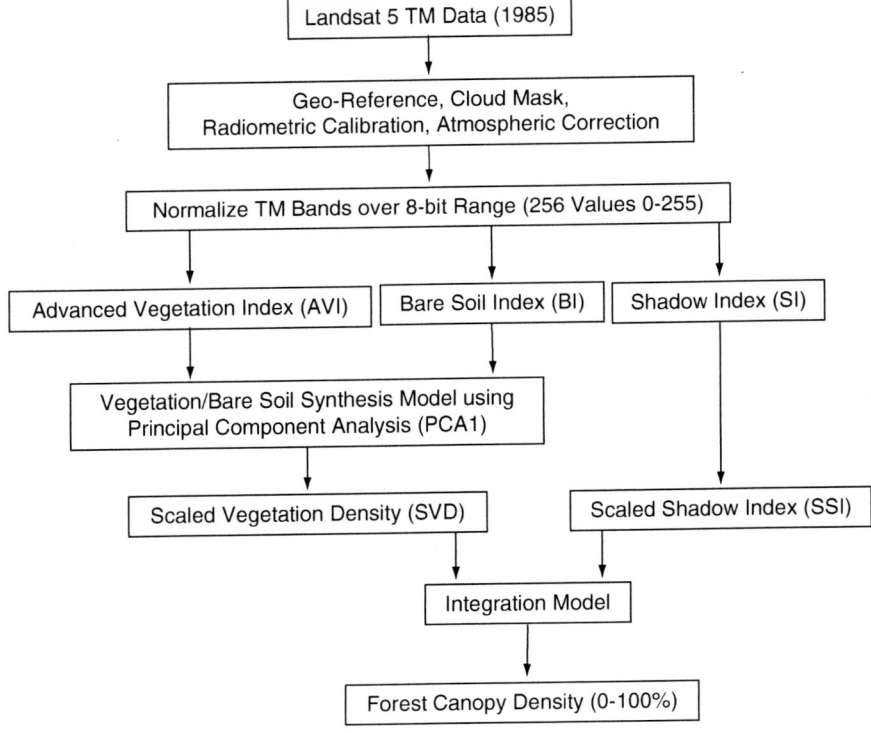

FIGURE 2.1. Forest canopy density processing flowchart.

density create differences in the amount of shadow present influencing reflectance. Scaled shadow index enhances the spectral differences between mature forests that have higher canopy shadow index values compared with that of younger forest stands (Jamalabad and Abkar 2004).

Shadow Index (SI): $[(256 - B_1) * (256 - B_2) * (256 - B_3)]^{1/3}$

where B_3 is TM band 3, B_2 is TM band 2, and B_1 is TM band 1.

Scaled Shadow Index (SSI): Shadow index (SI) scaled to values 0 to 100.

Vegetation Density (VDS) is produced using principal component analysis. AVI and BI (high negative correlation) are used as the model inputs.

Scaled Vegetation Density (SVD): First principal component of AVI and BI scaled to values 0 to 100.

Input parameters, SVD and SSI, share like characteristics of dimension and percentage scale units of density (Jamalabad et al., 2004) and are used to compute

Forest canopy density: $[(SVD * SSI + 1)^{1/2}] - 1$.

Figure 2.2 was generated from these data.

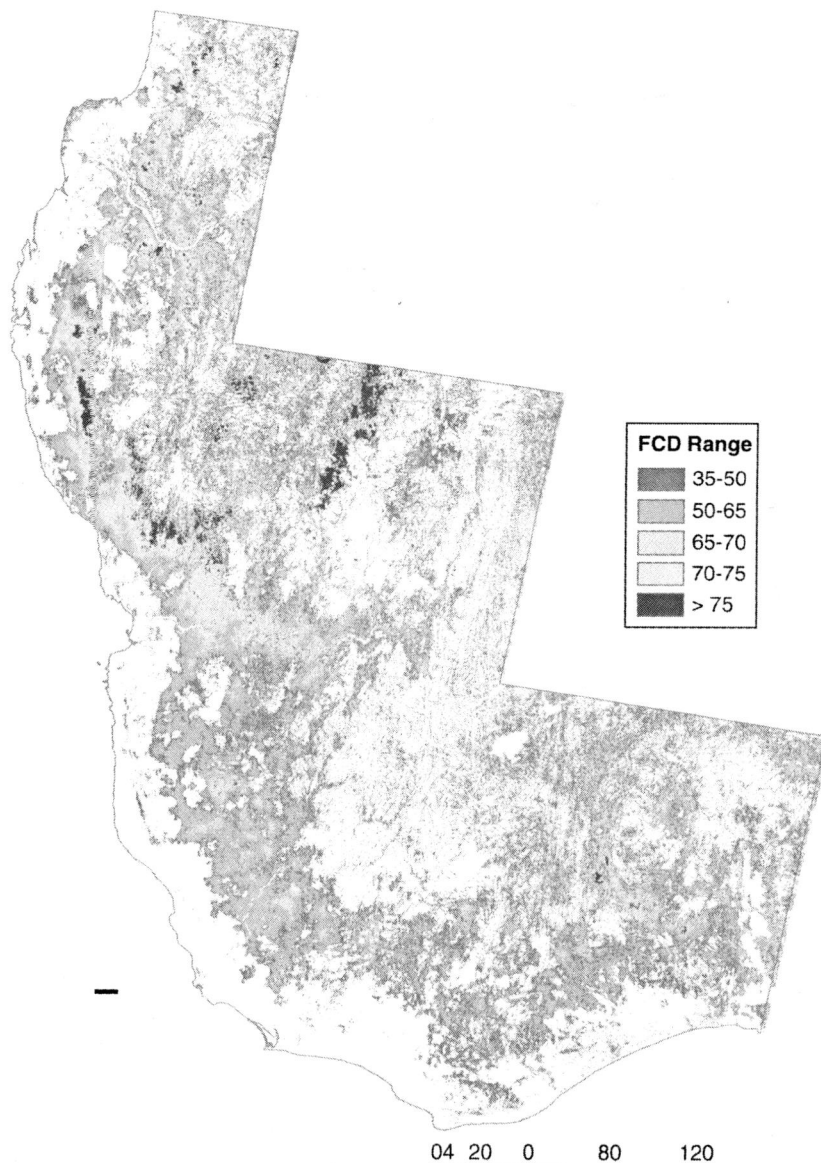

FIGURE 2.2. Stratified forest canopy density map derived from Landsat 5 Thematic Mapper images acquired January and February 1985. Pixels with forest canopy density (FCD) values less than 35 assigned a value of "0." [*See* Colour Plate I]

(This model also has the capacity to incorporate a *thermal index* (TI) to separate soil, particularly burn scars, from shadow other than that cast by trees, but this was not incorporated in this paper).

2.2.3. Linking Lemur Density to Forest Canopy Density

Mean FCD values were extracted from forest locations with known lemur densities. Using a curve-fitting software, TableCurve 2D, two relationships were examined: linear and nonlinear. A best-fitting line and a transition function, standard logistic (sigmoid), were applied to the data (Figure 2.3). Assuming that the lemur densities being employed are representative of those at the specific values of forest canopy density, the linear method underestimates lemur density at FCD values below 52% and greater than 72%, and overestimates lemur densities at FCD values greater than 53% and less than 72%. The transition function was employed to predict lemur density for the study area using FCD as the independent variable (Figure 2.4).

2.2.4. Methodology for Approximating Deforestation Between 1985 and 2000

An approximation of deforestation between 1985 and 2000 was calculated for our study area in an initial attempt to assess how this dynamic has affected habitat

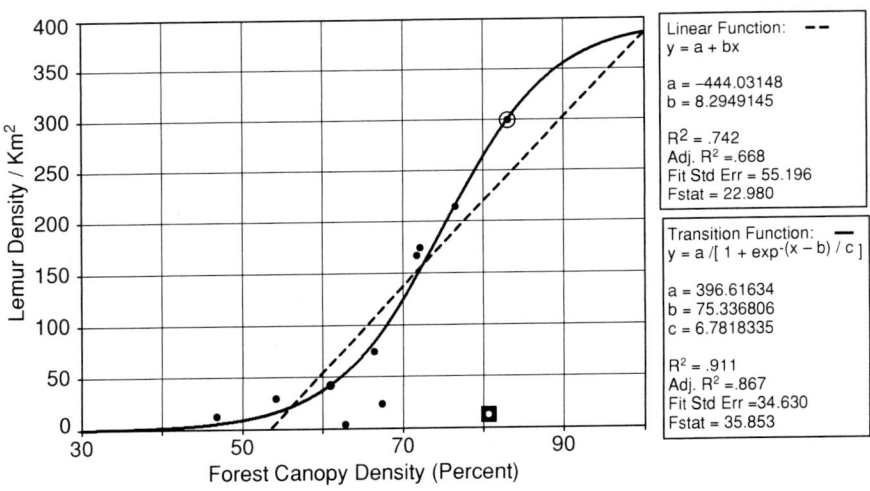

FIGURE 2.3. Plot of *Lemur catta* data with relationship functions, linear (dashed) and transition (solid). Beoloka (represented by a solid square) is an outlier and was not included as part of the data set in determining the prediction function but was added to illustrate the impact of external factors, in this instance hunting. Berenty (point with circle) is noteworthy in that the lemurs that reside within its boundaries are provisioned, likely producing the highest concentrations in this study area.

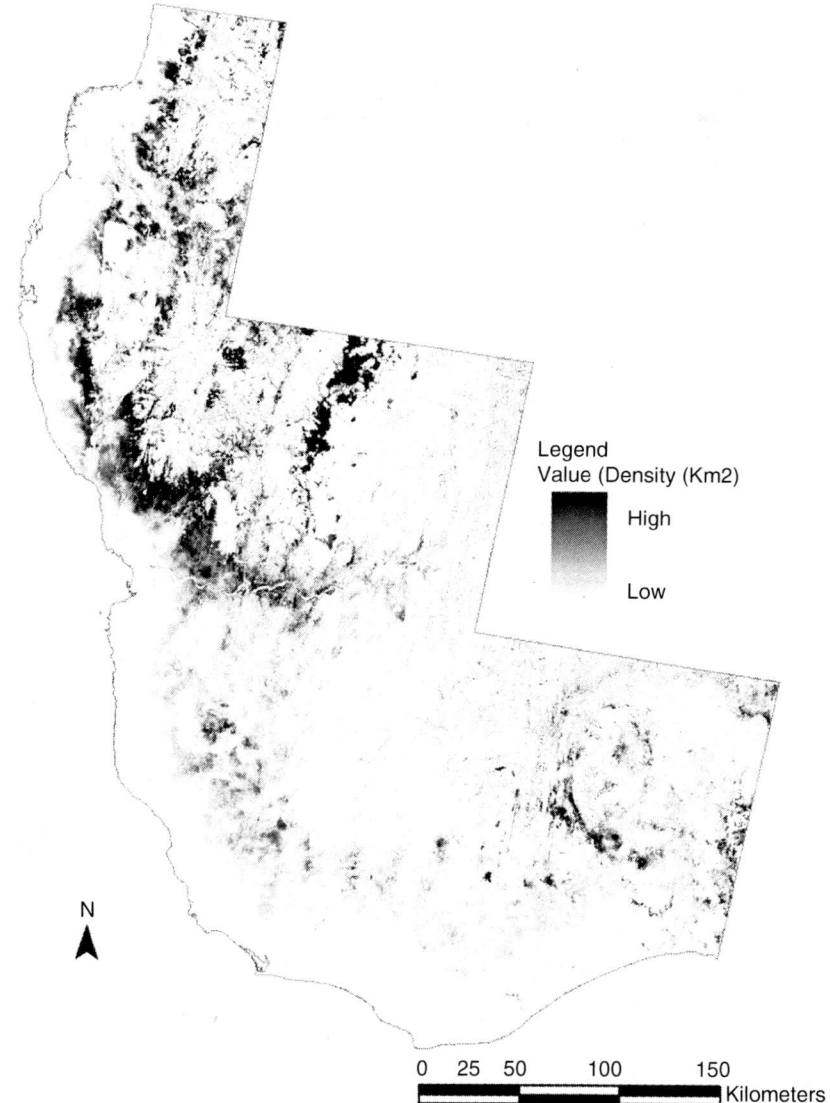

FIGURE 2.4. Lemur prediction map produced using the transition function and forest canopy density (FCD) as the independent variable.

extent and predicted populations. To identify deforested parcels, a multitemporal composite was employed. A three-layer image was constructed assigning band 3 (2000) to red, band 3 (1985) to green, and band 3 (1985) to blue. Band 3 is particularly sensitive to soil and exhibits high reflective properties in response to such. Pixels deforested between 1985 and 2000 appear red due to an increase in soil exposure; pixels reforested in that time frame appear cyan due to a decrease in soil exposure. Although a minimal amount of reforestation has taken place in this study area, the amount is negligible in comparison with that of deforestation and was not addressed in this analysis.

Deforested areas were identified in the multitemporal composite and areas of interest (AOIs) that delineated their extents used to extract spectral signatures, training samples, from the composite image. Training samples were then displayed over a frequency scatterplot, feature space image, with band 3 (2002) assigned to the Y-axis and band 3 (1985) assigned to the X-axis. An AOI delineating the cluster boundary of deforestation training samples was drawn on the feature space image. Pixels within this boundary were classified as deforested and those outside as stable. A binary image was created from the classified thematic image assigning deforested pixels a value of "0" and stable a value of "1." As a final step, the binary mask was applied to the 1985 forest canopy density map and pixels deforested between 1985 and 2000 given an FCD value of "0" (Figure 2.5). Habitat extents and ringtailed lemur numbers were recalculated for all FCD ranges.

2.2.5. *Assumptions and Potential Problems*

- The lemur density data were in units of km^2. Prior to calculating the lemur population for the study area, lemur densities in forests smaller in area than $1 km^2$ were normalized for forest extent. Predicted lemur density represents the potential population at a particular FCD value and an area of $1 km^2$. A linear relationship was assumed between lemur density and forest parcel size. The ratio of forest area to $1 km^2$ was applied to lemur density when the forest patch was less than $1 km^2$.
- Forest parcels were assumed to be homogeneous. No classification has been performed on the study area, and no distinction made between forest types. The mean FCD value was calculated for each forest parcel risking overestimating or underestimating FCD depending on the majority fraction and composition in a mixed forest.
- Unique relationships between FCD and lemur density may exist between forest types (gallery, dry, and xerophytic) and conditions (degraded and not). Additional data will reveal this.

2.3. Results

Lemur catta population densities are available from a number of sites (Table 2.1), and in Figure 2.3 we illustrate the relationships between lemur density and forest canopy density (FCD). As can be seen, we found an excellent curvilinear relationship between ringtailed lemur density and percent FCD ($R^2 = .91$).

In Table 2.2 (parts a and b), we give the amount of area represented by various levels of FCD over the entire geographic range in which suitable ringtailed lemur

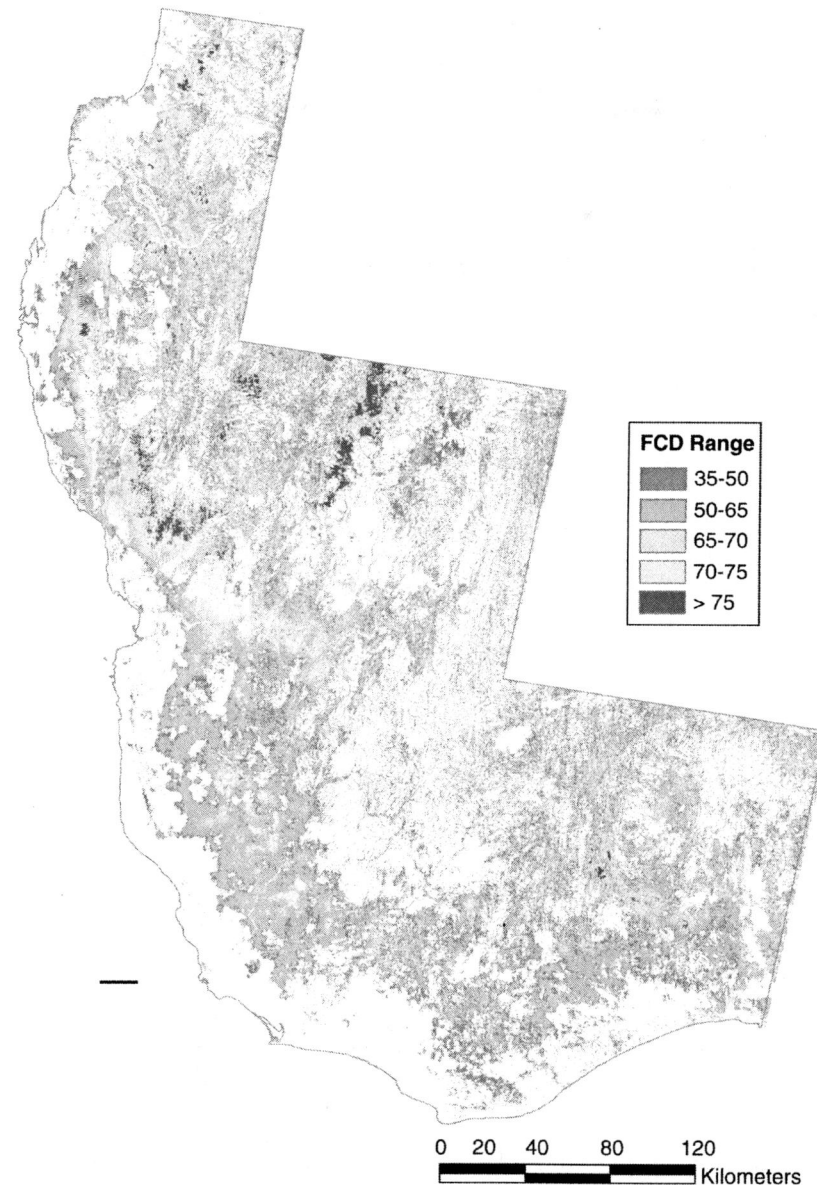

FIGURE 2.5. Stratified forest canopy density map derived from Landsat 5 Thematic Mapper images acquired January and February 1985. Pixels with forest canopy density (FCD) values less than 35 and those deforested between 1985 and 2000 assigned a value of "0." [*See* Colour Plate II]

TABLE 2.1. Known density figures and forest canopy density (FCD) for various sites.

Location	Lemur density (km^2)	FCD	References
Beza Mahafaly, Parcel 1 east	175	70.09	Sussman, 1991; Gould et al., 2003
Beza Mahafaly, Parcel 1 west	75	66.34	Sussman, 1991; Gould et al., 2003
Beza Mahafaly, Parcel 2	5	62.77	Sauther, pers. comm.
Beza Mahafaly, 1 km north	13	46.79	Sauther, pers. comm.
Beza Mahafaly, 1.5 km south	24	67.30	Sauther, pers. comm.
Beza Mahafaly, 1 km west	30	54.12	Sauther, pers. comm.
Beza Mahafaly, 1.5 km parallel to road south	42	60.87	Sauther, pers. comm.
Antserananomby	215	76.48	Sussman, 1974
Bealoka*	16	80.99	O'Conner, 1987
Berenty, Malaza west	300	83.02	Mertl-Millhollen et al., 1979; Jolly et al., 2002
Berenty, Malaza east	167	71.67	Budnitz and Dainis, 1974; Jolly et al., 2002

* Bealoka was omitted from analysis because it is a forest in which *Lemur catta* is hunted, but the forest has remained relatively intact. Therefore, the density is much lower than would be predicted for this FCD value. Also, new density figures are available form Berenty Malaza west (Jolly, this volume), but these figures are much higher than would be predicted for FCD values because some provisioning of food and water is available to these animals.

habitat is found. These areas representing each FCD range are illustrated in Figures 2.2 and 2.5. We estimate that in 1985, ringtailed lemurs occupied a total area of 27,248 km^2, representing approximately 27% of the total 100,000 km^2 area examined (Table 2, part a). The total number of ringtails at that time is estimated to have been a maximum of 933,162. By 2000, the total area occupied by the lemurs was 24,645 km^2 with a maximum of 751,251 ringtailed lemurs (Table 2, part b). Within this 15-year period, this represents a total loss of approximately 10% of suitable habitat and a 20% loss in the number of ringtailed lemurs (Table 2.3), an estimated loss of 180,000 individual lemurs. As explained in the "Methods" section, the actual numbers of ringtailed lemurs are likely overestimates because, in some forests such as Bealoka (Table 2.1), lemur densities will be lower than predicted for a particular FCD range due to hunting or other factors affecting the lemur population.

The loss of habitat varies at different ranges of FCD, and this directly relates to the relative number of ringtails affected (Table 2.2). This is related to two factors; the density of lemurs at lower FCD ranges is lower, and deforestation of lower quality forests is occurring at a much lower rate. Removal of deforested pixels between 1985 and 2000 results in an overall net loss of lemur habitat of 9.5% and a reduction in predicted lemur population of 19.5% (Table 2.3). The largest habitat extents, those at lower FCD values (35–50 and 50–65), experienced the smallest net loss, 3.9% and 9.1%, respectively, and a decrease in lemur population of 3.8% and 11.8%, respectively. Forests that are represented by the

Color Plate I

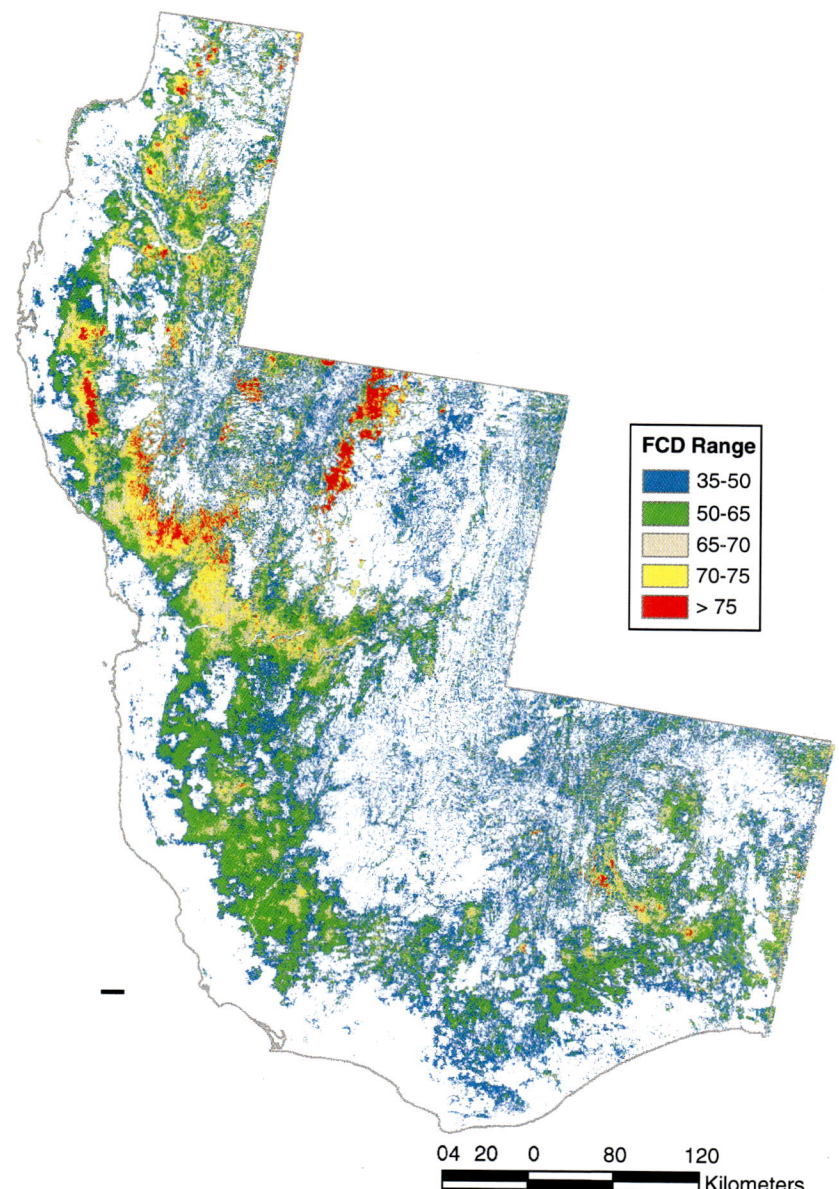

FIGURE 2.2. Stratified forest canopy density map derived from Landsat 5 Thematic Mapper images acquired January and February 1985. Pixels with forest canopy density (FCD) values less than 35 assigned a value of "0." [*See* page 21]

Color Plate II

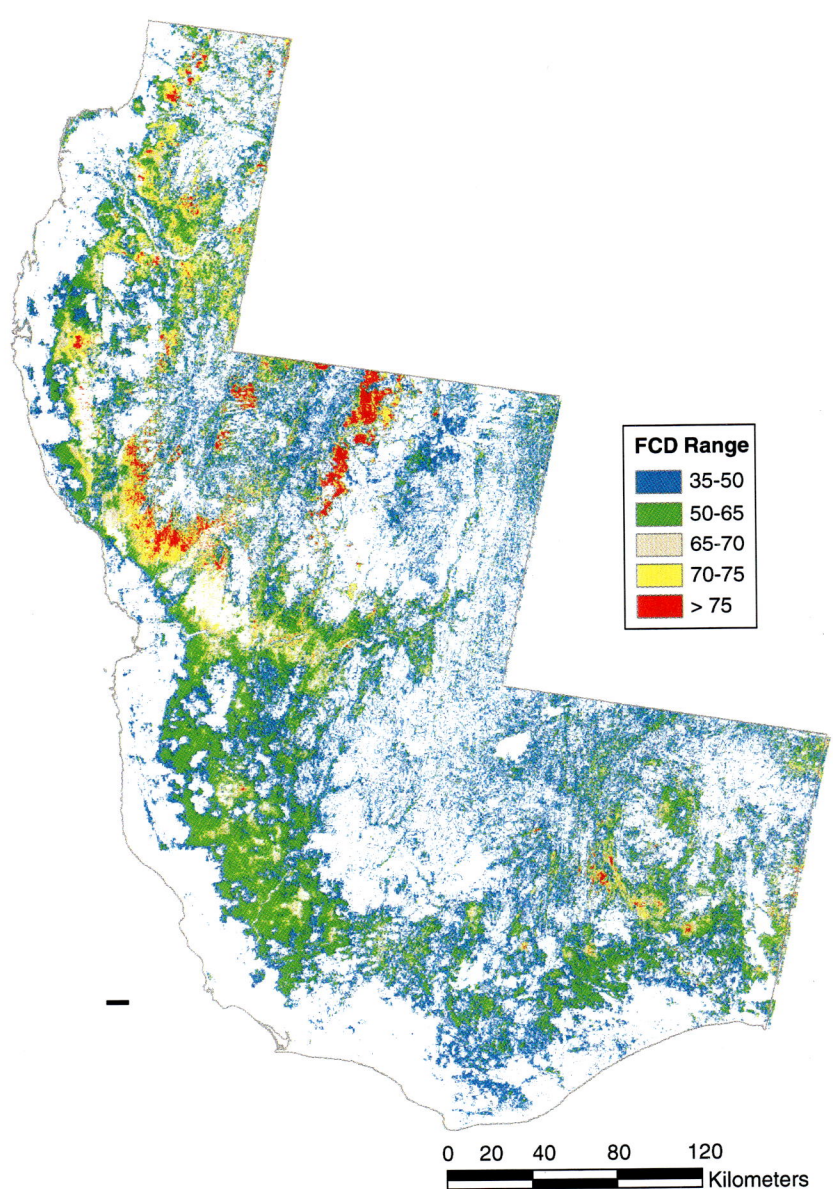

FIGURE 2.5. Stratified forest canopy density map derived from Landsat 5 Thematic Mapper images acquired January and February 1985. Pixels with forest canopy density (FCD) values less than 35 and those deforested between 1985 and 2000 assigned a value of "0." [*See* page 25]

TABLE 2.2 Area of total habitat occupied by ringtailed lemurs, percentage of total area occupied within the entire 100,000 km² research area, percentage of total ringtailed lemur habitat occupied, number of predicted lemurs, and average density of lemurs per FCD range for (a) 1985 and (b) 2000.

FCD Range	Area (km²)	% Study area	% Potential habitat	No. predicted lemurs	Avg. lemur density
(a.) 1985					
35 ≤ FCD ≤ 50	11,933	11.93	43.79	36,266	3.04
50 ≤ FCD ≤ 65	10,653	10.65	39.10	301,202	28.27
65 ≤ FCD ≤ 70	2548	2.55	9.35	238,003	93.41
70 ≤ FCD ≤ 75	1540	1.54	5.65	233,939	151.91
FCD > 75	574	0.57	2.11	123,752	215.6
Totals:	27,248	27.25	100.00	933,162	
(b.) 2000					
35 ≤ FCD ≤ 50	11,472	11.47	46.55	34,898	3.04
50 ≤ FCD ≤ 65	9682	9.68	39.29	265,593	27.43
65 ≤ FCD ≤ 70	1850	1.85	7.51	171,606	92.76
70 ≤ FCD ≤ 75	1189	1.19	4.82	181,975	153.05
FCD > 75	452	0.45	1.83	97,179	215.0
Totals:	24,645	24.65	100.00	751,251	

three highest ranges of FCD (65–70, 70–75, and >75) as well as the smallest in areas all experienced at least a 20% reduction in extent up to an amount in excess of 27%. Likewise, the lemur populations in these areas declined similarly, all reduced by values greater than 20%, as high as nearly 30%.

To illustrated this point, in 1985 we estimate that 337,468 ringtails existed in an area of 22,586 km² with FCD of lower than 65 (Table 2.2, part a), whereas by 2000, 300,400 ringtails inhabited 21,154 km² in forests within this FCD range (Table 2.2, part b). This represents a 9.4% loss in habitat and a 9% reduction in the number or ringtailed lemurs. By contrast, in 1985, 357,691 ringtails occupied 2,114 km² in forests with 70+ FCD. In 2000, 279,154 lemurs existed in 1,641 km² in forests with these canopy densities. Thus, the habitat at these higher FCD ranges was lost at a rate of more than 22%, and the rate of lemur population reduction was also 22%. Although the actual area deforested during these

TABLE 2.3. The percentage of ringtailed lemur habitat loss and population reduction between 1985 and 2000 at different forest canopy density (FCD) ranges and overall.

FCD range	Habitat loss	Population reduction
35 ≤ FCD ≤ 50	3.86	3.77
50 ≤ FCD ≤ 65	9.11	11.82
65 ≤ FCD ≤ 70	27.39	27.90
70 ≤ FCD ≤ 75	22.79	22.21
FCD > 75	21.25	21.47
35 ≤ FCD ≤ 100	9.55	19.49

15 years in regions with 70+ FCD was only 437 km^2, the reduction in the lemur population in these high-density forests is inordinately high because of the high density of ringtailed lemur populations in these forests.

Although the population density is much higher for ringtails living in areas with high FCD, the total area represented by low FCD forests within the ringtailed lemur habitat is very high. Of the total area occupied by ringtailed lemurs, more than 90% is in forests ≤70 FCD. This area was occupied by 575,471 ringtailed lemurs in 1985 and 472,097 in 2000, representing 62% of the population in 1985 and 63% in 2000. In 1985 and 2000, respectively, 35% and 40% of the ringtailed lemur population lived in areas ≤65 FCD. Almost all long-term research on ringtailed lemurs has been conducted in areas with FCD ≥70 FCD (i.e., Antserananomby, Berenty, and Beza Mahafaly Parcel 1) (Table 2.1). Thus, we know very little about the ringtailed lemur populations living in low-density forests, which represent the majority of their populated area.

2.4. Discussion

In this paper, we develop a method of measuring forest canopy density (FCD) using satellite images from 1985 and 2000. This methodology enables us to identify habitat capable of sustaining ringtailed lemur populations and the condition of that habitat. Furthermore, this parameter of forest condition is directly related to lemur density data, and we employ the function representing this relationship to predict population densities for all potential habitats.

The total habitat covered in this study is 100,000 km^2, of which we estimate approximately 27,000 km^2 was occupied by ringtailed lemurs in 1985 and 24,500 km^2 in 2000. This represents a 9.5% loss in habitat during that 15-year period. During that same period, we estimate that there were 933,162 ringtailed lemurs in 1985 and 751,251 in 2000, a loss of almost 20% of the population.

Between 1985 and 2000, there was a much higher rate of deforestation in areas with higher measures of FCD, those forests with richer and denser vegetation. Habitat loss in these areas ranged between 21% and more than 27%. Furthermore, even though these areas represent less than 5% of the total area inhabited by the ringtails, because the lemur population densities are so high, the loss in the number of the ringtailed lemurs was inordinately high in these high FCD regions, reaching 21% to 28% between 1985 and 2000.

In areas with less rich vegetation (lower FCD ranges), habitat loss was less than 10% and as low as 4% in areas with very sparse vegetation, such as extremely dry brush and scrub regions. The reduction of the ringtail population in these regions was also proportionately lower, between 4% and 9% in the regions with low FCD values (35–65). However, even if this is so, a great proportion of ringtailed lemurs inhabit these dryer forest regions, and the majority of ringtail lemurs (more than 60%) live in forests able to sustain population densities lower than those at <70 FCD values. We know next to nothing about the behavior and ecology of the ringtailed lemurs living in these types of habitats.

We realize that the numbers presented here are based on a number of broad and general assumptions and estimates. We have not related specific habitat types to particular FCD levels, and different habitats with similar FCD levels may support different densities of ringtailed lemurs, though our surveys indicate that this is not generally the case (Sussman et al., 2003). Furthermore, we have ringtailed lemur population density figures from very few research sites, and these are almost exclusively from areas with FCD values above 70. As mentioned above, many areas may exist in which forest canopy density reflects a potential carrying capacity higher than actually exists due to factors affecting the ringtailed lemur population but not the forest. The forest at Bealoka is a case in point, where the lemurs are hunted but the forest remains intact (O'Conner, 1987). We also know that forest areas closer to large villages usually contain fewer lemurs than predicted by our analysis. Finally, the size, dimensions, and topography of forest parcels and the distance between these parcels may affect ringtailed lemur densities and our estimates thereof.

In order to improve the estimates that we provide in this paper, we suggest that the following data need to be collected:

1. Lemur population densities in various habitats. In higher density areas, we need to know the densities of lemurs in small forest patches and where hunting has caused population loss. In low-density areas, where more than 60% of these lemurs live, we need basic data on home range size, group size, amount of overlap of ranges, and general population density. In fact, we need to know the basic ecology and behavior of these populations.
2. We need to collect specific forest measurements (e.g., DBH, height of vegetation, canopy diameter, branching height, canopy closure, species composition, leaf area index, etc.) in different habitat types in order to better relate satellite signatures to specific habitat types.
3. We need to be able to determine how the topography, size, and dimensions of forest patches and distance between patches and to water sources and settlements affect both our FCD measures and ringtailed lemur densities.

Given the fact that forest areas with high FCD values, mainly the gallery and continuous canopy forests, are being cut at a very high rate and that these forests sustain very high densities of ringtailed and other lemurs, as well as of other endangered flora and fauna, all efforts must be made to protect what little remains of these forests. The dryer regions of the south and southwest are not being deforested at such a rapid pace. However, ringtailed lemurs and other animals and plants that are adapted to these unique xerophytic conditions are found nowhere else on Earth and are endangered. We must learn how these species, including the ringtailed lemurs, adapt to these extremely harsh conditions. Further, we must appreciate the fact that these areas and their inhabitants are also threatened by habitat modification and destruction and need to be protected.

Acknowledgments. Field research was made possible by grants from the St. Louis Zoo Field Research for Conservation (FRC) Program and Washington University and by assistance from the School of Agronomy, University of Antananarivo. During our field work, we were assisted in the field by many personnel of the National Association for the Management of Protected Areas (ANGAP) and the World Wildlife Fund (WWF), Madagascar. Many professionals were particularly helpful to us, including B. Andriamihaja, Backoma, Christopher, M.H. Faramalala, the de Heaulme family, J.-J. Rainimiharantsoa, F. Ramiaramanana. A. Randriananrana, R. Rasary, J. Rasoarinanana, M. Sauther, N. Vogt, L. Waller, and D. Whitelaw. People in villages throughout the survey area also assisted us and provided their hospitality. We greatly appreciate all of this assistance. The initial set of satellite images used in this study were purchased with funding from the St. Louis Zoo Field Research for Conservation Program. Image selection and topographic map registration for this project were done by Sara Ivie and Mateus Batistella and were also funded through the St. Louis Zoo FRC. We appreciate the contributions of the Center for the Study of Institutions, Population, and Environmental Change (CIPEC), Indiana University, which is funded by National Science Foundation grant SBR9521918. We also appreciate the support of other CIPEC personnel. NSF, through CIPEC, also funded the purchase of the remaining satellite images used and transportation to Madagascar for Green and Vogt. Finally, we thank Alison Jolly for the initiating the symposium in Torino, Italy and this volume and for her diligence in getting the final project completed.

References

Budnitz, N., and Dainis, K. (1975). *Lemur catta*: Ecology and behavior. In: Tattersall, I., and Sussman, R. W. (eds.), *Lemur Biology*. Plenum, New York, pp. 219–236.

Cabido, M. R., and M. R. Zak. (1999). Vegetación del Norte de Córdoba. Secretaría de Agricultura, Ganadería y Recursos Renovables de Córdoba, Córdoba, Argentina.

Dirzo, R., and Sussman, R. W. (2002). Human impact and species extinction. In: Chazdon, R. L., and Whitmore, T. C. (eds.), *Foundations of Tropical Forest Biology: Classic Papers with Commentaries*. University of Chicago Press, Chicago, pp. 703–711.

Gould, L., Sussman, R. W., and Sauther, M. L. (2003). Demographic and life-history patterns in a population of ringtailed lemurs (*Lemur catta*) at Beza Mahafaly Reserve, Madagascar: A 15-year perspective. *Am. J. Phys. Anthropol.* 120:182–194.

Jamalabad, M. S., and Abkar, A. A. (2004). Forest canopy density monitoring, using satellite images. XXth ISPRS Congress, 12–23 July 2004, Istanbul, Turkey. Available at http://www.isprs.org/istanbul2004/comm7/papers/48.pdf.

Janzen, D. H. (1988). Tropical dry forests. The most endangered major tropical ecosystem. In: Wilson, E. O. (ed.), *Biodiversity*. National Academy Press, Washington, D.C., pp. 130–137.

Jolly, A. (1966). *Lemur Behavior: A Madagscar Field Study*. University of Chicago Press, Chicago.

Jolly, A., Dodson, A., Rasamimanana, H. M., Walker, J., O'Connor, S., Solberg, M., and Perel, V. (2002). Demography of *Lemur catta* at Berenty Reserve, Madagascar: Effects of troop size, habitat and rainfall. *Int. J. Primatol.* 23:325–353.

Koyama, N., Nakamichi, M., Oda, R., Miyamoto, N., and Takahata, Y. (2001). A ten-year summary of reproductive parameters for ring-tailed lemurs at Berenty, Madagascar. *Primates* 42:1–14.

Kramer, E. A. (1997). Measuring landscape changes in remnant tropical dry forests. In: Laurance. W. F., and Bierregaard, R. O., Jr. (eds.), *Tropical Forest Remnants: Ecology, Management, and Conservation of Fragmented Communities.* University of Chicago Press, Chicago, pp. 400–409.

Mertl-Millhollen, A. S., Gustafson, H. L., Budnitz, N., Dainis, K., and Jolly, A. (1979). Population and territory stability of the *Lemur catta* at Berenty, Madagascar. *Folia Primatol.* 31:106–122.

O'Connor, S. M. (1987). *The Effect of Human Impact on Vegetation and the Consequences to Primates in Two Riverine Forests, Southern Madagascar.* Ph.D. thesis, Cambridge University, Cambridge.

Rikimaru, A., and Miyatake, S. (1997). Development of forest canopy density mapping and monitoring model using indices of vegetation, bare soil, and shadow. Available at http:\www.gisdevelopment.net/aars/acrs/1997/ts5/index.shtmm.

Roy, P. S., Miyatake, S., and Rikimaru, A. (1996). Biophysical spectral response modelling approach for forest density stratification. Available at http://www.gisdevelopment.net/aars/acrs/1996/ts5/index.shtml.

Smith, A. P. (1997). Deforestation, fragmentation and reserve design in Western Madagascar. In: Laurance, W. F., and Bierregaard, R. O., Jr. (eds.), *Tropical Forest Remnants: Ecology, Management, and Conservation of Fragmented Communities.* University of Chicago Press, Chicago, pp. 415–441.

Sussman, R. W. (1974). Ecological distinction in two species of *Lemur.* In: Martin, R. D., Doyle, G., and Walker, A. C. (eds.), *Prosimian Biology.* Duckworth, London, pp. 75–108.

Sussman, R. W. (1977). Distribution of the Malagasy lemurs Part 2: *Lemur catta* and *Lemur fulvus* in southern and western Madagascar. *Ann. N. Y. Acad. Sci.* 293:170–184.

Sussman, R. W. (1991). Demography and social organization of free-ranging *Lemur catta* in the Beza Mahafaly Reserve, Madagascar. *Am. J. Phys. Anthropol.* 84:43–58.

Sussman, R. W., Green, G. M., Porton, I, Andrianasolondraibe, O. L., and Ratsirarson, J. (2003). A survey of the habitat of *Lemur catta* in southwestern and southern Madagascar. *Primate Conservation* 19:32–57.

Trejo, J., and Dirzo, R. (2000). Deforestation of seasonal dry forest: A national and local analysis in Mexico. *Biodiversity Conservation* 94:133–142.

3
Berenty Reserve: A Research Site in Southern Madagascar

ALISON JOLLY, NAOKI KOYAMA, HANTANIRINA RASAMIMANANA, HELEN CROWLEY, AND GEORGE WILLIAMS

3.1. Introduction

The forest reserves of Berenty Estate were established by the de Heaulme family in consultation with local Tandroy clans, beginning in 1936 when the de Heaulmes founded a sisal plantation beside the Mandrare River (Jolly, 2004). Some 5000 ha of spiny forest were felled, but 1000 ha remain as original forest reserves. The reserves comprise several different parcels, including a spiny forest parcel called Rapily and two large areas of gallery forest, Bealoka (100 ha) and the main Berenty Reserve (200 ha). These two gallery forest reserves were natural "islands" of extremely rich habitat formed by ancient oxbow lakes or an entire river arm. The forests are dominated by *Tamarindus indica,* the tamarind tree (Figures 3.1 and 3.2). Berenty has the semiarid climate of Madagascar's southern domain. Only along rivers with their high water tables can tamarind forest survive; elsewhere, there is the surreal succulent vegetation of Madagascar's spiny forest. (Figure 3.3).

Originally, the gallery forest was divided from the spiny forest by the steep banks of the old riverbed, easily traversed by lemurs, but with sharply different vegetation at top and bottom of the bank. Now the reserves are almost wholly isolated by sisal fields. The "islands" of gallery forest might seem too small to matter for conservation, but two overflights of the Mandrare Valley in 2004 showed that they are the only gallery forests remaining below the headwaters, except for two much smaller sacred forests near Ifotaka and a tract of tamarinds across the Mandrare River from Berenty that has little undergrowth and sparse canopy (Jolly, pers. obs.). Elsewhere there are isolated tamarind trees but no actual blocks of this forest type. Southern gallery forests are clearly one of the most threatened forest types of Madagascar (see Sussman et al., this volume).

3.2. Climate

The climate of southern Madagascar alternates hot wet summers, with temperatures above 40 °C at midday, and cold dry winters, when temperatures fall below

FIGURE 3.1. Berenty tamarind with the de Heaulme forest guards, 1963. Pencil drawing by Alison Mason Kingsbury.

10 °C at night. Rainfall varies erratically from 300 cm to 900 cm per year, if calculated in lemur-years beginning October 1, which group all of a wet season together (Figure 3.4). Conventional years, starting January 1, group the end of one wet season with the beginning of the next and so blur the degree of variation. Even lemur-years mask some of the variation, as in 1991–1992, when two thirds of the season's rain fell during a 3-day storm in January, with the drought bringing crop failure and human famine. El Niño years usually mean drought for the south of Madagascar as for southern Africa, but variation in latitude of winds may result in exceptionally wet El Niño years instead. Ringtailed lemurs, like the other plants and animals, adapt their breeding and growth to the alternation of wet and dry seasons (Figure 3.5), but their life-history strategies can only be understood in the light of recurrent catastrophic years (Gould et al., 1999; Wright, 1999; Richard et al., 2002; Jolly, this volume).

3.3. Berenty Habitat Zones

The main Berenty Reserve contains about 200 ha of gallery and scrub forest connected on the west to a corridor of spiny forest and on the east by a very narrow

FIGURE 3.2. The same tree in 2005. The lowest branch to the left has broken short, and small branches to the right have broken off, but the tree continues to flourish. Diameter at breast height in 2002 was 127 cm. Photo: A. Jolly.

interface to the 150-ha Akesson/Kaleta forest. Its four habitat zones embrace a fivefold difference in ringtailed lemur population density (Figures 3.6 and 3.7).

Starting from the north, the 40-ha lobe called Ankoba had been largely cleared for Tandroy local farms at the time the de Heaulmes settled at Berenty. The de Heaulmes attempted to grow crops in this zone. Old drainage ditches traverse the forest floor. However, after some years they gave up farming there, having found it was not suitable as sisal nursery fields. The original tamarind trees (*Tamarindus indica*) remained because neither Tandroy nor French cut down tamarinds. The de Heaulmes also planted an alley of *Pithecellobium dulce,* the monkey pod tree, "foreigner's tamarind" or *kilimbazaha*. These leguminous trees are excellent food for sifaka and lemurs, with protein-rich flowers and pods. They also function as nurse trees that shelter wild seedlings planted by feeding lemurs, unlike the tamarinds that inhibit seedlings (Blumenfeld-Jones, this volume). Ankoba is now a mature second-growth forest 50–60 years old, with canopy at 10–15 m and some emergent acacias to more than 20 m. Ringtail density is around 500/km^2; sifaka and brown lemurs are also extremely dense in Ankoba. Troops based

FIGURE 3.3. *Alluaudia procera* spires rise above other succulent and thorny plants in a spiny forest reserve of Berenty Estate. Oxcarts fetch water to provision villages many kilometers from the river. Photo: A. Jolly.

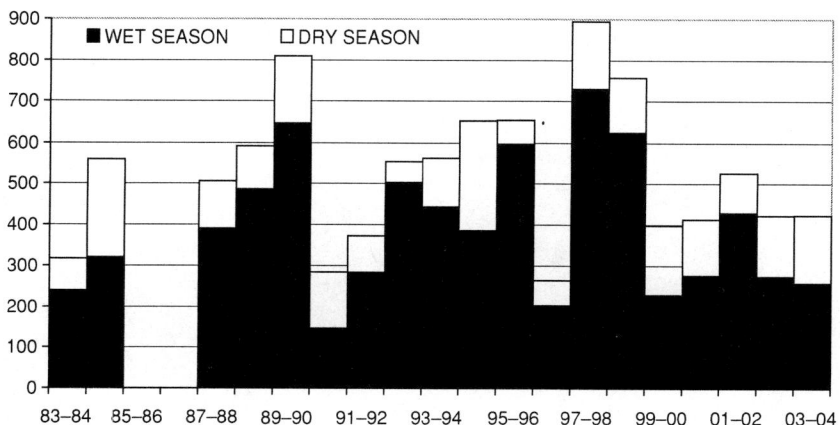

FIGURE 3.4. Rainfall at Berenty in millimeters for "Lemur-years." These begin October 1 as infants are born, showing the October–March wet season and succeeding April–September dry season. "Drought years" are those with wet season rainfall under 450 mm: that is, 1983–1984, 1984–1985, 1990–1991, 1991–1992, 1996–1997, and four of the five years since October 1999. Data for 1983–1985 courtesy S.M. O'Connor; data for 1987–2004 courtesy C. Rakotomalala.

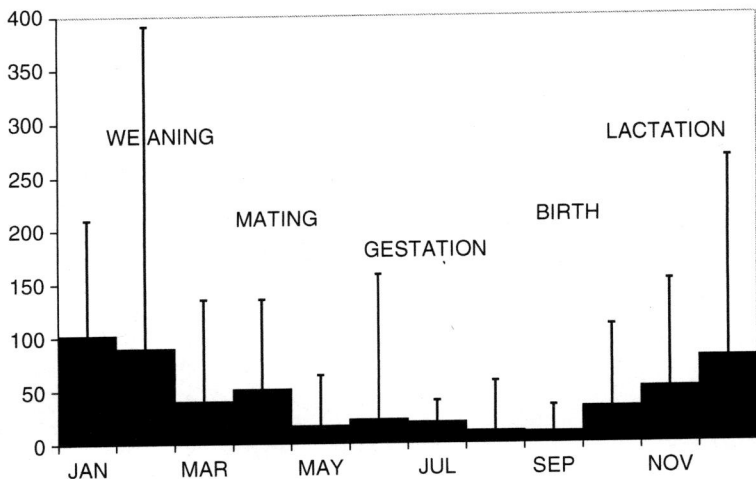

FIGURE 3.5. Mean monthly rainfall at Berenty in millimeters, with *Lemur catta* breeding seasonality. T-bars show maximum. Minimum in December and February is 5 mm, in other months 0 mm. Data for 1983–1985 courtesy S.M. O'Connor; data for 1987–2004 courtesy C. Rakotomalala.

within the forest range outward to the planted vegetable and fruit garden to the north and to open areas with introduced trees including neem (*Azadirachta indica*) and kantsa-kantsa (*Leucaena leucocephala*).

South of Ankoba lies the Malaza section of the forest. Malaza is the 100-ha area first chosen by Blumenfeld-Jones and Budnitz as their main study area (Budnitz and Dainis, 1975; Budnitz, 1978; Mertl-Millhollen et al., 1979; Blumenfeld-Jones, this volume) and studied by the Jolly team since 1989. Jolly's original study troop of 1963–1964 ranged within Malaza (Jolly, 1966). For the purposes of analyzing lemur demography and behavior, the 100 ha of Malaza are subdivided into four habitat zones: front, gallery, scrub, and spiny forest.

The "front" is a part of Malaza's western edge. This area has been inhabited since the 1940s: the modern tourist buildings derive from the original sisal factory. The lemurs of the front sleep and usually siesta in the gallery forest, spending the first and last hours of activity feeding on native trees. They then range for most of the day among planted trees and original tamarinds between the houses. They feed on neem (*Azadirachta indica*), *Cordia rothii*, Persian lilac (*Melia azedarah*), *Cassia* spp., *Eucalyptus* flowers, *Bougainvilea* buds, sisal flowers, and until 2005 *Leucaena* (Simmen et al., this volume; Soma, this volume; Crawford et al., this volume on the deleterious effects of leucaena). *Leucaena* is being removed from the front zone in 2005, though it will be left for study at the north end of Ankoba. Water is always available here. Since the growth of tourism in the 1980s and especially the 1990s, there is some garbage and offered food. Deliberate banana feeding increased from the 1980s until 1999, when it was

FIGURE 3.6. Air photo of Berenty, taken from the north. Tracks of the old river arm that embraces the reserve can be seen leading through the sisal toward the south (top) of the picture. The Ankoba lobe is a 50-year-old secondary forest. The main 100-ha Malaza lobe shades from the front with its introduced trees and tourists, through rich gallery forest and drier scrub. Above the ancient river bank lies a corridor of spiny forest. A cattle drove divides Malaza from the unstudied areas of Analamalangy and the Akesson-Kaleta Reserve. Courtesy of Barry Ferguson and the Centre Ecologique de Libanona, Taolagnaro.

banned. The ban is mostly successful, but lemurs still enter buses and bungalows to raid fruit the tourists planned on eating themselves.

East of the tourist front is lush gallery forest along the Mandrare River, dominated by tamarinds (*Tamarindus indica*), called *kily*, with emergent acacias (*Acacia rovumae*), called *benono* ("the many-nippled," from the thorn bases on their trunks.). Tamarind forest grades into open canopy forest, largely dominated by *Neotina isoneura*, which in turn grades into transitional brush and scrub as one moves south from the river. (For detailed vegetation maps, see Blumenfeld-Jones, this volume.) For the purposes of assigning lemur troops, we distinguish gallery from scrub, with the criterion that if ≥50% of the sky is covered, it is gallery. If more than 50% of the sky is open, it is scrub. The scrub zone still contains isolated tamarinds but with a variety of lower, thorny, and succulent species, including *Salvadorea angustifolia,* and various euphorbs, including the invasive smothering vine *Cissus quadrangularis.*

The transition from scrub to spiny forest is dramatically sharp. The reserve is bounded on the west by the ancient bank of an old river arm, about a 7-m nearly

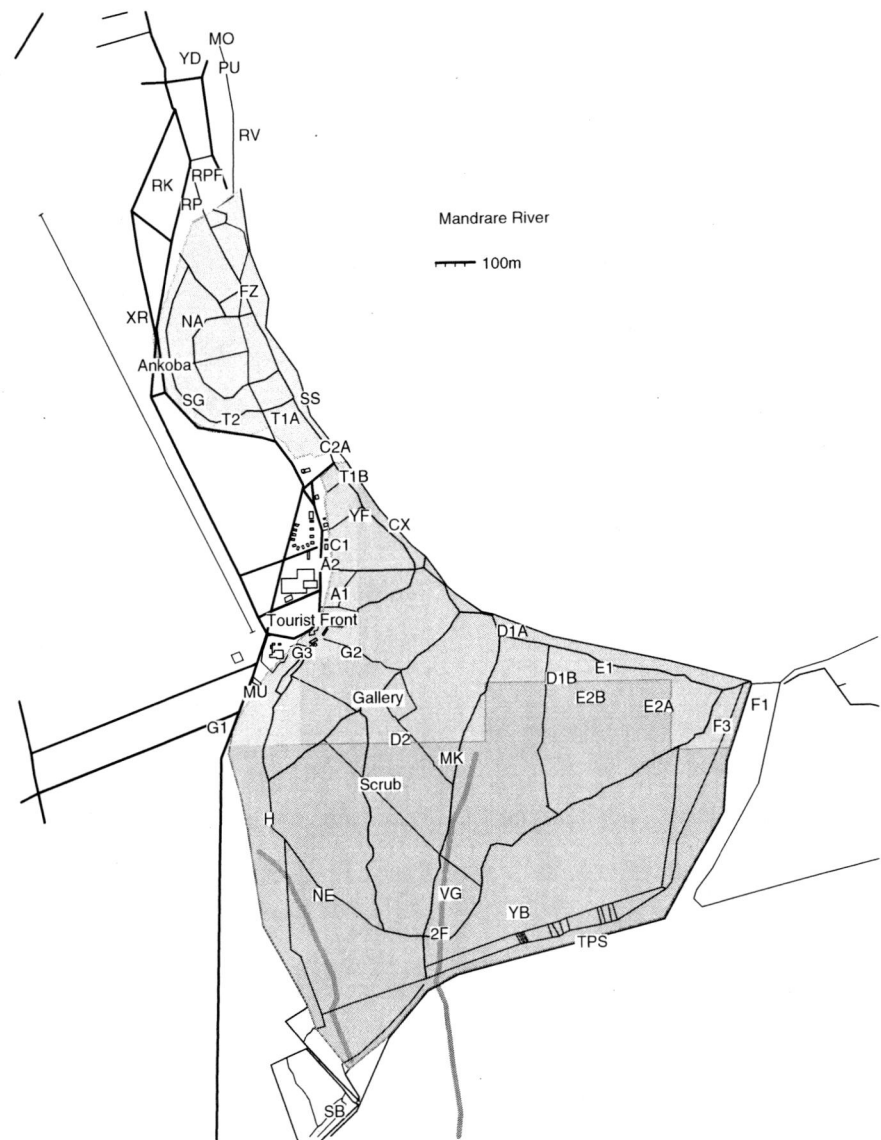

FIGURE 3.7. Ankoba and Malaza lobes of Berenty Reserve, showing habitat categories and the very close spacing of *Lemur catta* troops. Two-letter abbreviations are approximately at the center of the troop ranges. Ankoba from 2004 complete census, Malaza (Front, Gallery, and Scrub) from the 2000 complete census. Map prepared by G. Williams.

vertical scramble. Below lies scrub with some large tamarinds rooted in the water table of the old riverbed (Mertl-Millhollen, this volume; Blumenfeld Jones, this volume). At the top of the bank is thicket dominated by *Alluaudia procera* (fantiolotse) and other xerophytes.

Malaza's western boundary is a cattle drove to the river, with no undergrowth but a nearly continuous canopy of acacias. Beyond lies the approximately 60-ha parcel called Analamalangy. Again, this has gallery forest next to the river and scrub behind. However, a large part of the central section is covered by *Cissus quadrangularis*. This parcel has not been fully censused although ringtails, sifaka, and browns all live there.

A second cattle drove marks the boundary to the 150-ha Akesson-Kaleta reserve. This resembles Berenty, but with much more open undergrowth, because it was more recently subjected to cattle grazing.

None of the boundaries that humans assign to these zones are a barrier to lemurs. However, it is possible to distinguish troops whose core feeding areas fall in different habitats. Troops have quite regular day ranges, at least within a single season. We choose the daytime range as their habitat. "Front troops" range outside of the forest to the west for most of the day but sleep in adjacent closed-canopy gallery forest. "Gallery troops" spend the day near the river but may sleep further south in tamarind trees among the scrub. "Scrub" troops feed in scrub and spiny forest and also sleep there. "Ankoba troops" range both within Ankoba forest and in the human occupied land to west and north.

Any troop occasionally may make a 1- or even 2-km excursion out of its normal day range. Also, if key resources are only located outside its normal range in a given year or season, it may travel there (see Mertl-Millhollen, this volume). In areas as small as Ankoba or Malaza, the lemurs seem to be well aware of feeding locations and travel routes throughout the forest lobe. What keeps them bounded is the pressure of other groups (Pride, this volume).

3.4. Fauna

There are six species of lemurs at Berenty: *Propithecus verreauxi*, the white sifaka; *Lemur catta*, the ringtailed lemur; *Eulemur fulvus rufus x collaris*, hybrid brown lemurs (introduced in 1975); *Lepilemur leucopus*, the white-footed lepilemur; *Microcebus murinus*, the gray mouse lemur; and the newly identified *Microcebus griseorufus*, the gray-and-red mouse lemur (Rasoloarison et al., 2000). All of these live in all habitat zones except that the brown lemurs do not enter the spiny forest (so far), and the gray-and-red mouse lemur seems to be confined to spiny forest.

Berenty holds southern Madagascar's largest colony of the Madagascar giant fruit bat, *Pteropus rufus*, one of whose staple foods is sisal flowers from the surrounding fields (Long, 2002). Other mammals include *Setifer setosus*, the spiny tenrec; *Tenrec ecaudatus*, the large tenrec; *Microgale* sp., the shrew-like tenrec; *Eliurus myoxinus*, the Madagascar tree-rat; and *Viverricula indica*, the Indian civet. There are many *Rattus rattus*, the scourge of Malagasy small mammals (Crowley, 1995; Goodman, 1995). The forest is too small to hold *Cryptoprocta ferox*, the fossa, but domestic dogs and cats take its place as significant lemur predators.

Fifty-two species of resident birds have been recorded out of a total of 99 species seen, of which 41% are endemic to Madagascar (list compiled by M. Pigeon, in Goodman et al., 1997). The giant ground couas (*Coua gigas*) are particularly noticeable, as they are extensively trapped elsewhere. The two male color morphs of the Madagascar paradise flycatcher have been intensively studied in the Bealoka parcel of gallery forest to determine how they maintain their genetic polymorphism (Mulder et al., 2002). Lemur predators include *Polyboroides radiatus*, the harrier hawk; *Buteo madagascariensis*, the buzzard; and the endemic black kite, *Milvus migrans*.

Reptiles have been little studied at Berenty, but it is one of only two known localities for the burrowing snake *Pseudoxyrhophus kely*, the other one being at Mandena on the east coast in a wholly different type of forest. Tortoises and turtles include *Geochelone radiata*, the Madagascar radiated tortoise; *Pyxis arachnoides*, the spider tortoise; and the terrapin *Pelomedusa subrufa*. All three are widely distributed in the south, although the radiated tortoise is heavily trapped for illegal export. Two hundred confiscated radiated tortoises have been released by the Water and Forest Department in the Rapily spiny forest reserve parcel of Berenty Estate (Crowley, 1995).

No exhaustive plant list exists, although it is under active study. Simmen and colleagues have drawn up a list of plants eaten by *Lemur catta* (this volume), and Blumenfeld-Jones (this volume) describes the changing structure of Berenty gallery forest. There is the continued problem of invasive plants, especially the euphorb liana *Cissus quadrangularis*, and the toxic tree *Leucaena leucocephala* (Crawford, this volume).

3.5. Ringtailed Lemur Studies at Berenty

The first scientist to study in Berenty forest was Alison Jolly, in 1963–1964. The de Heaulme family welcomed her and inaugurated their tradition of hospitality to visiting scientists. There are now seven places reserved for students in the complex called "Naturaliste."

Peter Klopfer of Duke University visited in 1969. He then supervised a group of Duke PhDs at Berenty the early 1970s: Robert Sussman, Norman Budnitz, Kathryn Dainis (Blumenfeld-Jones), Lee McGeorge (Durrell), Jay Russell, and Anne Mertl (Millhollen). At the same time, Alison Richard, Pierre Charles-Dominique, and Marcel Hladik did shorter studies at Berenty in 1970 while concentrating on other sites. Jean-Jacques and Arlette Petter also visited.

With the political difficulties for foreigners in the late 1970s research slackened, beginning again in 1983 with a detailed ecological comparison of Berenty and Bealoka by Sheila O'Connor and a study of birds by Mark Pigeon. Jolly began studies with Earthwatch, notably including the first visit of Hantanirina Rasamimanana.

The recent era of intensive lemur work started in 1989 with the arrival of Naoki Koyama, of Kyoto University, and of Jolly and Rasamimanana. Since then, both the Japanese and the Anglo-Malagasy group have been present at Berenty for every birth season. PhDs and PhD candidates include Chiemi Saito, Ryo Oda, Shinichiro Ichino, Takayo Soma, Lys Rakototiana, and Ethan Pride. About 40 undergraduate and master's level students, almost half of them Malagasy from the Ecole Normale Supérieur, have done 2–6 month field projects at Berenty. The non-Malagasy include Japanese, French, Italians, British, Canadians, and Americans. Earthwatch has sent 120 short-term Earthwatch volunteers, and a few particularly dedicated volunteers have returned for several years of more intensive fieldwork.

George Williams has drawn up the base map of Berenty Reserve that comes with a program (MAP) for analyzing lemur day ranges, home ranges, and other desired data. Since 2000, Kathryn Blumenfeld-Jones and Anne Mertl-Millhollen have also returned to their 1970's study site.

In 1992–1994, Helen Crowley became the first manager of Berenty Reserve, funded by the Wildlife Trust and by the de Heaulmes. Her management plan remains the chief attempt to survey research done and to draw up recommendations for the future. Berenty as a small island of forest is threatened by edge effects, invasive species including the brown lemurs (Pinkus, this volume), lowered water table from many causes, and from the natural succession of tamarind trees as well as from the changing human context. Berenty Reserve will need continued active management in the future.

If someone had done a population and habitat viability study of Berenty 70 years ago when the reserve was founded, it would have seemed extremely unlikely that such a tiny fragment could survive the biological and political changes of the coming decades. However, in spite of the dilemmas posed by maintaining this small forest, there now seems every hope that lemur research—and the lemurs themselves—will continue to flourish at Berenty.

References

Budnitz, N., and Dainis, K. (1975). *Lemur catta:* Ecology and behavior. In: Tattersall, I., and Sussman, R. W. (eds.), *Lemur Biology*. Plenum, New York, pp. 219–236.

Budnitz, N. (1978). Feeding behavior of *Lemur catta* in different habitats. In: Bateson, P., and Klopfer, K. (eds.), *Perspectives in Ethology*, Vol. 3. Plenum Press, New York, pp. 85-108.

Crowley, H. M. (1995). *Berenty Reserve Management Plan*. Wildlife Preservation Trust International, Philadelphia.

Goodman, S. M. (1995). *Rattus* on Madagascar and the Dilemma of Protecting the Endemic Rodent Fauna. *Conservation Biology* 9(2):450–453.

Goodman, S. M., Pidgeon, M., Hawkins, A. F. A., and Schulenberg, T. S. (1997). The birds of southern Madagascar. *Fieldiana, Zoology* 187:1–132.

Gould, L., Sussman, R. W., and Sauther, M. L. (1999). Natural disasters and primate populations: The effects of a two-year-drought on a naturally occurring population of ringtailed lemurs *(Lemur catta)* in southwestern Madagascar. *Int. J. Primatol.* 20:69–85.

Jolly, A. (1966). *Lemur Behavior.* University of Chicago Press, Chicago.

Jolly, A. (2004). *Lords and Lemurs: Mad Scientists, Kings with Spears, and the Survival of Diversity in Madagascar.* Houghton Mifflin, Boston.

Long, E. (2002). *The Feeding Ecology of Pteropus rufus in a Remnant Gallery Forest Surrounded by Sisal Plantations in South-East Madagascar.* Unpublished PhD, University of Aberdeen, Aberdeen, UK.

Mertl-Millhollen, A. S., Gustafson, H. L., Budnitz, N., Dainis, K., and Jolly, A. (1979). Population and territory stability of the *Lemur catta* at Berenty, Madagascar. *Folia Primatol.* 31:106–122.

Mulder, R. A., Ramiarison, R., and Emahalala, R. E. (2002). Ontogeny of male plumage dichromatism in Madagascar paradise flycatchers. *J. Avian Biol.* 33:342–348.

Rasoloarison, R. M., Goodman, S. M., and Ganzhorn, J. U. (2000). Taxonomic revision of mouse lemurs *(Microcebus)* in the western portions of Madagascar. *Int. J. Primatol.* 21:963–1020.

Richard, A. F., Dewar, R. E., Schwartz, M., and Ratsirarson, J. (2002). Life in the slow lane? Demography and life histories of male and female sifaka (*Propithecus verreauxi verreauxi*). *J. Zool.* (London) 256:421–436.

Wright, P. C. (1999). Lemur traits and Madagascar Ecology: Coping with an island environment. *Yrb. Phys. Anthropol.* 42:31–72.

4
Beza Mahafaly Special Reserve: A Research Site in Southwestern Madagascar

ROBERT W. SUSSMAN AND JOELISOA RATSIRARSON

4.1. Lemur Studies at Beza Mahafaly

In the mid-1970s, R.W. Sussman received a late-evening phone call. "What can we do to save Madagascar's wildlife!?" the caller exclaimed. The caller was Edward (Ted) Steele, a member of the board of directors of Defenders of Wildlife (DOW), a conservation organization based in Washington, D.C. Mr. Steele had recently returned from Madagascar and had fallen in love with its animals, plants, and people (Steele, 1975). It just so happened that Sussman along with Alison Richard, then at Yale University, and Guy Ramanantsoa, then at the School of Agronomy, University of Madagascar, had been discussing the possibility of establishing a unique type of reserve somewhere in Madagascar—a reserve that would protect the flora and fauna, be used as a teaching and education center, and that would be accepted, integrated, and user friendly and provide developmental assistance to the neighboring local inhabitants. Sussman explained our vision to Mr. Steele, and he set up a meeting of board members of DOW with A.F. Richard and R.W. Sussman. The board was impressed with the idea; however, DOW did most of its work within the United States and was mostly involved in litigation. Therefore, one of the board members, Dr. Richard Pough, who was also a member of the board of directors of World Wildlife Fund, volunteered to present our ideas to WWF, and that organization agreed to fund the project. At that point, Professor Ramanantsoa began to survey areas in western and southern Madagascar in an attempt to find an undisturbed area with a diversity of flora and fauna that was relatively accessible. Another criteria was that the local inhabitants would be agreeable to and would actively participate in the project.

At Beza Mahafaly, Guy Ramanantsoa found a beautiful region that represented the dry forest habitats of southern Madagascar and local inhabitants who were conscious of the necessity to preserve this unique natural habitat. In July 1978, the Popular Consul of the local government of Beavoha (Commune de Beavoha) agreed to officially cede two noncontiguous parcels of forest to the Department of Water and Forests of the School of Agronomic Sciences (Ecole Supérieure des Sciences Agronomiques, Département des Eaux et Forêts; ESSA/Forêts), University of Madagascar (now University of Antananarivo). Thus, ESSA/Forêts began collaborative work with Yale University, Washington University, and with local, national, and international partners to establish the Beza Mahafaly Reserve. Research, education, and development projects were begun soon thereafter. On June 4, 1986, Beza Mahafaly was officially inaugurated as a Réserve Spéciale (Special Reserve) by official government decree No. 86-168. In 1989, WWF took over management of the reserve, and in November 1995, ESSA/Forêts officially became the principal operator and administrator of the Beza Mahafaly Project, with WWF continuing to be a major supporter through the refunding of debt program for the conservation of nature. Research, training and education programs, and local development projects have continued to flourish since ESSA/Forêts took over administration of the reserve. The site has hosted a multidisciplinary field course for fifth-year ESSA/Forêts students since 1986 (Ratsirarson, 2003). Additional support has been received from the Liz Claiborne and Art Ortenberg Foundation in collaboration with Yale University to involve the local community in research programs. Recently, in 2004, the management of Beza Mahafaly Reserve, like all the protected areas in Madagascar, has been transferred to ANGAP (National Agency to manage the network of protected areas in Madagascar), and the University of Antananarivo through ESSA/Forêts remains the main partner of ANGAP for research and training activities.

The Special Reserve of Beza Mahafaly is located 35 km to the northeast of Betioky-Sud, at 23 °41′60″ latitude south, and 44 °32′20″ and 44 °34′20″ longitude east (Figure 4.1). The reserve is made up of two noncontiguous parcels separated by 10 km. The first parcel (parcel no. 1) is characterized by a gallery forest dominated by *Tamarindus indica*. It covers an area of 80 ha of fenced and protected forest but is contiguous with a relatively small area (possibly another 200 ha) of unprotected gallery forest. This forest is located on the banks of the Sakamena River, a tributary of the Onilahy River, which is approximately 8–10 km north of the reserve. Southern Madagascar is characterized by a long dry season (<40 mm of rain/month) and a short wet season (>50 mm rain/month), although the amount of rain can vary throughout the year. Annual rainfall in the region of the reserve is about 750 mm of which 600 mm falls during the austral summer, November–March. The Sakamena River is normally dry during the long dry season. The wet season is also characterized by high ambient temperatures, averaging around 34 °C and reaching highs of 48 °C. Temperatures during the coolest months (July–August) usually range between 23 °C and 30 °C, but can fall to 3 °C at night. Annual temperatures average 25 °C (Sussman and Rakotozafy, 1994; Ratsirarson et al., 2001, Ratsirarson, 2003).

The gallery forest (parcel no. 1) is divided by marked transects whose paths intersect to form squares of 100 m × 100 m. This parcel was completely enclosed by barbed wire fence in 1979. Before this, it was exposed to cattle and goats and used by the local people for various resources, as is the surrounding forest currently. This gallery forest lies on flat terrain at an altitude of 100–200 m. Fenced parcel no. 1 is surrounded by similar but unprotected and somewhat degraded gallery forest on the north and south. To the east of the parcel is the Sakamena River. To the west is contiguous dry forest. The parcel is also bounded on the

4. Beza Mahafaly Special Reserve   45

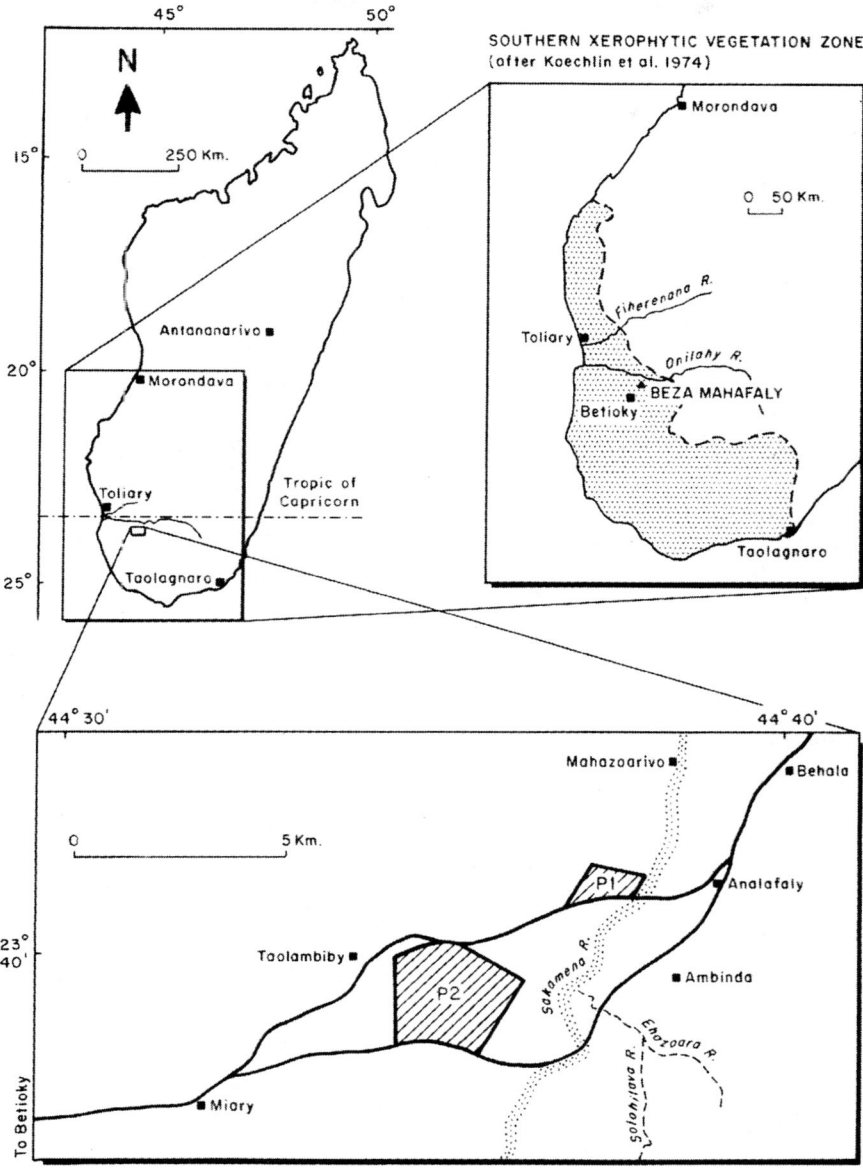

FIGURE 4.1. Location of the Beza Mahafale Special Reserve (P1, parcel no. 1; P2, parcel no. 2).

south by the dirt road that runs from Betioky to the reserve and on to the next small village of Analafaly about 2 km east. The reserve campsite and reception center is just south of the road adjacent to parcel no. 1. There is one large and another small wooden house, a museum, an office building, and a large open gazebo for courses and meetings. There is also open space for camping.

The forest represented in parcel no. 1 may be classified as western Malagasy dry deciduous forest (White, 1983). It has an average of 369 individual trees of ≥2.5 cm Diameter at Breast Height (DBH)/1000 m^2 (Sussman and Rokotozafy, 1994), which is typical of dry forests in continental Africa and the Neotropics (Gentry, 1993). In this parcel, vegetation varies according to the depth and moisture content of the soil. On drier soils away from the river there are fewer tall trees, but vegetation becomes denser. Distinctions between the canopy strata are obscured, and forest gradually passes into thicket. On more moist soils closer to the river, large *Tamarindus indica* trees are dominant. The proportion of trees over 10 cm DBH is similar in both microhabitats, and it is only in trees above 25 cm DBH that a distinction is seen between wet and dry soils.

On wet soils, the upper strata forms a closed canopy, mostly uniform in height (15–20 m). Members of the upper stratum are species whose trunks generally exceed 25 cm DBH and may attain 50 cm or more, especially on wet soils. The most common canopy species are *Tamarindus indica, Acacia rovumae, Euphorbia tirucalli*, and *Salvadorea angustifolia*. Other species of large trees include *Commiphora* spp., *Gyrocarpus americanus, Terminalia* spp., *Quivisianthe papinae*, and *Acacia bellula*. Most trees in the forest are small, constituting a middle stratum from about 2 to 15 m tall. The most common of these are *Azima tetracantha, Crateva excelsa, Gardenia* spp., *Gelonium adenophorum, Grewia* spp., *Rhigozum madagascariensis, Rhopalocarpus lucidus, Stereospermum variablile*, and *Tarenna pruinosum*. Only two species of tree are common throughout the forest: *Tamarindus indica* and *Azima tetracantha*. In general, those species found in both microhabitats are not equally distributed, and five of the most common species are found mainly on wet soils and eight mainly on dry soils.

In 25 identified transects, 25 plant families were represented with plants with Tiliaceae having the most species (15), followed by Burseraceae (7), Leguminoseae (7), and Euphorbiaceae (6). At least two families, Sphaeprosepalaceae and Didieriaceae, are endemic. Seventy-eight percent of 69 woody species ≥2.5 cm DBH in these transects were found to be native, and 26% of the 43 genera also were native (Sussman and Rokotozafy, 1994). Ratsirarson et al. (2001) found that, overall, this parcel contained approximately 120 species and 49 families of plants. However, half of the families were represented by a sole species.

The second parcel (parcel no. 2) is xerophytic, desert-like forest dominated by species adapted for the long dry season. This parcel is often referred to as spiny forest and is dominated by *Alluaudia procera* of the endemic family Didieriaceae. The second most common species is *Cedrelopsis grevei*. Other families represented are Burseraceae, Ptaeroxylaceae, Tiliaceae, Euphorbiaceae, and Combretaceae. The medium height of trees is 4.5 m, with a medium diameter of 6.5 cm (Ratsirarson et al., 2001). Parcel no. 2 is approximately 520 ha in size and is located southwest of parcel no. 1. This forest has been the subject of fewer studies than the gallery forest.

In a study of the phenology of Beza Mahafaly, Ratsirarson et al. (2001) found that most species lose their leaves during the long dry season, April to November. Most species produce fruit annually but there might be a massive production every other year in some species (such as *Azima tetracantha* and *Salvadorea angustifolia*). The flowering season is normally between October and February with a peak in December (Ratsirarson and Silander, 2003). In general, flowering corresponds with the rainy season, but flower buds for most species begin to appear at the end of the dry season when the plants no longer have leaves. *Tamarindus indica,* generally, flower for 8 months of the year (November–June), but flowers are often present on some individual trees throughout the year. During the driest and hottest portion of the year, between June and September, leaves, flowers, and fruit are all rare (Ratsirarson et al., 2001).

The region between the two noncontiguous parcels of the reserve is represented by transitional vegetation between the gallery and the xerophytic habitat, dominated by smaller trees (such as *Grewia* spp.) and shrub. This vegetation is more or less degraded because of intensive utilization as grazing land and for the collection of various forest products for food, medicines, building, and so forth. As stated above, the gallery forests surrounding parcel no. 1 are also degraded.

4.2. Fauna

There are four species of lemurs at Beza Mahafaly. The diurnal species are *Propithecus verreauxi,* the Verreaux's sifaka; and *Lemur catta,* the ringtailed lemur. The nocturnal species are *Lepilemur leucopus,* the white-footed lepilemur; and *Microcebus griseorufus,* the gray-and-red mouselemur. *Cheirogaleus medius,* the fat-tailed dwarf lemur, has been recorded within a kilometer of the reserve. *Microcebus murinus,* the gray mouselemur, is not found in the reserve (Godfrey and Rasoazanabary; pers. comm. to M. Sauther).

There are four species of bats found in the reserve: *Pteropus rufus, Hipposideros commersoni, Tadarida jugularis,* and *Taphozous mauritianus. Preropus rufus,* the giant fruit bat, is rare at Beza Mahafaly and does not have a colony within the reserve. The other three species are Microchiroptera. They are small, insectivorous bats. Another small, insectivorous bat, *Mormopterus fugularis,* has been identified in the pellets of the long-eared owl (*Asio madagascariensis).* Other small mammals include the tenrecs: *Echinops telfairi,* the pseudohedgehog; *Geogale aurita,* the large-eared tenrec; *Setifer setosus,* the spiny tenrec; and *Tenrec ecaudatus,* the large tenrec. The highest known density of the large-eared tenrec is at Beza Mahafaly (Stephenson, 2003). Rodents include two introduced species, the black rat, *Rattus rattus,* and the mouse, *Mus musculus.* Both of these species reproduce prolifically and are considered serious pests (Youssouf, 2004). The endemic Madagascar tree-rat, *Eliurus myoxinus,* is in the reserve but is rare.

Three species of carnivore exist at Beza Mahafaly. Two of these are introduced: *Viverricula indica* and the free-ranging, wild domestic-like cat, *Felis* spp. The latter species has larger and more pronounced ears than the domestic cat and shows some genetic differences (Goodman, pers. comm. to Ratsirarson et al., 2001).

The endemic fossa, *Cryptoprocta ferox,* is found here but is rare. The nocturnal, wild boar, *Potamochoerus larvatus,* lived in the reserve and was hunted for meat by the local people (Ratsirarson, 2003), but has not been observed in the reserve for many years (Sauther pers. comm.).

One hundred two species of birds representing 43 families have been observed at Beza Mahafaly. More than half of the families are represented by only one species. Of the 102 species, 27 are endemic and approximately 40 species are seen year-round in both parcels of the reserve (Ratsirarson et al., 2001). Besides being the home for many of these birds, Beza Mahafaly is a resting or breeding stop for many migrating birds. The Madagascar blue pigeon, *Alectroenas madagascariensis,* was observed in the reserve once in 1998.

The Beza Mahafaly Reserve is home to at least 15 species of snakes, 18 species of lizards, 2 species of tortoises and 1 freshwater turtle, and 1 species of crocodile (Ratsirarson et al., 2001). Three species of amphibian also are found here. Among the snakes, 13 species are of the family Colubridae. The families Boidae and Typlopidae are monospecific. Five families of reptiles are represented: Chamaeleonidae (12 species), Gekkonidae (7 species), Iguanidae (3 species), Cordylidae (12 species), and Scincidae (4 species) (Ratsirarson 2003). One species of Scincidae, *Amphiglossus splendidus,* was recently discovered in this region. It is a rare, semiaquatic species that was previously only known from the Fort Dauphin area. The Iguanidae, *Oplurus fierinensis,* is unique to the southwest of Madagascar. The Gekkonidae is very diverse with six genera represented by seven species. The two species of tortoise found in the reserve are *Geochelone radiata* and *Pelusios subniger,* and the freshwater turtle is *Erymnochelys madagascariensis* (Brockman, pers. comm.). The crocodile, *Crocodylus niloticus,* is seasonal along the Sakamena River. Amphibians are represented by three species from two families: Mantellidae (*Mantella* spp.) and Ranidae (*Ptychadena mascareniensis).* Little is known of these amphibians.

There also is a notable diversity of insects at Beza Mahafaly. This includes 105 species of lepidopterans from 16 families, 46 species of beetles from 17 families, and 28 species of hymenopterans from 9 families (Ratsirarson, 2003).

4.3. Lemur Studies at Beza Mahafaly

As part of the process of establishing the Beza Mahafaly Reserve, a student of ESSA, Randrianaivo Raymond, completed an inventory of the lemurs in the region. This became his Memoire de Fin D'Etude for the University of Madagascar (Randrianaivo, 1979). Shortly after the reserve was established, Ratsirarson Joelisoa conducted the first study of lemurs in the reserve-proper. He compared the ecology and behavior of ringtailed lemurs in the two different habitats and was the first to capture and collar lemurs in the reserve (Ratsirarson, 1987). Since that time, many Malagasy students have taken courses at the reserve, and many Memoire de Fin D'Etudes for Malagasy students mainly from ESSA/Forêts have been completed from research on lemurs as well as on other fauna and flora at the reserve. This research also includes socioeconomic studies of people neighboring the reserve (see a partial list of references in Ratsirarson et al., 2001).

As part of the accord between the three universities, Antananarivo, Washington, and Yale, cooperative projects sponsored by the two U.S. institutions began shortly after the reserve was founded (Rakotomanga et al., 1987). In 1984, Alison Richard began capturing and collaring *Propithecus verreauxi*, beginning a long-term project on the ecology, behavior, health status, and demography of individually identified animals living in groups in parcel no. 1 (e.g., Richard et al., 1991, 1993, 2002). In 1987, Sussman began a similar project on *Lemur catta* groups in parcel no. 1 (e.g., Sussman, 1991, 1992). These projects continue today, and a number of student theses, a great deal of research, and a new generation of research projects has resulted from studies on these identified populations (e.g., Sauther, 1992; Brockman, 1994; Gould, 1994; Kubzdela, 1997; Yamashita, 1998). Diane Brockman has continued to work with Alison Richard and her colleagues on the sifaka population. Michelle Sauther and Lisa Gould and their students and colleagues have extended the research on the ringtailed lemur population at Beza Mahafaly and the surrounding region, as can be seen in papers in this volume. In addition Malagasy students, in particular from ESSA/Forêt, continue to have an important role in lemur studies, particularly outside protected parcel no. 1 as well as inside the second parcel of the reserve (Ranarivelo, 1993; Ravelonjatovo, 1997; Raveloarisoa, 2000; Razafinjato, 2003; Randrianarisoa, 2005).

Much less research has been conducted on the other species of lemur living in the reserve. Short-term studies have been done on *Lepilemur* (Nash, 1998; Randriamboa, 1998; Ratsirarson and Emady, 2000) and on *Microcebus* (Rasoloarison, 2000), and a study of predation on *Microcebus* at the reserve has been completed (Goodman et al., 1993). As can be seen from papers in this volume, research on the ringtailed lemur in and around the Beza Mahafaly Reserve is beginning to cover many aspects of ecology, behavior, health status, and conservation in this region.

References

Brockman, D. K. (1994). *Reproduction and Mating System of Verreauxi's Sifaka, Propithecus verreauxi at Beza Mahafaly, Madagascar*. PhD thesis, Yale University, New Haven.

Gentry, A. H. (1993). Diversity and floristic composition of lowland tropical forest in Africa and South America. In: Goldblatt, L. P. (ed.), *Biological Relationships between Africa and South America*. Yale University Press, New Haven, pp. 500–546.

Goodman, S. M., O'Conner, S., and Langrand, O. (1993). A review of predation on lemurs: implications for the evolution of social behavior in small, nocturnal primates. In: Kappeler, P. M., and Ganzhorn, J. U. (Eds.), *Lemur Social Systems and their Ecological Basis*. Plenum Press, New York, pp. 51–66.

Gould, L. (1994). Patterns of Affiliative Behavior in Adult Male Ringtailed lemurs (*Lemur catta*) at the Beza Mahafaly Reserve, Madagascar. PhD thesis, Washington University, St. Louis.

Kubzdela, K. S. (1997). *Sociodemography in Diurnal Primates: The Effects of Group Size and Dominance Rank on Intra–group Spatial Distribution, Feeding Competition, Female Reproductive Success, and Female Dispersal Patterns in White Sifaka. Propithecus verreauxi verreauxi*. PhD thesis, University of Chicago, Chicago.

Nash, L. T. (1998). Vertical clingers and sleepers: Seasonal influences on the activities and substrate use of *Lepilemur leucopus* at Beza Mahafaly Special Reserve, Madagascar. *Folia Primatol.* 69(Suppl. 1):204–217.

Rakotomanga, P., Richard, A. F., and Sussman, R. W. (1987). Bezà Mahafaly: Formation et measures de conservation. In: Mittermeier, R. A., Rokotovao, L. H., Randrianasolo, V., Sterling, E. J., and Devitre, D. (eds.), *Prioités en Matières de Conservation de Espèces à Madagascar.* IUCN, Gland, pp. 41–44.

Ranarivelo, N. A. (1993). Etude de la variation locale et saisonnière du régime et du comportement alimentaire de *Propithecus verreauxi verreauxi* dans la première parcelle de la Réserve Spéciale de Beza Mahafaly. Mémoire de fin d'études. ESSA Département Eaux et Forêts, Université d'Antananarivo.

Randriamboa, R. (1998). Contribution à l'étude du comportement et de l' éthologie du *Lepilemur Leucopus* dans la Réserve Spéciale de Bezà Mahafaly. Mémoire de fin d'études. ESSA Eaux et Forêts, Université d'Antananarivo.

Randrianaivo, R. (1979) Essai D'Inventaire des Lemuriens de la Future Réserve de Bezà Mahafaly. Mémoire de fin d'études. ESSA Eaux et Forêts, Université d'Antananarivo.

Randrianarisoa, J. (2005). Étude comparative de la structure et comportement de la population de Propithecus verreauxi verreauxi à l'intérieur et à l'extérieur de la RS de Beza Mahafaly. Mémoire DEA en Sciences Forestières. ESSA Département Eaux et Forêts, Université d'Antananarivo.

Rasoloarison, R. M. (2000). Analyse de le diversité et densité des espèce de *Microcebus* à Madagascar. Thèse de Doctorat 3ème Cycle. Département Anthropologie, Université d'Antananarivo.

Ratsirarson, J. (1987). Contribution à l'étude comparative de l'Eco Ethologie de *Lemur catta* dans deux habitats different de la Réserve Spéciale de Bezà Mahafaly. Mémoire de fin d'études. ESSA Eaux et Forêts, Université d'Antananarivo.

Ratsirarson, J. (2003). Réserve Spéciale de Bezà Mahafaly. In: Goodman, S. M., and Benstead, J. P. (eds.), *The Natural History of Madagascar.* University of Chicago Press, Chicago, pp. 1520–1525.

Ratsirarson, J., and Emady, R. J. (2000). Predation de *Lepilemur leucopus* par *Buteo brachypterus* dans la forêt galerie de Beza Mahafaly. *Working Groups on Birds in the Madagascar Region Newsletter* 9:14–15.

Ratsirarson, J., and Silander J. A. (2003). Pollination ecology of plant communities in the dry forests of the southwest. In: Goodman, S. M., and Benstead, J. P. (eds.), *The Natural History of Madagascar.* University of Chicago Press, Chicago, pp. 272–275.

Ratsirarson, J., Randrianarisoa, J., Ellis, E., Emady, H. J., Efitroarany, Ranaivonasy, J., Razanajaonarivalona, E. H., and Richard, A. F. (2001). Béza Mahafaly: écologie et réalités socio-économiques. Recherches Pour Le Developpement, Série Sciences Biologiques, N° 18, Antananarivo, Madagascar.

Raveloarisoa, A. (2000). Contribution à l'étude De la préférence alimentaire du *Propithecus verreauxi verreauxi* de la RS de Beza Mahafaly (Parcelle I). Mémoire de DEA, Sciences Biologie Appliquées. Faculté des Sciences, Université d'Antananarivo.

Ravelonjatovo, S. A. (1997). Contribution à l'étude du comportement et de l'écologie de Sifaka dans la deuxième parcelle de la RS de Beza Mahafaly. Mémoire de fin d'études. ESSA Département Eaux et Forêts, Université d'Antananarivo.

Razafinjato, A. P. 2003. Étude du comportement du *Lemur catta* (Linnaeus, 1758) en dehors de la première parcelle de Beza Mahafaly. Mémoire de fin d'études. ESSA Département Eaux et Forêts, Université d'Antananarivo.

Richard, A. F., Rakotomanga, P., Schwartz, M. (1991). Demography of *Propithecus verreauxi* at Beza Mahafaly: sex ratio, survival, and fertility, 1984–1988. *Am. J. Phys. Anthropol.* 84:307–322.

Richard, A. F., Rakotomanga, P., and Schwartz, M. (1993). Dispersal by *Propethicus verrauxi* at Beza Mahafaly, Madagascar. *Am. J. Primatol.* 30:1–20.

Richard, A. F., Dewar, R. E., Schwartz, M., and Ratsirarson. J. (2002). Life in the slow lane? Demography and life histories of male and female Sifaka (*Propithecus verreauxi verreauxi*). *J. Zool.* 256:421–436.

Sauther, M. L. (1992). *The effect of Reproductive State, Social Rank and Group Size on Resource Use among Free-ranging Ringtailed Lemurs (Lemur catta) of Madagascar.* PhD thesis, Washington University, St. Louis.

Steele, E. (1975). Needed: virtue and money. *Defenders of Wildlife.* 50(2):90.

Stephenson, P. J. (2003). Lipotyphla (ex Insectivora): *Geogale aurita,* large-eared tenrec. In: Goodman, S. M., and Benstead, J. P. (eds.), *The Natural History of Madagascar.* University of Chicago Press, Chicago, pp. 1265–1267.

Sussman, R. W. (1991). Demography and social organization of free-ranging *Lemur catta* in the Beza Mahafaly Reserve, Madagascar. *Am. J. Phys. Anthropol.* 84:43–58.

Sussman, R. W. (1992). Male life history and intergroup mobility among ringtailed lemurs (*Lemur catta*). *Int. J. Primatol.* 13:395–413.

Sussman, R. W., and Rakotozafy, A. (1994). Plant diversity and structural analysis of a tropical dry forest in southwestern Madagascar. *Biotroica* 26:241–254.

White, F. (1983). The vegetation of Africa, a descriptive memoir to accompany the UNESCO/AETFAT/UNSO Vegetation Map of Africa. UNESCO, Paris.

Yamashita, N. (1998). Functional dental correlates of food properties in five Malagasy lemur species. *Am. J. Phys. Anthropol.* 106:169–188.

Youssouf, J. (2004). Bioécologie de *Rattus rattus* dans la Réserve Spéciale de Beza Mahafaly. Mémoire DEA, Département des Sciences Biologiques, Faculté des Sciences, Université de Toliary.

Part II

Ringtails and Their Forests: Feeding and Ranging

5
Plant Species Fed on by *Lemur catta* in Gallery Forests of the Southern Domain of Madagascar

BRUNO SIMMEN, MICHELLE L. SAUTHER, TAKAYO SOMA, HANTANIRINA RASAMIMANANA, ROBERT.W. SUSSMAN, ALISON JOLLY, LAURENT TARNAUD, AND ANNETTE HLADIK

5.1. Introduction

In this paper, we provide an overview of the feeding trends of *Lemur catta*, the ringtailed lemur, including a checklist of all plant species and plant items known to be ingested by this prosimian species in three different forests of southwestern and southern Madagascar. Ringtailed lemurs have been mainly studied in gallery forests including riverine forest, closed canopy forest, and drier habitats with opened forests and scrub as distance perpendicular to the river increases. There is little published information on food species eaten consistently in other areas throughout the distribution range of this species such as in the dry spiny forest (Didiereaceae and arborescent Euphorbiaceae forests) or in montane areas up to 2000 m [see however, the study of Rakotoarisoa (1999) in the Andringitra massif; and Goodman et al. (this volume)]. The checklist provided here thus emphasizes plant species from the gallery forests and associated drier areas, with few observations made in spiny forests and mixed dry forest and bush.

The three field areas considered here are respectively the southwestern sites of Antserananomby and the Beza Mahafaly Special Reserve, and in the south the Berenty Private Reserve. In these areas, groups of ringtailed lemurs have been studied intensively, in some cases over several years, and/or are under current studies. It should be noted that dietary differences described below between ringtailed lemurs of Berenty and those of Beza Mahafaly appear to a great extent linked to the availability of many more introduced plant species in the first site and still some human-derived food and water. Detailed information on geographical locations and plant species composition of the study sites can be found in Jolly (1966), Sussman (1974), O'Connor (1988), and Sussman and Rakotozafy (1994). Food plants collected and dried as herbarium samples have been identified by botanists from the Missouri Botanical Garden and the Muséum National d'Histoire Naturelle, Paris (France), in collaboration with the Parc Zoologique et Botanique de Tsimbazaza, Antananarivo.

5.2. Food Plants, Food Items, and Seasonal Variations

Ringtailed lemurs are considered mainly frugivorous/folivorous primates that can shift their diet toward either leaves or fruits as one main food category in different seasons. They have been observed feeding on ripe and unripe fruits, young and mature leaves, leaf stems, flowers, unripe seeds, and dead wood; additionally, they ingest parts of termite galleries as well as small pieces of earth and they prey on invertebrates and on vertebrates on rare occasions (Jolly, 1966; Sussman, 1974, 1976; Budnitz and Dainis, 1975; Sauther, 1992, 1998; Yamashita, 2002; Simmen et al., 2003; Soma, this volume; S. Ichino, pers. comm.). *Lemur catta* apparently shows adaptations to feeding on poor-quality leaves or leaves that are rich in secondary metabolites (see Simmen et al., this volume; Ganzhorn, 1986). With a somewhat enlarged haustrated caecum (Campbell et al., 2000), ringtailed lemurs harbor an intestinal symbiotic flora that is assumed to facilitate leaf fermentation and that, to some extent, may help detoxify foods. Geophagy is quite frequently observed and may be a behavioral response to coping with toxic foods, leading to the neutralization of leaf tannins through adsorption by clay (Johns and Duquette, 1991).

In Antserananomby, where groups have been studied during the late dry season, ringtailed lemurs feed on at least 23 species. In Beza Mahafaly and Berenty where lemur feeding behavior and diet has been repeatedly investigated relative to both feeding strategy and social organization as well as reproductive state, at least 61 and 109 plant species, respectively, with a wide variety of plant food items, are included in the diet (Table 5.1). At Beza Mahafaly, 40 plant species are used for leaves, 28 species for fruits and 16 species for flowers. At Berenty, 82 plant species are used by lemurs for leaves, 40 for fruits and 38 for flowers.

In terms of the number of plant species or food items used, the diet of *Lemur catta* thus appears relatively diverse. However, only a few species within this food repertoire actually play a major role in any season, once dietary proportions are accounted for. Such a feeding pattern would correspond, overall, with an unselective, opportunistic feeding behavior: the diet reflects to a great extent the composition and structure of the gallery forest, a relatively low-diversity environment in which less than 15 tree species account for most of the total basal area (Berenty: O'Connor, 1988; Simmen and Tarnaud, unpublished results; see also Sussman and Rakotozafy, 1994, for Beza Mahafaly). This also reflects the availability of introduced plant species at Berenty, especially for groups foraging near the tourist area.

Tamarindus indica (vernacular name: *kily*) has long been recognized as a keystone resource (sensu Terborgh, 1986) in gallery forests inhabited by ringtailed lemurs (Jolly, 1966; Sussman and Rakotozafy, 1994; Sauther, 1998; Blumenfeld-Jones, this volume; Mertl-Millhollen et al., this volume). At Beza Mahafaly, *Tamarindus indica* is an important food resource as it is the only species that is used throughout the year (Figure 5.1). At Berenty, kily ripe pods are used intensively during the late dry–early wet season, when females give birth and lactate, and may still be consistently ingested during the wet season in favorable years.

TABLE 5.1. List of plant species and items consumed by ringtailed lemurs in Berenty, Antserananomby, and Beza-Mahafaly.

Plant family	Species	Food items (Berenty)	Food items (Antserananomby)	Food items (Beza-Mahafaly)
Acanthaceae	*Justicia glabra* K. Koenig ex Roxb			yl
	Mimulopsis sp.		l	
	Ruellia anaticollis R. Ben.	yl, ml		yl, ml
	Thunbergia convolvulifolia (R. Ben.) Bak.	ml		
Agavaceae	*Agave angustifolia* Haw.*	stem, flb		
	Agave sisalana Perrine ex Engelm.*	stem, flb		
	Sansevieria sp.	F		
Aizoaceae	*Mollugo* sp.	ml		
	Unidentified	l		
Amaranthaceae	*Achyranthes aspera* L.	flb, l, stem	flb, l, stem	ml
	Aerva javanica Jussieu	ml		
Anacardiaceae	*Mangifera indica* L.*	F		
	Operculicarya cf. decaryi H. Perr.*	f		
	Sclerocarya birrea (Sond.) H. Perr.*	fl, flb, yl, ml, F	l	
Annonaceae	*Annona* sp.*	F		
Aristolochiaceae	*Aristolochia aurita* Duch.			ml
Asclepiadaceae	*Cynanchum nodosum* (Jum. & H. Perrier) Desc.			F, fl
	Gonocrypta grevei Baill.	ml		ml, yl
	Marsdenia cordifolia Choux			F, fl, ml, yl, st
	Pentopetia androsaemifolia Decne.	l, ml, stem		ml, yl
	Secamone sp.			ml
	cf. Secamone uncinata Choux	l		
Asteraceae	*Bidens* sp.			ml
	Senecio sp.	ml		
	Tridax procumbens L.	ml		
Bignoniaceae	*Fernandoa madagascariensis* (Bak.) Gentry*	flb		ml, st
	Tecoma stans Griseb.*	flb		
Boraginaceae	*Cordia caffra* Sond.	yl, ml, fl, flb, F, f		
	Cordia sinensis Lam.*	yl, ml, fl, flb, F, f		
	Ehretia sp.	yl, l, F		
Burseraceae	*Commiphora* sp.	yl		
Cactaceae	*Cereus* sp.*	flb, fl, F		
	Opuntia vulgaris Mill.*	stem, F, sap		
Caricaceae	*Carica papaya* L.*	F		
Capparaceae	*Capparis chrysome* Bojer			F, yl, ml, st
	Capparis sepiaria L.	ml, yl, flb, fl, F, f		
	Crateva cf excelsa Bojer	F, f, fl, flb, yl, p		fl, yl
	Maerua filiformis Drake	yl, ml, fl, stem		F, fl, yl
	Maerua nuda Scott-Elliot			F

(*Continued*)

TABLE 5.1. List of plant species and items consumed by ringtailed lemurs in Berenty, Antserananomby, and Beza-Mahafaly—*Cont'd.*

Plant family	Species	Food items (Berenty)	Food items (Antserananomby)	Food items (Beza-Mahafaly)
Casuarinaceae	*Casuarina* sp.*	ml		
Celastraceae	*Maytenus* sp.	ml		
	Maytenus linearis (L.f.) Marais			fl
Combretaceae	*Combretum albiflorum* (Tul.) Jongkind	ml, yl, fl		
	Combretum sp.	F		
	Combretum sp.			ml
	Terminalia mantaly H. Perrier		l, flb, sap	
Commelinaceae	*Commelina* sp.	ml		
Convolvulaceae	*Hildebrandtia* sp.			ml, yl
	Ipomoea cairica (L.) Sweet	yl, ml		fl, yl, ml
	Ipomoea mojangensis Vatke			fl
	Metaporana parvifolia (K. Afz.) Verdcourt	yl		yl, ml
Crassulaceae	*Kalanchoe beharensis* Drake*	ml		
Cucurbitaceae	*Corallocarpus grevei* (Keraudren) Keraudren			F, ml, yl
	Seyrigia sp.			F
	Zehneria sp.	l		
Didiereaceae	*Alluaudia dumosa* (Drake) Drake	fl		
	Alluaudia humbertii Choux*	ml		
	Alluaudia procera Drake*	yl, ml, fl		
	Didierea trollii Capuron & Rauh*	yl, fl		
Euphorbiaceae	*Acalypha* sp.	l		fl, yl
	Acalypha sp.		l	
	Alchornea sp.		flb, l	
	Antidesma madagascariense Lam.	F, f, ml		
	Antidesma petiolare Tul.			F
	Chamaesyce aff. hirta (L.) Millsp.	yl, ml, fl		
	Croton sp.	ml		
	Euphorbia tirucalli L.	fl		st
	Flueggea sp.			ml
	Pedilanthus tithymaloides Den.*	yl		
	Phyllanthus casticum Willem.	yl, fl, f		
	Phyllanthus sp.	fl, flb, F, f, ml, yl		
Fabaceae Caesalpinioideae	*Caesalpinia pulcherima* (L.) Sw.*	ml, fl		
	Delonix regia (Bojer ex Hook.) Raf.*	ml		

5. Plant Species Fed on by *Lemur catta*

	Senna siamea (Lam.) Irwin*	fl, flb		
	Senna spectabilis (DC.) Irwin & Barneby*	fl		
	Senna sp.	l		
	Tamarindus indica L.	f, F, ml, yl, fl, flb, w, stem, sap	l, F, stem, w, sap	f, F, ml, yl, fl
Mimosoideae	*Acacia rovumae* Oliv.	yl, ml, fl	l, sap	
	Acacia sp.			ml
	Acacia sp.		l	
	Albizia polyphylla E. Fourn.	l, ml		
	cf. *Entada* sp.	ml		
	Leucaena leucocephala (Lmk.) De Wit*	sd, yl, ml, fl, f		
	Pithecellobium dulce (Roxb.) Benth.*	yl, F, f, sd, fl		
Papilionoideae	*Abrus precatorius* L.			ml
	Crotalaria sp.			F, yl, ml
	Crotalaria sp.	yl		
	Lablab boivinii (Drake) R. Vig.			F, yl
	Rhynchosia sp.			ml
	Unidentified		l	
Flacourtiaceae	*Flacourtia ramontchi* L'Hér.	F	F, l	F
	Physena sessiliflora Tul.	yl, ml		F
Hernandiaceae	*Gyrocarpus americanus* Jacq.	yl		fl, yl
Hippocrateaceae	*Hippocratea* sp.			yl, ml
	Loeseneriella sp. 1	ml		
	Loeseneriella sp. 2	yl		
Hydnoraceae	*Hydnora esculenta* Jum. & Perr.	(F)		
Icacinaceae	*Apodytes dimidiata* E. Mey.	F, ml		
Liliaceae	*Aloe vahombe* Dec.*	yl, ml		
	Aloe cf *capitata* Bak.*	ml		
	Gloriosa superba L.			fl, st, ml
Loganiaceae	*Strychnos madagascariensis* Poiret	F		
Lythraceae	*Lawsonia inermis* L.	F, ml		
Malpighiaceae	*Microsteia* sp.	ml		
Malvaceae	*Abutilon pseudocleistoganum* Hochr.	ml		ml
	Hibiscus sp.*	flb, ml		
	Sida rhombifolia L.	ml		
Meliaceae	*Azadirachta indica* Jussieu*	yl, ml, F, fl		
	Melia azedarach L.*	yl, ml, flb, F		
	Quivisianthe papinae Baill.	ml, fl, F	fl, yl	fl
Menispermaceae	*Cissampelos pareira* L.	yl		
	Cissampelos sp.	l		
	Rhaptonema cf. *swinglei* Kundu & Guha	yl		
Moraceae	*Ficus* cf. *grevei* Baill.	F		
	Ficus cf. *pachyclada* Bak.	F		
	Ficus cf. *polita* Vahl	F		
	Ficus sp.		F, l	
	Ficus sycomorus L.	l, F	F	
	Ficus sp.		l, F	

(*Continued*)

TABLE 5.1. List of plant species and items consumed by ringtailed lemurs in Berenty, Antserananomby, and Beza-Mahafaly—*Cont'd.*

Plant family	Species	Food items (Berenty)	Food items (Antserananomby)	Food items (Beza-Mahafaly)
Musaceae	*Musa* sp.*	F		
Myrtaceae	*Eucalyptus* sp.*	fl, flb		
Nyctaginaceae	*Boerhavia diffusa* L.	yl, ml		
	Bougainvilea spectabilis Willd.*	yl, ml, fl		
	Commicarpus commersonii Cav.	yl, ml	l, stem	yl, ml
Oleaceae	*Noronhia seyrigii* Perr.	f, F		F
Papaveraceae	*Argemone mexicana* L.*			fl, ml, st
Passifloraceae	*Adenia* sp.		l	
Poaceae	*Panicum maximum* Jacq.	ml, yl		
	Unidentified 1	ml		
	Unidentified 2	ml		
	Unidentified 3	ml		
Polygonaceae	*Polygonum* sp.	ml		
Portulacaceae	*Talinella dauphinensis* Scott-Elliot			F, yl, ml, st
Rhamnaceae	*Gouania* sp.			ml
	Scutia myrtina (Burm.) Kwz			F
	Ziziphus jujuba Mill.	l		
Rubiaceae	*Catunaregam spinosa* (Thunb.) Tirveng.			F
	Enterospermum sp.	F		
	Enterospermum sp.	F, fl, flb, ml, yl		
	Enterospermum pruinosum			F
	Paederia grandidieri Drake	ml	l	fl, yl, ml
	Paederia sp.		l	
	Tricalysia sp.	yl, ml, p, fl		
Rutaceae	*Cedrelopsis grevei* Baill.	yl		F, fl, yl, ml
Salvadoraceae	*Azima tetracantha* Lam.	F, f, flb, fl, yl, ml, stem		F, yl, ml
	Salvadora angustifolia Turril	yl, ml, stem		F, fl, yl
Sapindaceae	*Cardiospermum halicacabum* L.	l		
	Neotina isoneura (Radlk.)	F, f, yl, ml		
Sterculiaceae	*Byttneria voulily* Baill.			ml
Tiliaceae	*Grewia saligna* Baill.	f, fl, flb		
	Grewia calvata Baker			F, fl
	Grewia humbertii Capuron			F
	Grewia grevei Baill.			F, fl, yl
	Grewia leucophylla Capuron			F
	Grewia triflora (Bojer) Walp.			F, yl
	Grewia sp.		l, stem, flb	

Family	Species				
Ulmaceae	*Celtis bifida* J.-F. Leroy	f, F, fl, flb, yl, ml			
	Celtis philippensis Blanco	F, f, flb, fl, ml			
Verbenaceae	*Clerodendrum* sp.				ml
	Lantana camara L.*	F, f			
	Vitex beraviensis Vatke		F		
	Vitex sp.				F
Violaceae	*Rinorea greveana* H. Bn	f, F, flb, fl, ml			
Vitaceae	*Cissus microdonta* (Baker) Planch.				F
Unidentified	Unidentified	1			

The list includes a few plant species that are typically found in the dry spiny forest but which are eaten by lemur groups foraging in both gallery forest and edge or open areas. The asterisk (*) indicates either ornamental and introduced species or naturalized species. Food items are defined as ripe fruit pulp and seed (F), unripe fruit (f), mature leaf (ml), young leaf and leaf bud (yl), leaf of unknown maturity (l), petiole (p), tip of stem (stem), flower (fl), floral bud (flb), ripe seed (Sd), unripe seed (sd), wood (w), and exudate (sap). Items that are probably eaten are figured in brackets.

During the dry season, immature pods of *Tamarindus* are a main fruit source and, together with *Tamarindus* leaves, form the staple food.

According to phenological studies carried out at Beza Mahafaly (Sauther, 1998; Yamashita, 2002), there appears to be a number of other plant species that serve as keystone species. In addition to kily fruits and leaves, the fruits of *Enterospermum pruinosum* (Rubiaceae) may be important as they are available during the dry season. Patches of *Salvadora angustifolia* (Salvadoraceae) can be shared by several groups for their fruit at times when individuals may be energetically constrained (birth season and early lactation; Sauther, 1998). In any case, ringtailed lemurs appear to use resources as they become available so that the food species can change dramatically from month to month. As a result, only two or three species and plant parts make up the major percentage of the diet at this study site for any one month, as shown in Figure 5.1. Beside resources mentioned above, fruits of *Talinella dauphinensis* (Portulacaceae) and *Grewia* spp. (Tiliaceae) are important in the wet season. *Hildebrandtia* spp. (Convolvulaceae), *Talinella dauphinensis* (Portulacaceae), *Justicia glabra* (Acanthacaeae), *Rynchosia* sp. (Fabaceae), *Secamone* sp. (Asclepiadaceae), and *Commicarpus commersonii* (Nyctaginaceae) are among plant species most widely used as leaf sources, some of them accounting for a large part of feeding observations during the dry season (Figure 5.1). Flowers of *Quivisianthe papinae* (Meliaceae) are a major food source during the birth season but are available only for a limited period. Introduced species that are critical fall-back foods during the dry season include *Argemone mexicana* (Papaveraceae), which is only eaten at that time.

Fruit species predominantly fed on by lemurs at Berenty are *Tamarindus indica* (Caesalpiniaceae), *Rinorea greveana* (Violaceae), *Cordia sinensis* and *Cordia caffra* (Boraginaceae), *Celtis philippensis* and *Celtis bifida* (Ulmaceae),

62   B. Simmen et al.

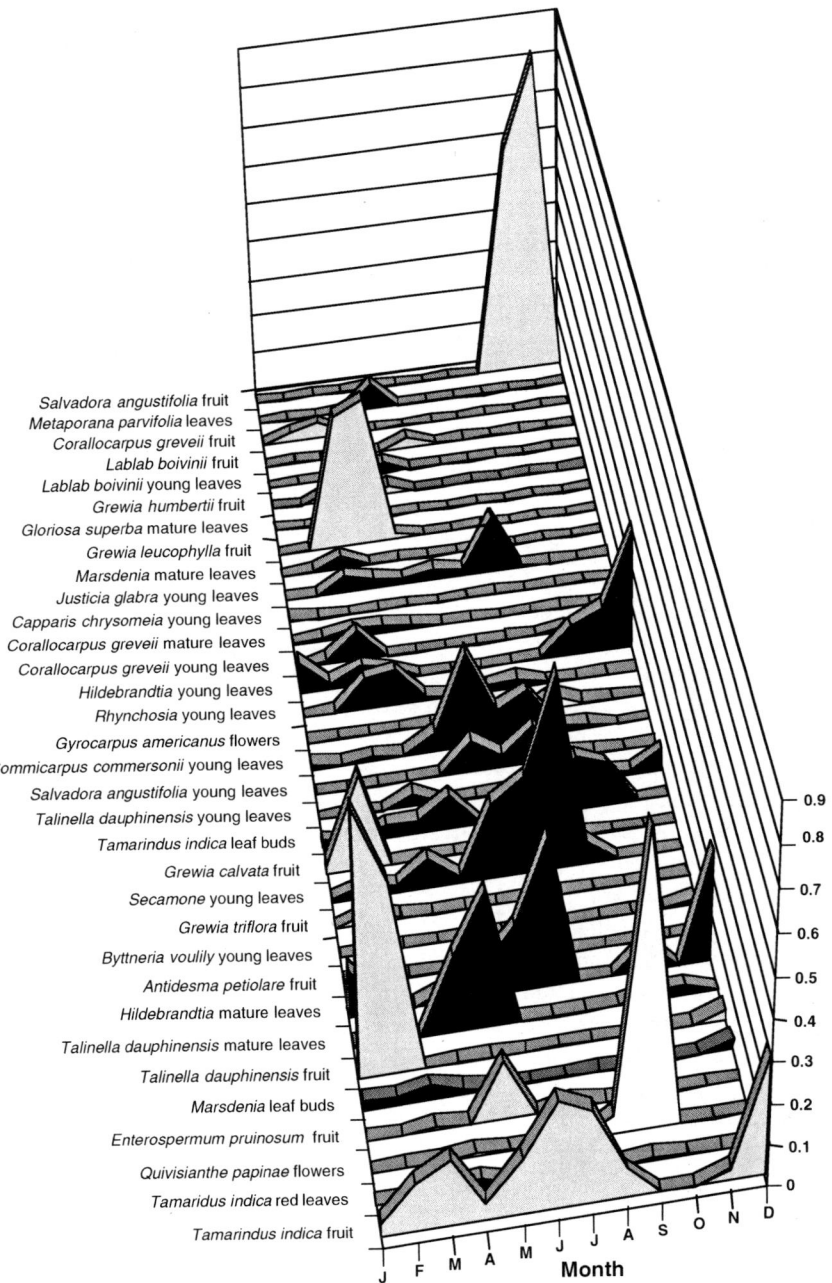

FIGURE 5.1. Seasonal food use in *Lemur catta* by percentage species/part plotted by month, as recorded at Beza Mahafaly. Food categories are indicated as fruit (gray area), leaf (black area), and flowers (white area).

Neotina isoneura (Sapindaceae), *Crateva* sp. (Capparaceae), and *Azadirachta indica* (Meliaceae). *Cordia sinensis* and *Azadirachta indica* are introduced ornamental species growing at the edge of the forest and are important fruit resources for lemur groups foraging there. There are no published long-term records of variations of fruit supplies at Berenty, but recurrent observations indicate that all these species provide ripe fruits with partly overlapping fruiting peaks throughout the wet season. This pattern nevertheless appears somewhat irregular or even disrupted during drought years. Fruits available during the wet season allow increased energy intake and coincides with the need for high-energy input in lactating females and high-quality diet in juveniles at weaning (see Rasamimanana, this volume). This also favors accumulation of body energy reserves before entering the mating season and for coping with the subsequent dry season of food scarcity. Phenological data recorded over two separate years at Beza Mahafaly provide evidence of greater resources when females are lactating and especially weaning their infants but reduced food availability during gestation (Sauther, 1998; Yamashita, 2002).

In recent years, many lemur groups in Berenty shifted from kily mature leaves to leaves of an introduced leguminous tree, *Leucaena leucocephala* (Mimosaceae) during the dry month (see Soma, this volume). Several troops have even been observed sharing patches of *Leucaena leucocephala* without conflict, but this tolerant behavior related to leaf consumption seems restricted to the dry season. During the early lactating period, for instance, conflicts systematically occur for *Rinorea* fruits between neighboring groups (Soma, unpublished observation). Other main dry season foods at this site are mature leaves of *Azadirachta indica* and *Opuntia* rackeets, fruits of *Cordia sinensis*, and flowers of *Senna siamea* and *Eucalyptus* sp., which all are introduced species available to troops foraging at the edge of the forest. Such groups also spend a high proportion of their feeding time ingesting the protein-rich flushes of new leaves of the ornamental *Bougainvilea* (Nyctaginaceae), another introduced plant. Feeding on this highly rewarding food is observed throughout the year and may be especially important for lactating females (Rasamimanana, 2004). Most of the above-mentioned tree species are very common in Berenty according to previous investigations of the forest structure. This would largely explain the occupation by large troops of very small home ranges, as compared with similarly sized primate species.

Lemur catta has been observed seasonally (between October and December) visiting the very dry forest in the reserve at Faux Cap and feeding on seasonal fruit from species of the genera *Phyllanthus*, *Capurodendron*, and *Poupartia* (Reserve Naturalist, Randrianananrana, pers. comm. to Sussman et al., 2003). As a matter of fact, there is evidence that *Lemur catta* is able to survive in sub-desertic areas (see Goodman et al., this volume). At Beza Mahafaly, the terrestrial and arboreal habits of this species allow groups to exist within open areas beside closed canopy forest and to include large amounts of herbaceous species in their diet. Compared with Berenty, Beza Mahafaly appears to contain a greater number of herbaceous lianas that are very important, including *Hildebrandtia* spp. (Convolvulaceae). At this site, shrubs, herbs, and low-level lianas contribute 62% of time spent feeding on leaves during the wet season (Sauther, 1992), stressing the importance of these plant biological types for lemur groups inhabiting open areas. Forests or scrub areas subjected to grazing by cattle contain a less diverse composition of herbs and lianas than protected areas (Sussman and Rakotozafy, 1994). Given the importance of these plants in the diet, disappearance of these species could be a source of concern. At Antserananomby, *L. catta* ate herbs or grasses (mainly herbs) for about 15% of the time during the late dry season (*Achyranthes aspera*, *Mimulopsis* spp., and *Commicarpus commersonii*). Use of herbaceous species and vines is not frequent in Berenty where the ecotone forest/scrub has been largely modified, leading to the disappearance of a specific flora for troops whose home ranges include the tourist front (Pinkus, 2004). Instead, introduced plants available at the edge of the forest are largely used by such groups, accounting, in some cases, for up to 73% of the time spent feeding in some months (see Soma, this volume). Dietary shifts toward introduced masting resources such as *Azadirachta indica* and other introduced plants may have been to a large extent responsible for increased populations of *Lemur catta*, compared with other forest areas undergoing low-level anthropogenic effects. However, plant introduction may also have contributed to the appearance of serious diseases unknown prior to their use, as with *Leucaena leucocephala*, known to affect cell division among other major lethal symptoms in nonruminant mammals and some folivorous species (see Crawford et al., this volume). Reasons why groups changed their feeding habits from kily mature leaves to *Leucaena* leaves are under investigation. They may involve discovery of a new valuable food source, reduction of food provisioning especially for troops with high contact with tourists, or competition with introduced *Eulemur fulvus* that now forms a large population with considerable dietary overlap with *Lemur catta* in the closed canopy forest (Simmen et al., 2003; Pinkus, 2004, this volume; Crawford et al., this volume).

Forest heterogeneity due to natural and/or anthropogenic effects, patchy distribution of several plant species, and asynchronous phenological patterns is associated with diet differences between troops (Sauther, 1992; Rasamimanana, 1993; Pinkus, 2003). At Beza Mahafaly, there are differences in mean food weights ingested between troops, and this appears to be a reflection of the phenological availability of resources within each group's home range (Sauther, 1992). Intertroop variations of the diet may be pronounced at Berenty, where groups tend to defend well-defined home ranges and rarely range far away from their normal range to feed on shared patches of food resources. Some groups also benefit from foods available from the large patches of introduced plants at this site and eventually from food provisioning by tourists (although this has been considerably reduced in last years).

5.3. Conclusion: Drawbacks and Perspectives

Interpreting the role of female dominance in *Lemur catta* in terms of priority of access to foods has been complicated by the fact that different studies failed to demonstrate clear variations of diet quality between dominant and subordinates—dietary differences nevertheless occur between males and females during lactating and mating seasons (Sauther 1993, 1994; Rasamimanana, 2004, this volume). From a methodological standpoint, however, it should be noted that the maturity of fruits or of leaves eaten has been commonly inferred from external coloration or toughness but that these characteristics can be misleading (especially for dull-colored fruits; see also van Roosmalen, 1984). It is not rare, indeed, to observe lemurs in a feeding tree carefully choosing fruits through olfactory cues, focusing on a few fruits that apparently share external characteristics similar to avoided fruits (e.g., *Azadirachta indica*). Furthermore, some fruits consumed may be categorized as unripe on the basis of their color and their immature seeds despite seeds being surrounded by an edible fleshy pulp (e.g., in *Neotina isoneura*). It follows that operational criteria such as categorizing selected fruits into ripe versus unripe categories (or immature versus mature leaves) limits one's ability to assess individual energy budget and its variation between individuals. In this respect, the subtlety of food-related sensory abilities in *Lemur catta* is investigated in the chapter of Simmen et al. (this volume) showing that this species probably discriminates close taste or olfactive stimuli.

It has been argued that the timing of reproduction in this species is tuned to the seasonal variation of food availability, and especially of particular plant species, in a quite predictable environment (e.g., Sauther, 1998). The late dry season–early wet season indeed globally corresponds with a period of increased new leaf and flower availability, followed by fruiting peaks in successive months, whereas the dry season is a period of food scarcity. At Beza Mahafaly, there are also periodic droughts, which have been associated with a marked decline in the population, indicating that in the short-term, this synchrony between food resources and reproduction can have a dramatic effect on survival. However, for these ringtailed lemurs, the population rebounded quickly and were at predrought numbers within 6 years (Gould et al. 1999, 2003). This suggests that ringtailed lemurs are able to maintain themselves in seasonal environments that include periodic droughts (Gould et al., 1999, 2003). It is also possible that many physiological and behavioral traits of *Lemur catta* have been shaped in relation to supra-annual variations of food production, as proposed for other Malagasy prosimians (e.g., Wright, 1999; Jolly et al., this volume).

To date, we lack phenological data for the dry spiny forests, an ecosystem that constitutes one major habitat of *Lemur catta*, as well as for more northern areas of the distribution range of these animals. If populations of ringtailed lemurs living at high altitudes spend more than 75% of their feeding activity on fruits in both the winter and the summer season (Rakotoarisoa, 1999), does this pattern correspond with a predictable or unpredictable environment? In the future, it will be highly profitable to contrast years of high food availability with drought years in different habitats (see Gould et al., 1999). This involves studying the effects of El Niño Southern Oscillation on plant reproduction in southern Madagascar and plant responses to cyclones. That lemur populations may actually never reach or

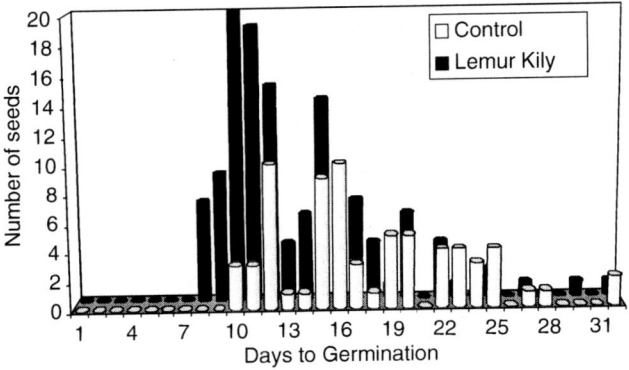

FIGURE 5.2. Germination rate of control versus passed seeds of *Tamarindus indica* by *Lemur catta*. The global germination success is 41% for control seeds versus 75% for passed seeds (after Sauther, unpublished data).

stay for long periods at an equilibrium can be better understood integrating long-term implications of these abiotic factors.

Finally, the role of ringtailed lemurs as seed dispersers has been only briefly investigated (Sauther, unpublished data). Ringtailed lemurs may play an important role in the germination of one of their keystone species, the kily. Tests indicate that germination success is significantly increased and germination occurs much more quickly when the seeds of this plant pass through the lemur's digestive tract (Figure 5.2). The range of other plant species whose seeds are dispersed by ringtailed lemurs as well as the fate of dispersed seeds have not been investigated. Undamaged seeds are found in the feces, and one would expect that the probability of a seedling to establish in a favorable environment after seeds transited by the digestive tract is increased, as in other zoochorous dispersal systems. Nevertheless, that lemurs will prefer to feed on fruits of exotic species (and swallow their seeds) over native species fruiting at the same time (e.g., *Azadirachta indica* versus *Crateva* and *Celtis* spp.) may have large consequences in the long term on the regeneration of gallery forest subjected to anthropogenic effects. Seed dispersal by lemurs within the guild of frugivores then needs to be studied further to improve the management of remaining gallery forests.

Acknowledgments. We thank the de Heaulme family for providing facilities and for interest in the study and conservation of the Berenty gallery forest. We are grateful to the Ministère des Eaux et Forêts for giving permission to study lemur ecology and their forest habitat. We also thank Joel Ratsirarson, Jo Ajimy Enafa Efitroaromy, and Edidy Ellis for their long-standing assistance for our work at Beza Mahafaly.

References

Budnitz, N., and Dainis, K. (1975). *Lemur catta*: ecology and behavior. In: Tattersall, I., and Sussman, R. (eds.), *Lemur Biology*. Plenum Press, New York, pp. 219–235.

Campbell, J. L., Eisemann, J. H., Williams, C. V., and Glenn, K. M. (2000). Description of the gastrointestinal tract of five lemur species: *Propithecus tattersalli, Propithecus verreauxi coquereli, Varecia variegata, Hapalemur griseus*, and *Lemur catta. Am. J. Primatol.* 52:133–142.

Ganzhorn, J. U. (1986). The influence of plant chemistry on food selection by *Lemur catta* and *Lemur fulvus*. In: Else, J. G., and Lee, P. C. (eds.), *Primate Ecology and Conservation, 2*. Cambridge University Press, Cambridge, pp. 21–29.

Gould, L., Sussman, R. W., and Sauther, M. L. (1999). Natural disasters and primate populations: the effect of a 2-year drought on a naturally occurring population of ring-tailed lemurs (*Lemur catta*) in Southwestern Madagascar. *Int. J. Primatol.* 20: 69–84.

Gould, L., Sussman, R.W., and Sauther, M.L. (2003). Demographic and life-history patterns in a population of ring-tailed lemurs (*Lemur catta*) at Beza Mahafaly, Madagascar: A 15-year perspective. *Am. J. Phys. Anthropol.* 120:182–194.

Johns, T., and Duquette, M. (1991). Detoxification and mineral supplementation as functions of geophagy. *Am. J. Clin. Nutr.* 53:448–456.

Jolly, A. (1966). *Lemur Behavior. A Madagascar Field Study*. The University of Chicago Press, Chicago.

O'Connor, S. (1988). Une revue des différences écologiques entre deux forêts galeries, une protégée et une dégradée, au centre sud de Madagascar. In: Rakotovao, L., Barre, V., and Sayer, J. (eds.), *L'équilibre des écosystèmes forestiers à Madagascar. Actes d'un séminaire international*. UICN, Gland, pp. 216–227.

Pinkus, S. (2004). *Impact of an introduced population of Eulemur fulvus on a native population of Lemur catta at Berenty Reserve, Southern Madagascar*. Master's thesis. University of British Columbia, Vancouver, Canada.

Rakotoarisoa, S. V. (1999). *Contribution à l'étude de l'adaptation de* Lemur catta *Linnaeus, 1758, aux zones sommitales de la Réserve Naturelle Intégrale d'Andringitra*, DEA Anthropologie, Université d'Antananarivo, Madagascar.

Rasamimanana, H. (2004). *La dominance des femelles makis (Lemur catta): quelles stratégies énergétiques et quelle qualité de ressources dans la réserve de Berenty, au sud de Madagascar?* PhD diss., Thesis of the National Museum of Natural History, Paris, France.

Rasamimanana, H. R., and Rafidinarivo, E. (1993). Feeding behavior of *Lemur catta* females in relation to their physiological state. In: Kappeler, P. M., and Ganzhorn, J. U. (eds.), *Lemur Social Systems and Their Ecological Basis*. Plenum Press, New York, pp. 123–133.

van Roosmalen, M. G. M. (1984). Subcategorizing foods in Primates. In: Chivers, D. J., Wood, B. A., and Bilsborough, A. (eds.), *Food Acquisition and Processing in Primates*. Plenum Press, New York, London, pp. 167–175.

Sauther, M. L. (1992). *Effect of Reproductive State, Social Rank and Group Size on Resource Use Among Free-Ranging Ringtailed Lemurs (Lemur catta)*. PhD thesis, Washington University, St. Louis.

Sauther M. L. (1993). Resource competition in wild populations of ringtailed lemurs (*Lemur catta*): Implications for female dominance. In: Kappeler, P. M., and Ganzhorn, J. (eds.), *Lemur Social Systems and Their Ecological Basis*. Plenum Press, New York, pp. 135–152.

Sauther, M. L. (1994). Changes in the use of wild plant foods in free-ranging ringtailed lemurs during lactation and pregnancy: Some implications for hominid foraging strategies. In: Etkin, N. L. (ed.), *Eating on the Wild Side: The Pharmacologic, Ecologic, and Social Implications of Using Noncultigens.* University of Arizona Press, Tucson, pp. 240–246.

Sauther, M. L. (1998). Interplay of phenology and reproduction in ring-tailed lemurs: Implications for ring-tailed lemur conservation. *Folia Primatol.* 69(Suppl. 1):309–320.

Simmen, B., Hladik, A., and Ramasiarisoa, P. L. (2003). Food intake and dietary overlap in native *Lemur catta* and *Propithecus verreauxi* and introduced *Eulemur fulvus* at Berenty, Southern Madagascar. *Int. J. Primatol.* 5:949–968.

Sussman, R. (1974). Ecological distinctions in sympatric species of *Lemur*. In: Martin, R. D., Doyle, G. A., and Walker, A. C. (eds.), *Prosimian Biology.* Duckworth, London, pp. 75–108.

Sussman, R. W. (1976). Ecologie de deux espèces coexistantes de lémur: *Lemur catta* et *Lemur fulvus rufus. Bull. Acad. Malg.* 52:175–191.

Sussman, R. W., and Rakotozafy, A. (1994). Plant diversity and structural analysis of a tropical dry forest in southwestern Madagascar. *Biotrop.* 26:241–254.

Sussman, R. W., Green, G. M., Porton, I., Andrianasolondraibe, O. L., and Ratsirarson, J. (2003). A survey of the habitat of *Lemur catta* in southwestern and southern Madagascar. *Primate Conserv.* 19:32–52.

Terborgh, J. (1986). Keystone plant resources in the tropical forest. In: Soulé, M. E. (ed.), *Conservation Biology. The Science of Scarcity and Diversity.* Sinauer Associates, Sunderland, Mass., pp. 330–344.

Wright, P. C. (1999). Lemur traits and Madagascar ecology: Coping with an island environment. *Ybk. Phys. Anthrop.* 42:31–72.

Yamashita, N. (2002). Diets of two lemur species in different microhabitats in Beza Mahafaly Special Reserve, Madagascar. *Int. J. Primatol.* 23:1025–1051.

6
Tamarind Recruitment and Long-Term Stability in the Gallery Forest at Berenty, Madagascar

KATHRYN BLUMENFELD-JONES, TAHIRIHASINA M. RANDRIAMBOAVONJY,
GEORGE WILLIAMS, ANNE S. MERTL-MILLHOLLEN, SUSAN PINKUS,
AND HANTANIRINA RASAMIMANANA

6.1. Introduction

Habitat loss is one of the main causes of species decline and extinction in Madagascar, as elsewhere. Evidence suggests that *Lemur catta*, one of Madagascar's most well-known lemurs, may be more endangered than was previously thought, due to the rapid degradation of the southern dry forests (Sauther et al., 1999). Gallery forest, in particular, is disappearing at an alarming rate. Along the Mandrare River, there are only four actual gallery forest patches (Jolly, pers. comm.). The largest is the Berenty–Kaleta complex, about 350 ha of forest and scrub, of which less than half is true gallery forest. There are three other smaller patches of forest, a few very degraded tracts, and some files of tamarind trees. That is all there is. These last remnants of gallery forest are worth protecting, both for their own sake as a highly endangered forest type and for the lemurs that depend on them.

The Malaza forest at Berenty is a 100-ha gallery forest fragment along the Mandrare River, which has been protected from outside disturbance since 1936. The forest is dominated by *Tamarindus indica* and provides a refuge for six species of lemurs. As early as 1980, it was noted that mature tamarind trees were dying in the Malaza forest and that there seemed to be little recruitment of young tamarinds (Jolly et al., 1982). *Lemur catta* is heavily dependent on a few keystone plant species that are able to provide food during times of unusual environmental stress, such as drought (Sauther, 1998). *Tamarindus indica* is such a keystone species for *Lemur catta* at Berenty. Its leaves, flowers, and fruit provide a year-round food source. Additionally, the broad, dense canopies of mature tamarinds are preferred by *Lemur catta* troops for sleeping and resting and provide protection from aerial and ground predators (Mertl-Millhollen et al., 2003). At Berenty, *Lemur catta* biomass is greater in the riverfront forest than in other parts of the reserve, correlating with the higher density of tamarinds in this patch of forest (Budnitz and Dainis, 1975). If tamarind regeneration in the Malaza forest is not keeping pace with the death of old trees, then the lemur population is likely to decline as well. As Jolly (1986) has pointed out, to save the lemurs one will have to save the trees.

In this paper, we examine the change in the Malaza forest over 30 years and use the results of a recent study of tamarind recruitment to speculate on the future of the Berenty gallery forest. We address the following questions:

1. Has the Malaza gallery forest maintained stability during the past 30 years?
2. What is the spatial and age distribution of young tamarinds (<30 cm diameter at breast height [dbh]) at Berenty?
3. How is tamarind recruitment related to environmental factors?
4. Is tamarind recruitment sufficient to replace aging adult trees?

Forest communities are naturally dynamic, as plant populations increase or decrease over the short-term. However, the conservation of small forest fragments is often based on the assumption that the protected community will maintain stability in its present location through a long-term balance of new recruitment, growth, and mortality (Felfili, 1997). Successful conservation of gallery forest fragments will depend not only on protecting the habitat exactly as it exists today but also on understanding and protecting the dynamics of its regeneration.

6.2. Three Decades of Change in the Malaza Forest at Berenty

The Malaza forest at Berenty is composed of three different forest types that are easily distinguished by casual observation (Budnitz and Dainis, 1975). Continuous-canopy forest exists only along the riverbank, where there is a permanent supply of groundwater. It is dominated by tall *Tamarindus indica* (20–25 m in height) and other large tree species, such as *Celtis* and *Neotina*. A large middle section of the reserve is characterized by a more open forest where the canopy is discontinuous, and open spaces may be filled with nearly impenetrable tangles of thorny vines, such as *Capparis sepiaria* and the invading succulent *Cissus quadrangularis*. Here the dominant trees may be shorter tamarinds (averaging 16 m) or other large, spreading trees such as *Neotina isoneura* or *Acacia rovumae*. A third type, the brush and scrub forest, is found in the driest areas of the reserve. Thorny bushes, succulents, and small trees such as *Azima tetracantha* and *Salvadora angustifolia* dominate this habitat. Although there is no canopy in the brush and scrub, tamarinds are scattered throughout, either singly or in small groups.

We use data gathered in 1973, 1995, and 2000 to describe change in the forest over the past 22–27 years. We first examine the Malaza reserve as a whole to determine if there has been a general trend toward a drier, more open forest during the 22 year period from 1973 to 1995. We then analyze in more detail a single 100 × 100 m plot located in the closed canopy riverfront forest to detect changes in species composition during a 27-year period (1973 to 2000) in the most intensively studied portion of the reserve.

6.2.1. Methods

In 1973, Blumenfeld-Jones and Budnitz made canopy maps of large sections of each of the above three forest types. We considered canopy cover to be a better estimate of potential food resources than stem count for *Tamarindus indica*, a year-round dietary staple for *Lemur catta*. From a lemur-eye view, canopy volume also determines what aerial pathways and sleeping sites are available. Transects were laid down at 50-foot intervals going north to south and west to east. While walking the transects, the crowns of all canopy-level trees were drawn onto a grid superimposed on the reserve map. Total canopy cover was calculated for the three forest types (Budnitz, 1978). In 1973, 88% of the area labeled continuous canopy had tall tree canopy overhead, approximately 50% of the open forest was covered by canopy, and 12% of the brush and scrub had tree canopy, indicating that the three habitats observed when walking forest paths differed by canopy cover.

Primatologists working at Berenty have found the above three habitat types useful in describing differences in population dynamics, resource quality and availability, feeding behavior, and ranging patterns (Budnitz and Dainis, 1975; Jolly et al., 2002; Mertl-Millhollen et al., 2003). Using these three forest types as a foundation, we developed six distinct vegetation categories that were both easy to identify and described the more subtle habitat variations and differences in species composition that we found on the detailed canopy maps. The vegetation zones were numbered in order, from the community with the highest water need (1) to the most xeric (6).

1. *Closed canopy tamarind forest* is dominated by large *Tamarindus indica* and the canopy is continuous (>75% cover).
2. *Open Neotina–tamarind forest* averages about 55% canopy coverage, consisting primarily of *Tamarindus* or *Neotina* with a few *Acacia* and *Crateva*.
3. *Open tamarind parkland* is an open forest of almost 100% tamarind. In 1973, the understory was small shrubs and open grassy areas.
4. *Open Acacia-scrub forest* is dominated by *Acacia* spp., followed by *Tamarindus indica*, a scattering of *Quivisianthe,* and a few large banyans. The undergrowth begins to thicken here into a tangle of shrubs and vines.
5. *Brush and scrub* is characterized by the sudden appearance of a more xeric vegetation type, including *Quivisianthe, Azima tetracantha,* and *Salvadora angustifolia* with a low-growing thorny understory. There is no true canopy, although *Tamarindus* is scattered throughout.
6. *Brush and scrub–spiny forest transition* is the same as (5) above but true desert trees, such as *Allaudia procera* and *Aloe* spp., begin to appear marking a transition to the true spiny forest that lies outside the study area.

In 1981, Howarth et al. (1986) independently did a comparable study, observing and mapping seven similar vegetation zones in this same forest. Field work by Pinkus in 1995 resulted in a more detailed map of the entire Malaza forest, using both the 1973 vegetation types and 1981 the vegetation zones of Howarth and colleagues. We compare Pinkus' map to Blumenfeld-Jones' 1973 map to illustrate macrolevel changes in the Berenty Reserve over 22 years. The Pinkus map was spot checked in 2000 to make certain that the vegetation categories were consistent between studies. Based on this census, we combined groups 5 and 6 above into one category, (5) brush and scrub, and created transition zones where vegetation did not strictly conform to one of the five types.

A stem count was done in each vegetation zone to check the validity of these categories. The sample was a permanently labeled set of 451 trees along the forest trails. All trees above 20 cm dbh with canopy touching a vertical plane from the trails edge were included. Results of this count (Table 6.1) verify the species composition of the above categories.

6.2.2. Results

6.2.2.1. Change in the Overall Pattern of Vegetation

In general, the vegetation types were quite consistent between the two maps, and most of the reserve could easily be categorized in both years. The two maps are presented in Figures 6.1a and 6.1b. The area (as a percent of entire reserve) of each vegetation type is given in Table 6.2.

Both maps clearly show two vegetation gradients from higher water use to low water use vegetation, suggesting a more complex model of groundwater than simply distance from the river. The north to south gradient is more prominent in the western half of the reserve and correlates with the current route of the Mandrare

TABLE 6.1. Number (and percent) of individual trees >20 cm dbh in each forest type.

	Closed tamarind	Open tamarind–*Neotina*	Open tamarind parkland	*Acacia* scrub	Brush and scrub
Total transects (m)	1000	1400	425	1450	875
Total stems	123	127	44	104	53
Stems/100 m	12.3	9.1	9.5	7.2	6.0
Total stems (percent by vegetation type)					
Tamarindus indica	43 (35)	27 (27)	27 (61)	25 (18)	18 (34)
Neotina isoneura	17 (14)	46 (36)	0	5 (4.5)	0
Rinorea greveana	12 (10)	20 (16)	0	2 (7)	0
Celtis philippensis	6	11	0	0	0
Acacia spp.	16 (13)	6 (5)	2 (4.5)	32 (24.5)	6 (11)
Crateva spp.	9 (7)	4 (3)	3 (7)	9 (13)	2 (4)
Albizia spp.	5	7	1	5	1
Celtis bifida	5	3	1	4	3
Quivisianthe papinae	1 (1)	2 (1.5)	1 (2)	0	10 (19)
Azima tetracantha	0	0	5	2	2
Salvadora	0	0	0	4	4
Euphorbia spp.	0	0	0	3	5
Other	9	1	4	13	2

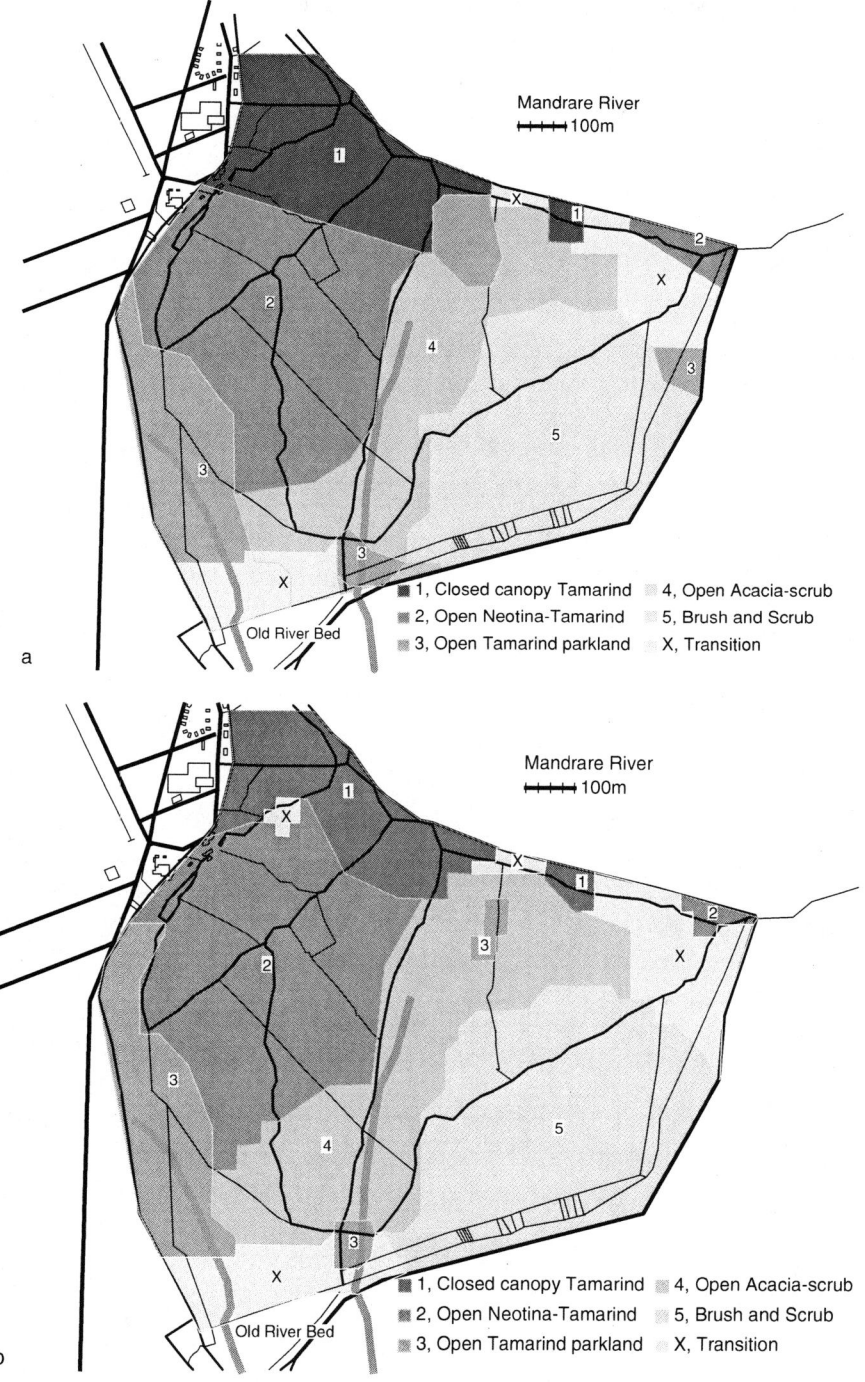

FIGURE 6.1. (a) Map of vegetation zones in Malaza forest, 1973. (b) Map of vegetation zones in Malaza forest, 1995.

TABLE 6.2. Percent area of each vegetation type in 1973 and 1995.

Year	Closed tamarind (%)	Open tamarind–Neotina (%)	Open tamarind parkland (%)	Acacia scrub (%)	Brush and scrub (%)	Transition (%)
1973	14.3	26.9	10.7	13.9	23.3	10.8
1995	10.7	24.9	9.4	20.8	23.6	10.6
Change	−3.6	−2.0	−1.3	+6.9	+0.3	−0.3

River. The west to east gradient is most likely associated with the path of an old river, visible both in air photos, and as a depression at ground level. Williams plotted the position of this old riverbed (see Figure 6.1), and our five vegetation types appear to follow its outline closely. In particular, the *Tamarindus* parkland (zone 3) seems to thrive along the banks of this old river channel. As might be expected, the vegetation type requiring the most water (closed-canopy tamarind forest) is found only in the northwest corner where these two gradients converge.

The maps in Figure 6.1 depict two patterns of change that seem related to these two gradients. The west to east pattern of vegetation appears to be fairly stable. The boundaries that parallel the old riverbed have changed very little in 22 years. This includes the drier half of the reserve, primarily *Acacia* forest and brush and scrub. The 1995 boundaries of the *Acacia* scrub forest (zone 4) and the brush and scrub (zone 5) are nearly identical to those recorded in 1973. The dry riverbed does not seem to have much impact on present-day forest reproduction and maintenance at Berenty.

On the other hand, the north–south vegetation zone boundaries paralleling the current river have shifted north toward the riverbank, particularly in the western half of the reserve. This has resulted in a sizeable reduction in the amount of closed canopy forest (zone 1) bordering the river and an increase in the amount of *Acacia* scrub (zone 4) at the southern edge of the reserve, furthest from the river. Additionally, three small patches of forest embedded in *Acacia* scrub along the eastern edge of the current river had all decreased in size by 1995. This pattern is consistent with an overall drying of the Malaza forest due to changing dynamics in the current river. Because the closed canopy *Tamarindus* forest was most affected by this drying trend, we looked at this piece of forest in more detail.

6.2.2.2. Differences in Canopy Cover and Species Composition in the Closed Canopy Forest (1973 to 2000)

In 2000, a 100×100 m plot was laid out in the center section of the closed canopy riverfront forest (zone 1). The canopy was mapped using the techniques described above, and a comparison was made to a map of the same plot drawn in 1973. The results are shown in Table 6.3 and Figures 6.2a and 6.2b.

TABLE 6.3. Percent canopy cover.

Canopy species	Closed canopy Plot 1 1973 (%)	Closed canopy Plot 1 1995 (%)	Open Neotina–tamarind Plot 2 1995 (%)
Tamarindus indica	50.7	20.9	15.8
Neotina isoneura	11.2	16.6	14.7
Celtis philippensis	4.7	2.1	0
Acacia spp.	2.2	2.1	0
Albizia spp.	0.4	0.5	1.8
Quivisianthe papinae	0	0.3	0.7
Other	0.8	1.7	1.9
Total canopy cover	75.5	42	34.8

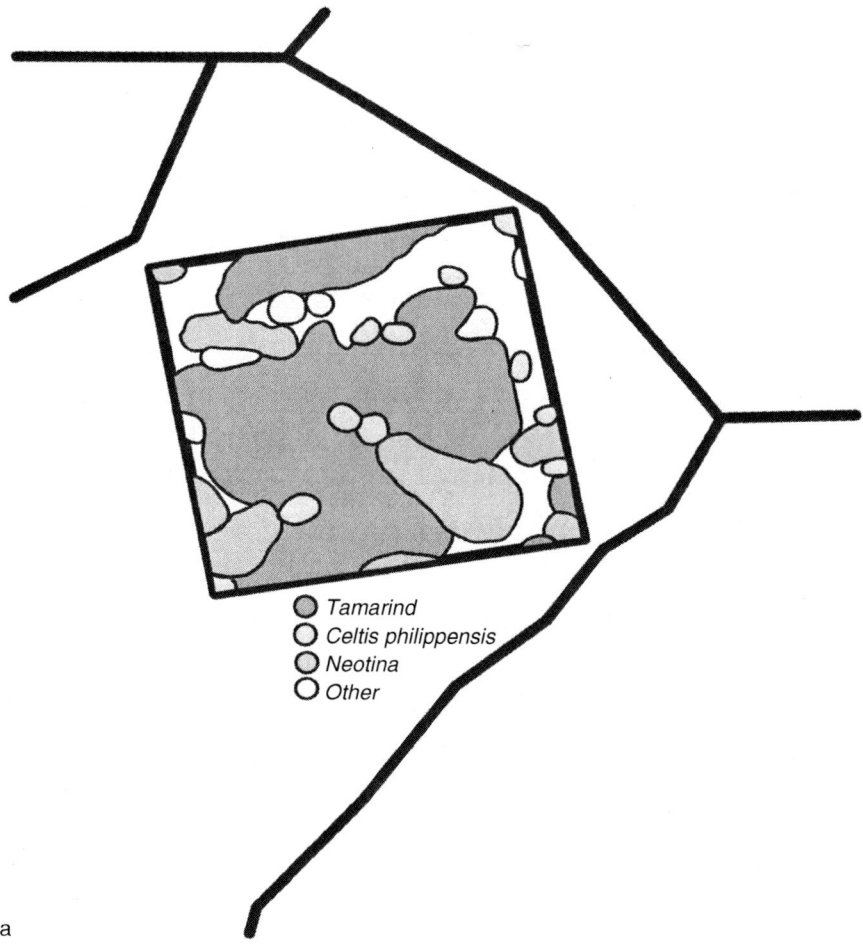

FIGURE 6.2. (a) Canopy map of 100 × 100 m plot in closed canopy forest, 1973.

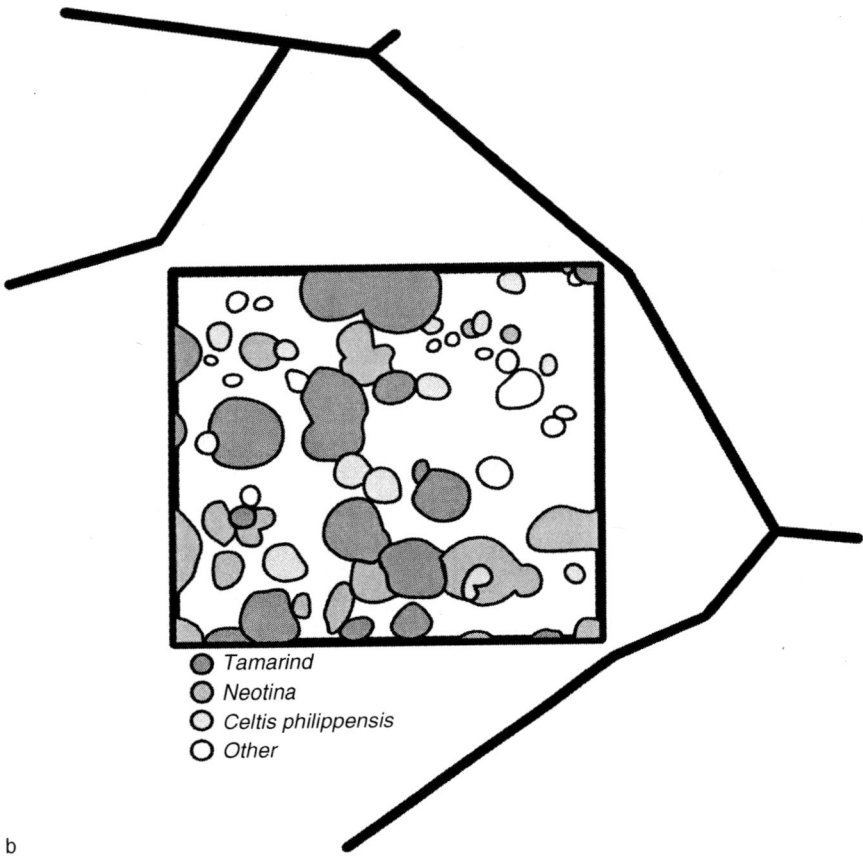

FIGURE 6.2. (*Continued*) (b) Canopy map of same 100 × 100 m plot in closed canopy forest, 2000.

Table 6.3 shows a rather remarkable decrease in both the overall amount of canopy cover and the amount of tamarind cover from 1973 to 2000. A comparison with the other tree species present indicates that *Tamarindus indica* is being disproportionately affected by the environmental processes at work in this part of the reserve. The loss of nearly half of the overall canopy cover from 1973 to 2000 can be accounted for by the loss of tamarind cover. A similar 100 × 100 plot located in the Open *Tamarindus–Neotina* forest (zone 2) shows a tamarind population that has remained stable at about 16–17% for the past 27 years (Budnitz, 1978). A comparison of all three columns in Table 6.3 indicates that the 2000 plot located in what is today considered to be the closed canopy forest bears more resemblance to the open *Tamarindus–Neotina* plot than it does to the closed canopy plot in 1973 in both total area covered and species composition.

To summarize, the most notable change in the Berenty Reserve during the past three decades is the decrease in the amount and quality of the tamarind dominated closed-canopy forest. This change appears to be associated with the location of the current river, but the exact water dynamics are not known at this time. It has been well documented that large tamarinds are dying in the Berenty Reserve (both above and see Koyama et al., this volume), but the extent to which *Tamarindus indica* is regenerating in this forest is less well known. We next present a 2000 study focused on this issue.

6.3. Tamarind Regeneration

Although anecdotal evidence suggests that *Tamarindus indica* is not reproducing in the Malaza forest, there has been little study of the problem. One notable exception is Miles' 1995 study of tree regeneration at Berenty (Miles, 1996). She concludes that tamarind seedling production is quite high, but there is also a concomitantly high seedling mortality rate, with very few seedlings surviving to reach 30 cm in height. Dead seedlings appeared to have dried out, and some may have been preyed on by lemurs although this was not measured. Tamarind seedlings were uniformly distributed throughout her 13 study plots (primarily in the closed and open canopy forests), and abundance was not correlated with soil type or canopy cover. Seedling density was found to decrease with distance from paths and from both current and past riverbeds. In the following study, we examine *Tamarindus indica* regeneration from 2000 to 2004.

6.3.1. Methods

Data for this study were collected in September 2000 and September 2004. We recorded tamarind reproduction in every area of the forest. All tamarinds less than 30 cm dbh were sampled along 11,025 m of trails and transects covering all vegetation zones in the reserve (Table 6.4).

Trails consisted of large paths, approximately 3 to 4 m wide, and small paths that measured about 1 m in width. Interior transects were straight lines through the middle of the forest and did not intersect with paths. Interior transects were either 50 or 100 m long and 10 m wide. In addition to the above sampling method,

TABLE 6.4. Distribution and total length (in meters) of regeneration transects.

Transect type	Closed tamarind (m)	Open tamarind–*Neotina* (m)	Open tamarind parkland (m)	*Acacia* scrub (m)	Brush and scrub (m)	Transition (m)
Large trails	925	1575	150	900	500	700
Small trails	200	1075	750	750	1450	300
Interior lines	1100	650	—	—	—	—
Total	2225	3300	900	1650	1950	1000

we also walked areas away from the transects, looking for successful tamarind regeneration in areas not sampled. None was found, which leads us to believe that our study may actually overrepresent the proportion of young tamarinds in the reserve.

All tamarinds under 30 cm dbh and within 5 m of either side of a transect or trail edge were mapped, tagged, and measured in 2000. Plants were assigned to one of five size classes. Plants between 1 and 2 m tall were classed as seedlings. Plants between 2 m tall and 5 cm dbh were labeled saplings, and those from 5 to 10 cm dbh were designated as poles. Larger young trees were classed as either 10–20 or 20–30 dbh. Seedlings under 1 m tall are not included in this particular sample because our focus was on successful regeneration.

In 2004, the sample was remeasured, and new recruits into the seedling category were added. We measured distance from the current and past riverbeds and also looked at two environmental factors not included in Miles' study: edge effects and distance from mature tamarinds. Because roads passing through forests are said to affect the microclimate much like forest edges and treefall gaps (Turner, 1996), we compared the distribution of recruits along 4-foot-wide forest trails with the distribution along small paths and line transects to see if *Tamarindus indica* seedlings responded to edge effects. We also measured the distance from each tamarind in our regeneration sample to the nearest tamarind over 30 cm dbh and mapped the tamarind canopy cover over each young recruit.

6.3.2. Results

6.3.2.1. Distribution of Young Tamarinds by Habitat Type

In 2000, we located 249 tamarinds <30 cm dbh along the trails and transects in Table 6.4. We examined the distribution of these plants both by age class and by habitat type (Table 6.5).

TABLE 6.5. Distribution of young *Tamarindus indica* size classes.

Tamarind size class	Closed tamarind	Open tamarind–*Neotina*	Open tamarind parkland	*Acacia* scrub	Brush and scrub	Transition	Total
Seedling (1 to 2 m)	15	41	1	1	1	4	63
Sapling (2 m to 5 cm dbh)	45	26	0	2	0	16	89
Pole (5 to 10 cm dbh)	19	8	0	4	3	8	42
10 to 20 cm dbh	16	8	2	0	3	8	37
20 to 30 cm dbh	7	2	1	2	1	5	18
Total (stems/1000 m^2)	102 (4.6)	85 (2.6)	4 (0.4)	9 (0.5)	8 (0.4)	41 (4.1)	249

Size class counts ranged from a low of 18 trees in the largest class (20–30 cm dbh) to a high of 89 in the sapling class. In general, frequency decreased as size increased. The overall density of young tamarinds was highest in the two vegetation zones bordering the bank of the current river (closed canopy forest and transition), indicating an increase in habitat favorability as the river is approached. The open *Tamarindus–Neotina* forest along the path of the old riverbed had an intermediate density of regenerating tamarinds, whereas the dry outer regions had very little regeneration at all.

Tamarind recruitment was definitely patchy in all habitats. We measured the distance between each individual in our sample and the nearest young tamarind to it and found that 96.5% of our sample had another young tamarind stem within 15 m of it. In 80% of the cases that nearest young tamarind was less than 5 m away. Clusters of 5 to 10 young tamarinds in a single location were not uncommon, suggesting that there are specialized recruitment niches that particularly favor seedling survival.

6.3.2.2. Survival and Growth

In the 2004 re-census, we found that 15 of our 249 sample plants had died, including 6 seedlings, 6 saplings, 2 poles, and 1 small tree. The riverbank had eroded away from the roots of the two poles, and the small tree had been crushed when a larger tree fell during a windstorm. It was not obvious why the smaller plants had died. During this same time, 29 new tamarinds in the sample area exceeded the 1-m-tall size limit and were added to the study.

Growth rates between 2000 and 2004 varied by size category and location. In general, the smaller seedlings and saplings grew quite slowly. A few near the river increased as much as 0.4 cm dbh a year, but most (80%) had an annual increase of less than 0.25 cm a year. Growth decreased with distance from the river, and many seedlings and saplings in the drier areas of the forest failed to grow at all, maintaining the same size and shape during the 4 years. Poles increased dbh by an average of 0.35 cm per year (range 0–1.1 cm) and small trees by an average of 0.85 cm per year (range 0.05–1.5 cm). The variable rates of growth made it difficult to estimate the age of a plant by size. It is possible that a zero-growth sapling in a dry habitat could be the same age as a much larger tree growing in a more favorable environment.

6.3.2.3. Edge Effects

The altered microclimate at forest edges has direct effects on the forest community. Edges have been found to differ from forest interiors in light availability, temperature, relative humidity, and soil moisture, and seedling recruitment is sometimes enhanced at forest edges (Turner, 1996; Kellman et al., 1998; Laurance et al., 1998). We compared tamarind recruitment along 4-m-wide trails to that along small paths and interior line transects to see if there was a response to edge phenomena (Table 6.6).

TABLE 6.6. Effect of trail width on tamarind regeneration.

Transect type (width)	Total length of transects (m)	Number of tamarinds (less than 30 cm dbh)	Ratio of trees to transects
Large trails (3 to 4 m)	2500	156	.062
Small trails (1 m)	1275	8	.006
Interior lines (0)	1850	21	.011

Tamarind regeneration along wide trails was 6 to 10 times greater than along the same length of narrow path or interior transect. We conclude that trail edges are more favorable habitats for tamarind recruitment than interior forest sites.

6.3.2.4. Distance from Mature Tamarinds

A set of intriguing recent studies by Parvez et al. (2003, 2003a, 2004) has found biologically active growth regulators present in the roots, leaves, bark, and seeds of mature *Tamarindus indica*. Water-soluble exudates of these plant parts had a strong inhibitory effect on the growth of a variety of agricultural crops and weed species. They suggest that this likely contributes to maintaining the weed-free environment around the base of adult tamarinds. No mention is made of what effects these chemicals might have on the growth of tamarind seedlings. We examined the distribution of young tamarinds with respect to the location of large trees to see if large tamarinds might, in fact, be limiting their own reproductive success (Table 6.7).

In the three smallest size classes, 88% of the sample was located at least 15 m from the nearest adult tamarind, and 65.5% were more than 25 m away. A direct measure of the overhead canopy cover revealed that 97% of tamarind seedlings and saplings did not grow under tamarind canopy, the area likely to retain the highest concentration of allopathic chemicals. Miles (1996) reported no correlation between mean canopy cover (of undifferentiated species) and successful tamarind regeneration, which suggests that the lack of regeneration under large tamarinds is due to factors other than competition or decreased light level. Recruits above 10 cm dbh were more evenly distributed with respect to larger tamarinds, possibly because their age cohort included trees above 30 cm dbh.

TABLE 6.7. Distance from young tamarinds to nearest mature tamarind (>30 cm dbh).

Distance (m)	Seedlings	Saplings	Poles	10 to 20 cm dbh	20 to 30 cm dbh
0 to 5 m	1	1	3	8	2
5 to 15 m	5	9	2	6	7
15 to 25 m	16	17	5	4	1
>25 m	34	56	24	10	4

6.4. Discussion

6.4.1. Three Decades of Change in the Malaza Forest

There have been some rather striking changes in the Malaza forest during the past three decades, especially in the closed canopy forest that borders the bank of the Mandrare River. This small parcel of *Tamarindus*-rich forest, slightly more than 14 ha in 1973, was reduced to nearly 10 ha by 1995, a decrease in area of 25% over 22 years. There also have been changes in the forest structure and species composition. In 1973, this 10 ha parcel was a true continuous canopy forest dominated by large spreading *Tamarindus indica*. Jolly (1966) described this piece of forest as giving "the impression of a 500-year-old oak forest." Today this same 10-ha parcel resembles more closely the neighboring open *Tamarindus–Neotina* forest than it does the dense tamarind forest of the 1960s and 1970s.

The Malaza forest in general appears to be in transition toward a drier forest type where tamarind is present but perhaps not as uniformly dominant along the current riverbank. The open forests along the bed of the old river channel on the western edge of the reserve are becoming more impenetrable, the understory thick with tangles of small shrubs and vines. In places where we had little trouble following lemurs in 1973, today one must crawl on hands and knees through thorny scrub to keep them in view. *Allaudia procera*, a signature species of the spiny desert, limited to the southeast edge of the reserve in 1973, is now seen across the trail several hundred meters further north.

The loss of *Tamarindus indica* canopy along the river stands out as a critical difference between the 1973 forest and the one we see today. It has changed the qualitative character of the forest and potentially reduced the amount of resources available to the lemurs that depend on this small fragment of isolated forest. Whether these changes are a relatively short-term perturbation or the mark of a more irreversible long-term trend depends, in part, on the extent to which *Tamarindus indica* is able to regenerate in this forest.

6.4.2. Regeneration of Tamarindus indica

In 2000, *Tamarindus indica* recruits were found clustered in small groups in the closed canopy, open canopy, and transitional forests. Tamarind regeneration appears to be limited to the riverine environments near the bank of the Mandrare River and along the bed of the dry river channel and is probably associated with the height of the water table or specific soil types found there. However, distribution is still patchy within these habitats, indicating that *Tamarindus indica* has an even more specialized regeneration niche.

Seedling growth and survival was more successful along the edges of 4-foot-wide trails than along either small paths or transects through the undisturbed center of the forest. Possible reasons for this include increased light and routes taken by seed dispersers. Because newly germinated seedlings are found in abundance throughout the forest, we attribute the higher survival rate along wide paths to increased light availability. But even along wide trails in favorable river habitat, *Tamarindus indica* recruits thrive only in the few scattered patches not already colonized by mature tamarinds.

We examined the possibility that adult tamarinds produce a hostile chemical environment under their canopy that inhibits the growth of many species, including their own seedlings. Our findings, that 97% of the sample tamarind recruits did not have tamarind canopy overhead and that 88% of the sample was at least 15 m from the nearest large tamarind, are consistant with this hypothesis. Additionally, recruits were never found in the gaps surrounding large dead tamarinds, indicating that the inhibitory effect of these chemicals may persist long after the trees have disappeared from the landscape.

The decrease in the amount of closed canopy forest and the slow, but persistent death of mature tamarinds at Berenty, when coupled with the patchy distribution of new recruits is cause for concern. Young tamarinds are probably not present in sufficient numbers or distributed widely enough to maintain this forest in its current form. Even an adequate number of seedlings will not turn back the decline of this forest if they are bound to the few scattered sites that are suitable for tamarind regeneration. Regeneration that is limited to a specialized niche, requiring water, light, and soil that is free of tamarind roots and litter, is unlikely to produce either the numbers or breadth of distribution necessary to replace the loss of older tamarinds in the closed canopy forest.

6.4.3. *Adaptation to a Changing Environment*

The closed canopy forest at Berenty, dominated by huge, stately tamarinds in the past, does not appear to be a stable climax forest. One would hardly expect otherwise, existing as it does on the banks of the wide and sometimes powerful Mandrare River. Large rivers are always changing, shifting channels, changing course, alternatively flooding during high-water and turning into mud flats during low, building new flatland while eroding old banks. A gallery forest, dependent on a constantly shifting river, must be specially adapted to this mobile, highly unstable environment. There are some hints in the Malaza forest of how this process has occurred in the past.

The Berenty gallery forest has a life history that is intimately connected to the Mandrare River. The character of the forest bears a strong relationship to the contours of both the current river and an old riverbed, leaving a footprint of what this forest might have been like in the past, perhaps hundreds of years or more ago. Returning to Figure 6.1, we can imagine that the forest that once existed along the banks of the old riverbed might have been similar to today's closed canopy forest. Remnants of a tamarind-dominated forest still hug these old riverbanks, but the soil is dry now and the canopy is discontinuous.

Further east, in the drier brush and scrub, tamarinds are found in small groves scattered among low brushy vegetation sometimes dotted with *Allaudia* and other intruders from the spiny desert. The spatial patterning of these groves is not unlike today's patchy distribution of tamarind recruits near the current river.

Could these groves be the outcome of ancient regeneration niches, supporting a final few clusters of young tamarinds, as the river arm and forest around it begin to dry out? Looking further up the old bank and into what was once spiny desert, single large tamarinds can be found, standing in sharp profile to the sisal fields that now surround them. These old trees have persisted, even as the river changed course and everything around them turned to desert. Were they once part of a small grove? A lone seedling at the outer edge of sustainable habitat? What seems certain is that tamarinds did not suddenly appear in the desert, but that the forest they were once a part of gradually disappeared around them as the river shifted and the land dried out.

We suggest that the transition in progress at the Berenty Reserve is a natural process. The gallery forest is a community that is in equilibrium with a dynamic river. It survives as a spatial and temporal mosaic of patches in different successional stages, including dense riverfront forest, open canopy forest, and brush and scrub. Within this mobile system, the riverfront forest does not remain rooted to one spot on the bank, but slowly, over perhaps centuries, is able to follow the course of a changing river. The precarious status of Madagascar's gallery forests today is likely due to relatively recent human activity that has interfered with the natural regeneration of these forests. Sussman and Rakotozafy (1994) have called the southern gallery forest one of the most endangered forest types in Madagascar, due to overgrazing and cutting of the forest. Even in the protected environment at Berenty, it appears that cultivation along the riverbank has prevented the natural regeneration of gallery forest, probably for many years now.

At Berenty, however, there are hints that natural regeneration processes may reappear if farming is permanently relocated away from the lower riverbank. Cultivation was stopped in the reserve in the 1990s, and by 2004 the beginnings of a new forest was appearing at the bottom of the riverbank where crops once grew. A wide band of reeds now separates the river from its bank, and flood-tolerant trees, such as *Ficus* and *Pithecellobium*, grow on the soil trapped behind them. If the soil continues to build and dry below the riverbank, conditions there might someday be excellent for new tamarind recruitment. With a permanent source of water, light at the edge of the forest, and soil free from tamarind roots and litter, this may be where gallery forest will exist in the future if given adequate protection.

Conceptualizing the gallery forest and river as parts of a single dynamic system calls for a different approach to conservation and management. The tamarind-dominated gallery forests of southern Madagascar exist in discontinuous, narrow bands at the edges of rivers and streams and appear to be unable to regenerate continuously on the same small parcel of land. Although gallery forests may be hundreds of years old, efforts to preserve these small fragments by protecting the land on which they stand will ultimately fail as the river shifts course and the existing forest nears the end of its life cycle. That is perhaps what we are seeing in the Malaza forest at Berenty. But there are also hopeful signs at Berenty. Within 10 years of stopping cultivation on the lower banks, large trees stand where sweet potatoes once grew. It might even be possible to accelerate these regeneration processes with the use of appropriate management practices to enhance the capture of soil at the river's edge and facilitate the colonization of the lower bank with gallery forest species. To be successful, the conservation of southern gallery forests must be grounded in a thorough understanding of the dynamics of its natural regeneration. Our data suggest that long-term conservation success at Berenty will probably not come in maintaining today's small patches of forest but rather in stabilizing and protecting the areas on the lower riverbank where forest regeneration is possible.

Acknowledgments. We give a special thanks to Alison Jolly who has brought us all together and, with extraordinary amounts of enthusiasm, knowledge, and support, made this study possible, and to Richard Jolly who tagged trees and trails while helping us to see the big picture. Thanks also to the many Berenty researchers who have variously encouraged, inspired, helped, entertained, and in so many other ways enriched our visits to the field. We are most grateful to the de Heaulme family for protecting the gallery forest and for their gracious support and hospitality.

References

Budnitz, N. (1978). Feeding behavior of *Lemur catta* in different habitats. In: Bateson, P. P. G., and Klopfer, P. H. (eds.), *Perspectives in Ethology*, Vol. 3. Plenum, New York, pp. 85–108.

Budnitz, N., and Dainis, K. (1975). *Lemur catta:* Ecology and behavior. In: Tattersall, I., and Sussman, R. W. (eds.), *Lemur Biology*. Plenum, New York, pp. 219–236.

Connell, J. H., and Green, P. T. (2000). Seedling dynamics over thirty–two years in a tropical rain forest tree. *Ecology* 81(2):568–584.

Felfili, J. M. (1997). Dynamics of the natural regeneration in the Gama gallery forest in central Brazil. *Forest Ecol. Management* 91:235–245.

Howarth, C. J., Wilson, J. M., Adamson, A. P., Wilson, M. E., and Boase, M. J. (1986). Population ecology of the ring-tailed lemur, *Lemur catta*, and the white sifaka, *Propithecus verreauxi verreauxi*, at Berenty, Madagascar, 1981. *Folia Primatol.* 47:39–48.

Jolly, A. (1966). *Lemur Behavior*. University of Chicago Press, Chicago.

Jolly, A (1986). Lemur survival. In: Benirschke, K. (ed.), *The Road to Self-Sustaining Populations*. Springer-Verlag, New York, pp. 71–96.

Jolly, A., Dobson, A., Rasamimanana, H. M., Walker, J., O'Conner, S., Solberg, M., and Perel, V. (2002). Demography of *Lemur catta* at Berenty Reserve, Madagascar: Effects of troop size, habitat, and rainfall. *Int. J. Primatol.* 23(2):327–353.

Jolly, A., Oliver, W. L. R , and O'Connor, S. M. (1982). Population and troop ranges of *Lemur catta* and *Lemur fulvus* at Berenty, Madagascar: 1980 Census. *Folia Primatol.* 39:115–123.

Kellman, M., Tackaberry, R., and Rigg, L. (1998). Structure and function in two tropical gallery forest communities: Implications for forest conservation in fragmented systems. *J. Appl. Ecol.* 35:195–206.

Laurance, W. F., Ferreira, L. V., Rankin-De Merona, J. M., Laurance, S., Hutchings, R. W., and Lovejoy, T. E. (1998). Effects of forest fragmentation on recruitment patterns in Amazonian tree communities. *Conservation Biol.* 12(2):460–464.

Mertl-Millhollen, A. S., Moret, E. S., Felantsoa, D., Rasamimanana, H., Blumenfeld-Jones, K. C., and Jolly, A. (2003). Ring-tailed lemur home ranges correlate with food abundance and nutritional content at a time of environmental stress. *Int. J. Primatol.* 24(5):969–985.

Miles, L. (1996). *Regeneration in a Dry Forest Reserve*. MSc thesis, University College, London.

O'Conner, S. M. (1987) The effect of human impact on vegetation and the consequences to primates in two riverine forests, southern Madagascar. PhD thesis, Cambridge University, Cambridge.

Parvez, S. S., Parvez, M. M., Nishihara, E., Gemma, H., and Fujii, Y. (2003). *Tamarindus indica* L. leaf is a source of allelopathic substance. *Plant Growth Regul.* 40(2):107–115.

Parvez, S. S., Parvez, M. M., Fujii, Y., and Gemma, H. (2003). Allelopathic competence of *Tamarindus indica* L. root involved in plant growth regulation. *Plant Growth Regul.* 41(2):139–148.

Parvez, S. S., Parvez, M. M., Fujii, Y., and Gemma, H. (2004). Differential allelopathic expression of bark and seed of *Tamarindus indica* L. *Plant Growth Regul.* 42(3):245–252.

Rasamimanana, H. (1999). Influence of social organization patterns on food intake of *Lemur catta* in the Berenty Reserve. In: Rakotosamimananana, B., Rasamimanana, H., Ganzhorn, J. U., and Goodman, S. M. (eds.), *New Directions in Lemur Studies*. Kluwer Academic, New York, pp. 173–188.

Rasamimanana, H. R., and Rafidinarivo, E. (1993). Feeding behavior of Lemur catta females in relation to their physiological state. In: Kappeler, P. M., and Ganzhorn, J. U. (eds.). *Lemur Social Systems and their Ecological Basis*. Plenum, New York, pp. 123–134.

Sauther, M. (1998). Interplay of phenology and reproduction in ring-tailed lemurs: Implications for ring-tailed lemur conservation. *Folia Primatol.* 69(Suppl.1):309–320.

Sauther, M. L., Sussman, R. W., and Gould, L. (1999). The socioecology of the ringtailed lemur: Thirty-five years of Research. *Evol. Anthropol.* 8:120–132.

Sussman, R. W., and Rakotozafy, A. (1994). Plant diversity and structural analysis of a tropical dry forest in Southwestern Madagascar. *Biotropica* 26(2):241–254.

Turner, I. M. (1996). Species loss in fragments of tropical rain forest: A review of the evidence. *J. Appl. Ecol.* 33:200–209.

7
Home Ranges of Ringtailed Lemur Troops and the Density of Large Trees at Berenty Reserve, Madagascar

NAOKI KOYAMA, TAKAYO SOMA, SHINICHIRO ICHINO, AND Y. TAKAHATA

7.1. Introduction

Recently, primatologists have tried to analyze the complicated relationships between ecological factors and social systems of primates. Several socioecological models have been proposed (Nakagawa, 1998; Sterck, 1999). However, few long-term studies have been carried out both on wild primate and plant populations.

Some authors have argued about the importance of kily trees (*Tamarindus indica*) belonging to the family Leguminosae, as a staple food species for ringtailed lemurs (Sauther, 1992; Rasamimanana and Rafidinarivo, 1993). In Berenty Reserve, Madagascar, kily alone accounted for 34.9% of all feeding time (Soma, 2003), although this species is thought to have originated from the savanna of tropical Africa (Hotta, 1989).

At Berenty Reserve, socioecological studies of ringtailed lemurs (*Lemur catta*) have been conducted for 11 years from 1989 to 2000, and some of the results have been published (Koyama et al., 2001, 2002). Due to social changes such as troop divisions and evictions, the number of troops in the study area increased. Consequently, the home range size of a troop decreased (Koyama et al., 2002). Since 1982, we have been measuring and monitoring large trees exceeding 50 cm in diameter at breast height (DBH) within the area of 30.4 ha including our main study area of 14.2 ha.

In this paper, we analyze the following subjects: (1) changes in home range size of ringtailed lemur troops within the main study area of 14.2 ha for 11 years; (2) location, number, and density of large trees exceeding 50 cm in DBH both within the main study area (14.2 ha) and within the broader study area (30.4 ha); (3) changes in the number of kily trees per ringtailed lemur. We discuss the availability of foods with regard to the current conditions of ringtailed lemurs inhabiting the study area.

7.2. Study Area and Methods

7.2.1. Vegetation

This study was carried out at Berenty Reserve in southeastern Madagascar (Koyama et al., 2001, 2002). Annual rainfall in this area is about 580.6 mm (mean for the period 1989–1998). The natural vegetation types of this area are roughly grouped into (1) canopy forest dominated by *Tamarindus indica*, (2) open forest, (3) brush and scrub forest, and (4) subdesert forest, most of which has been cleared and replaced by plantations of sisal, *Agave rigida* (Budnitz and Dainis, 1975). The amount of rainfall varies from year, ranging from 911.2 mm (in 1998) to 226.0 mm (in 1991) (Figure 7.1). Severe droughts occurred in 1983 and 1991, which may have affected the population of ringtailed lemurs (Jolly et al., 2002; Koyama et al., 2002). On the other hand, the amount of precipitation showed no consistent correlation with the time (Spearman rank coefficient = 0.426, p = 0.061 > 0.05). Most rain falls between November and February, and little rain falls during July, August, and September (Figure 7. 2).

Based on the vegetation types, the Berenty Reserve is classified into several areas: (1) Ankoba, largely regrown forest with non-native trees from cleared ground; (2) tourist front, a part of the western boundary of the reserve, studded with tourist bungalows; (3) gallery forest, a natural forest with canopy covering more than 50% of the sky; and (4) scrub forest, a drier natural forest with more than 50% open sky (Jolly et al., 2002). Our main study area (ca. 14.2 ha) was located in the center of this reserve, corresponding with the "tourist front" and "gallery forest."

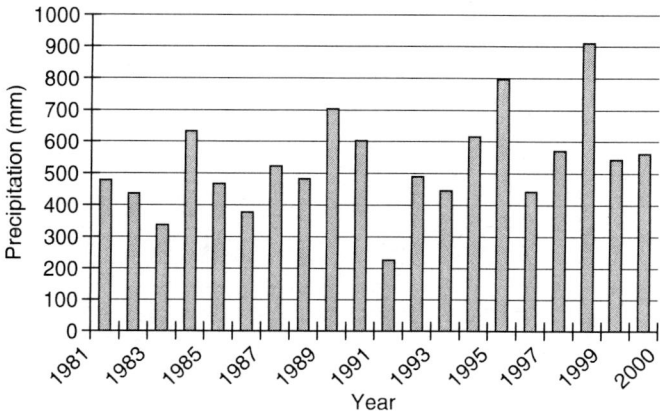

FIGURE 7.1. The annual rainfall recorded from 1981 to 2000.

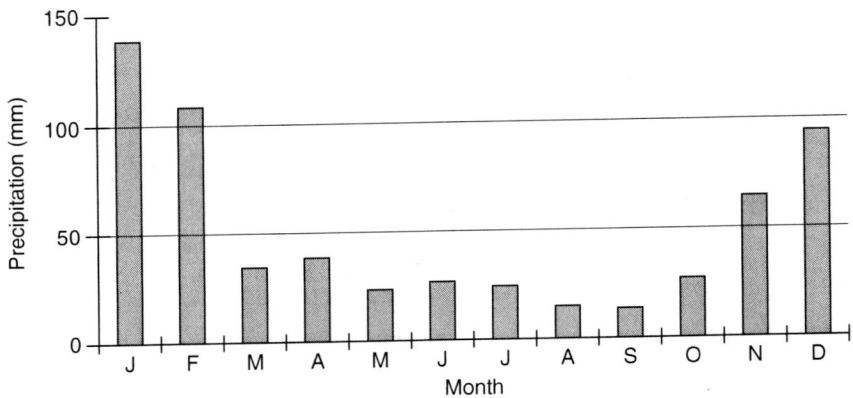

FIGURE 7.2. Mean monthly precipitation from 1989 to 2000.

7.2.2. Studies of Home Ranges

In a main study area of 14.2 ha, we have been monitoring the size of ringtailed lemur troops and their home ranges since 1989 (Figure 7.3). Home range of a troop is defined as an area that it utilized frequently and does not include areas belonging to the other troops. It means that we do not include the area where a troop makes "excursions" from its usual range. We divided the home ranges of ringtailed lemurs into quadrants, each approximately 25 × 25 m (1/16 ha).

7.2.3. Studies of Large Trees

Since 1982, we have been measuring the size of large trees exceeding 50 cm in DBH. We made a map (scale about 1/866) of the broader study area covering 30.4 ha, which included our main study area of 14.2 ha. It consisted of 487 quadrants (Figure 7.4). Locations and numbers of each species of tree were plotted on this map. The size of large trees including kily trees was measured in 1982, 1989, 1990, and 1991. In 2000, Soma and Ichino remeasured the size of kily trees. Hereafter, we refer to the populations of kily trees as the 82 cohort, 89 cohort, and so forth.

The total number of quadrants surveyed was 487 (1/16 ha × 487 = 30.4 ha). Excluding 108 kily trees that died or were not measured, or for which our data are incomplete, we used the data for 205 kily trees to calculate the annual increase of diameter for each tree. Using the mean value of annual growth, we estimated the year of birth, the age of trees when they reached the size of 50 cm in DBH, and the age in the year 2000.

In 2000, we also estimated the availability of fruits and leaves of kily trees by using the abundance scores. The abundance scores were calculated scoring 0 (no fruits or leaves on the tree), 1 (fruits or leaves yielding 1–33% of full harvest),

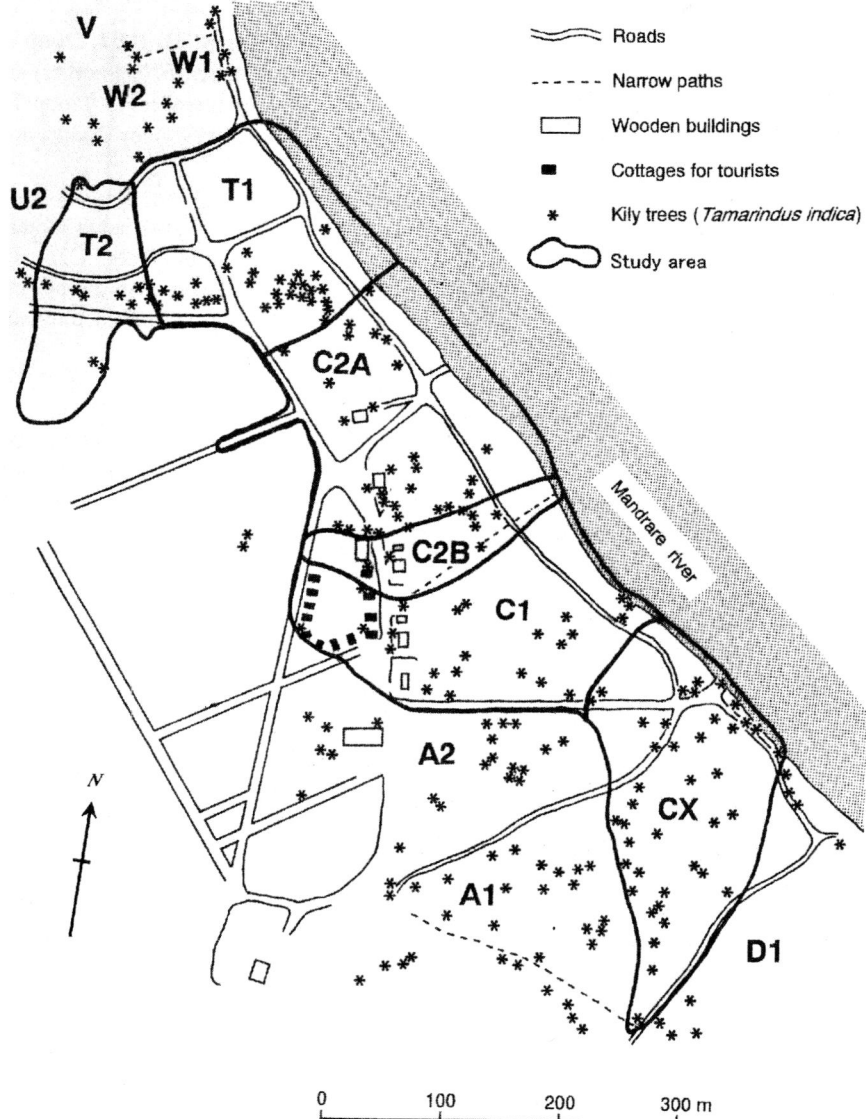

FIGURE 7.3. Home ranges of six troops within main study area in 2000.

2 (fruits or leaves yielding 34–66% of full harvest), and 3 (fruits or leaves yielding 67–100% of full harvest) to each kily tree.

Differences between groups were examined for statistical significance using the non-parametric tests, such as Mann–Whitney U test. Correlations between variables were examined by Pearson regression analysis. A p value less than 0.05 denoted the presence of a statistically significant difference.

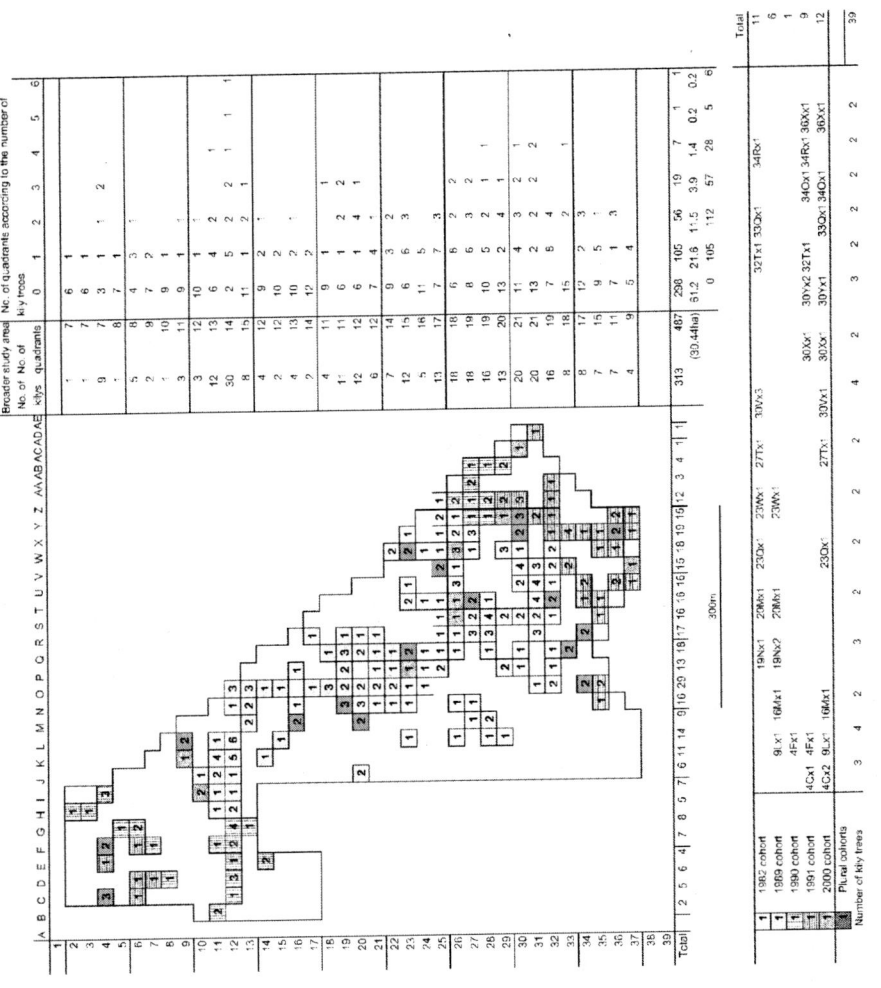

FIGURE 7.4. Location, year of study, and the number of kily trees by quadrants (n = 313).

7.3. Results

7.3.1. The Study Populations of Lemurs and Large Trees

In the main study area of 14.2 ha, we have been studying the changes of the home range sizes of ring-tailed lemur troops for 11 years from 1989 to 2000. At the beginning of our study in September 1989, there were three troops (T, B, and C) inhabiting this area. Due to social changes (troop divisions and evictions, etc.), the number of troops increased to six (T1, T2, C2A, C2B, C1, and CX) (see Figure 1 of Koyama et al., 2002). The number of animals also increased from 63 to 89 (see Table 1 of Koyama et al., 2002), and home range size of each troop also changed. Figure 7.3 shows the home ranges of six troops in September 2000.

We also measured the large trees in the broader study area of 30.4 ha, in 1982, 1989, 1990, and 1991. There were 475 large trees belonging to 14 species, 11 genera and 9 families within the broader study area of 30.4 ha (Table 7.1). They were three species of Moraceae, three species of Leguminosae, two species of Ulmaceae, and one species each of Verbenaceae, Capparidaceae, Rubiaceae, Violaceae, Meliaceae, and Sapindaceae. The most abundant species was kily (n = 289), the second benono (*Acacia rovumae*) (n = 74), the third voleli (*Neotina isoneura*) (n = 66), and these three species alone accounted for 90.3% of all large trees. Population density of large trees was 15.6 per ha and that of kily was 10.3 per ha.

7.3.2. Population of Kily Trees in the Broader Study Area of 30.4 Hectares

7.3.2.1. Cohorts and DBH

In the following section, we analyze the distribution and density of kily trees. Figure 7.4 shows the distribution of tamarind trees exceeding 50 cm in DBH measured in the 1982, 1989, 1990, 1991, and 2000 studies. The cohort measured

TABLE 7.1. Scientific name and the number of 14 species of trees exceeding 50 cm in DBH (4 cohorts).

Scientific name	Family	Vernacular name	n
Acacia rovumae	Leguminosae	benono	74
Albizia polyphylla	Leguminosae	halombolo	6
Celtis bifida	Ulmaceae	bemavo	5
Celtis philippensis	Ulmaceae	tsilikatsifaka	6
Clerodendrum emirnense	Verbenaceae	ambifotse	1
Crateva excelsa	Capparidaceae	keleon	2
Enterospermum pruinosum	Rubiaceae	mantsak	1
Ficus coculifolia	Moraceae	adabo	6
Ficus grevei	Moraceae	adabonara	1
Ficus megapoda	Moraceae	fihamy	6
Neotina isoneura	Sapindaceae	voleli	66
Rinorea greveana	Violaceae	tsatsake	6
Tamarindus indica	Leguminosae	kily	289
Quivisianthe papinae	Meliaceae	valiandro	6
Total			475

in 1982 (82 cohort) consisted of 124 trees, the 89 cohort 73 trees, the 90 cohort 41 trees, and the 91 cohort 51 trees (Table 7.2).

Figure 7.5 shows the distribution of DBH for the 1982, 1989, 1990, and 1991 cohorts. There was a significant difference in DBH among these cohorts (Kruskal–Wallis test, $H = 16.740$, $DF = 3$, $p = 0.0008$), which may be because the 1989 cohort contained smaller individuals than other cohorts: median of DBH was 78.8 cm in the 1982 cohort, 68.5 cm in the 1989 cohort, 83.3 cm in the 1990 cohort, and 85.4 cm in the 1991 cohort. There was no significant difference among the 1982, 1990, and 1991 cohorts ($H = 1.876$, $DF = 2$, $p = 0.3914$).

For the pooled data of the four cohorts, there was no significant correlation between the DBH and the distance from the Mandrare River (Kruskal–Wallis test, $H = 1.361$, $DF = 5$, $p = 0.929$) (Figure 7.6). However, note that there was no larger kily exceeding 130 cm in DBH in the area 200–300 m from the river.

In 2000, there were 24 kily trees newly measured. Of these, 5 trees had escaped our notice at the time of initial measuring in 1982, and they should have exceeded 50 cm in DBH in 1982. Other ones ($n = 19$) were probably under 50 cm in DBH at the time of initial measuring, and some years later they reached 50 cm in DBH. Out of 19, we measured the size of DBH on 14 trees, which were actually under 50 cm in DBH at the time of initial measuring. Thus, 313 individuals grew in the broader study area of 30.4 ha in total.

7.3.2.2. Distribution Pattern

Out of 487 quadrants, there were 297 quadrants (61.0%) with no kily trees, 107 quadrants with 1 kily, 55 quadrants with 2 kilys, 19 quadrants with 3 kilys, 7 quadrants with 4 kilys, and 1 quadrant each with 5 and 6 kilys. The index of dispersion (= variance ÷ average) = $1.503 > 1.0$. While there was a significant difference from the expectation of Poisson (random) dispersion (Kolmogorov–Smirnov test, $D = 0.08399$, $p < 0.01$), no significant difference existed from the expectation of negative binomial (clumped) dispersion (chi-square = 6.024, $DF = 3$, $p > 0.10$). Thus, these data suggested the clumped dispersion pattern of this species.

It is uncertain what environmental factors affect the dispersion pattern of kily trees. For example, in the area undisturbed by human activities, the individual density did not correlate with the distance from the Mandrare River (12.4 trees/ha within 100 m from the river, 16.2 trees/ha in the distance 100–200 m from the river; 11.2 trees/ha in the distance 200–300 m from the river).

TABLE 7.2. Mortality and survivorship of each cohort.

Cohort	n	Dead	Mortality(%)	Live	Survival rate (%)
1982	124	42	33.9	82	66.1
1989	73	12	16.4	61	83.6
1990	41	2	4.9	39	95.1
1991	51	8	15.7	43	84.3
Total	289	64	22.1	225	77.9

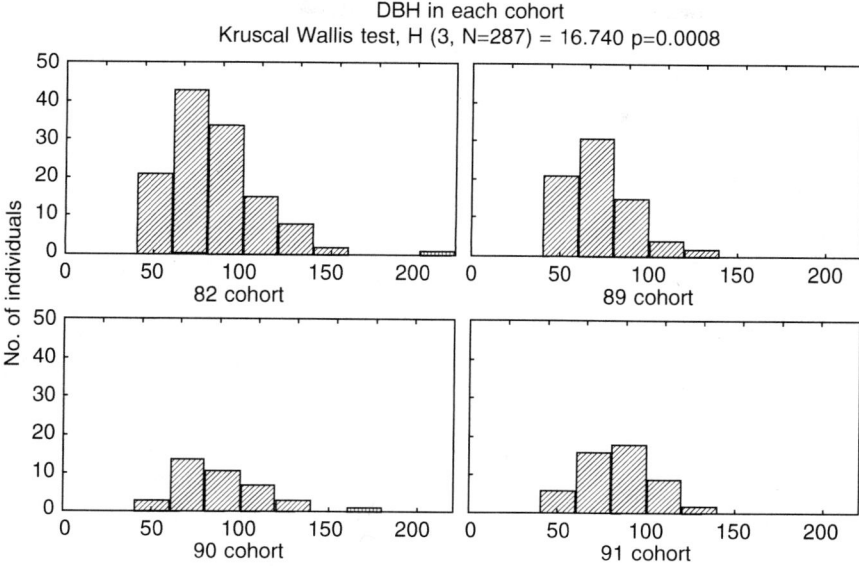

FIGURE 7.5. Distribution of DBH in the 1982, 1989, 1990, and 1991 cohorts.

7.3.2.3. Population Dynamics

If the age of an individual tree was positively correlated with DBH, the distribution pattern of DBH suggested this population was falling into a decline in each cohort (see Figure 7.5). In fact, out of 294 (=289 + 5) trees that had grown in the study area in 1982, 1989, 1990, or 1991, 64 trees had died, and only 19 trees joined into this population (Table 7.2). Figure 7.7 shows the locations of dead kily trees within main study area.

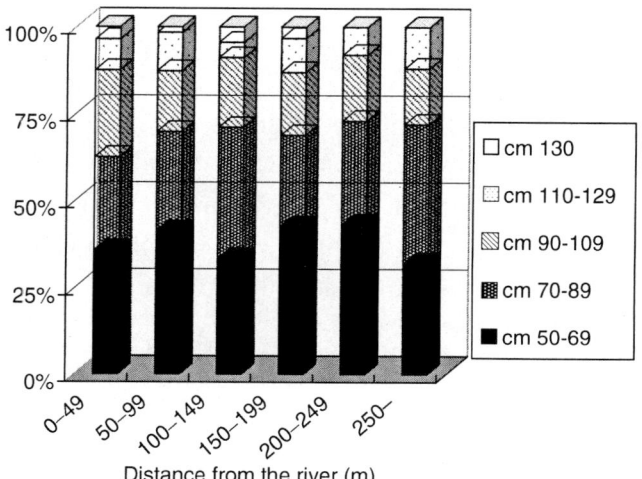

FIGURE 7.6. The correlation between the DBH and the distance from the Mandrare River.

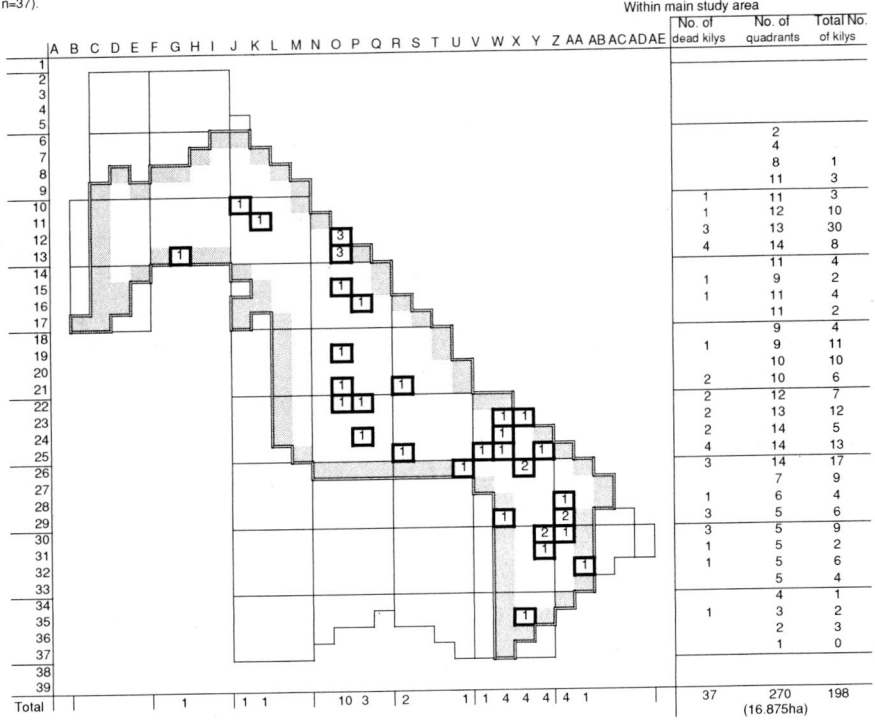

FIGURE 7.7. Locations of dead kily trees within main study area (n = 37).

Figure 7.8 shows the correlation between survival rate and intervals of research for four cohorts. The survival rate linearly correlated with interval (r = −0.894, p = 0.041 < 0.05). There was no significant difference in the distance from the Mandrare River between the dead and surviving trees (Mann–Whitney U test, z = 0.9230, p = 0.356, n_1 = 225, n_2 = 64). On the other hand, there was significant difference in the DBH between the dead and surviving trees (Mann–Whitney U test, z = 2.820, p = 0.0048 < 0.005). Apparently, surviving rate decreased in the larger trees with DBH of 90 cm or more (Figure 7.9).

7.3.2.4. The Estimation of Birth Year Based on the Growth Rate

Out of the 289 kily trees measured in 1982, 1989, 1990, and 1991, 64 trees died, and 2 were not measured. Furthermore, the measurement of 18 trees was incomplete. Thus, the remaining 205 trees were measured twice at different times. Figure 7.10 shows the correlation of DBH and its annual increase for each kily tree. Mean annual increase of DBH was 0.45 cm. If a kily increased its diameter at the rate of 0.45 cm per year, it would take 112 years to reach 50 cm in DBH.

Using these data, we estimated the year of birth, the year when a tree reached the size exceeding 50 cm in DBH, and the age in the year 2000 for 311 kily trees

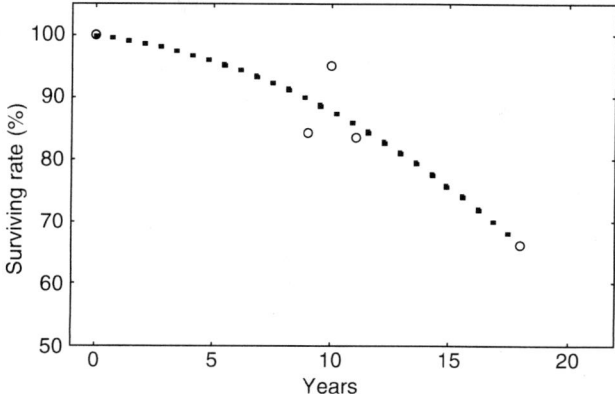

FIGURE 7.8. The correlation between survival rate and intervals of researches for the 1982, 1989, 1990, and 1991 cohorts.

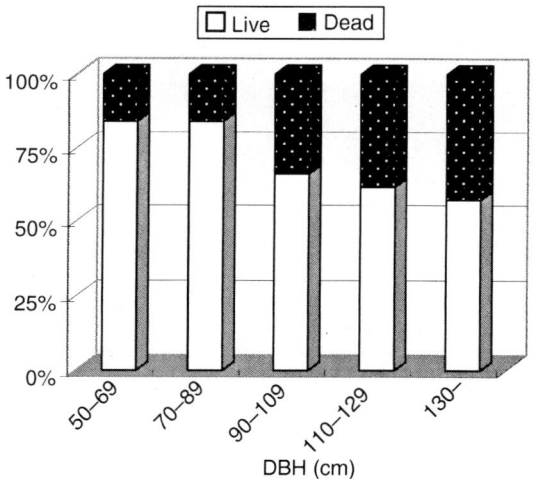

FIGURE 7.9. DBH and survival rate for all cohorts.

(2 unmeasured kilys were excluded) (Figure 7.11). For example, the year of birth for the largest kily (no. 19), with a diameter of 216.8 cm in 2000, was calculated according to the following formula. Estimated year of birth = 2000 − (216.8 ÷ 0.45) = 1518. The length of time it reached the size of DBH 50 cm = 50 cm ÷ 0.45 cm/year = 112 years. Estimated year of the tree when it reached the size exceeding 50 cm in DBH was 1630 (1518 + 112), and the estimated age in the year 2000 was 482 (= 2000 − 1518). When plotted the number of surviving trees by decade of birth, Figure 7.11 seems to suggest two things. First, there is erratic survival by decade, which may be random, or may reflect the erratic climate. Second, there is somehint that the number surviving to reach 50 cm is decreasing in recent decades. On the other hand, there was no correlation between annual growth and the distance from the river (Figure 7.12).

96   N. Koyama et al.

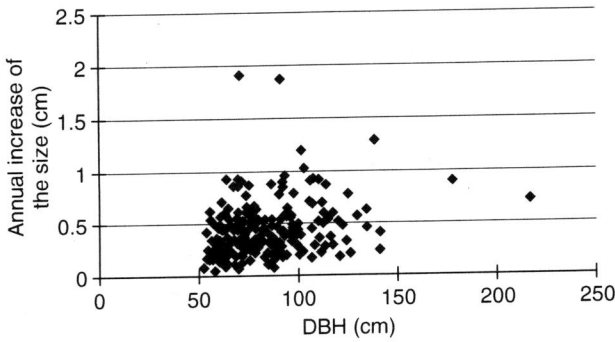

FIGURE 7.10. DBH and annual increase of the size (n = 205).

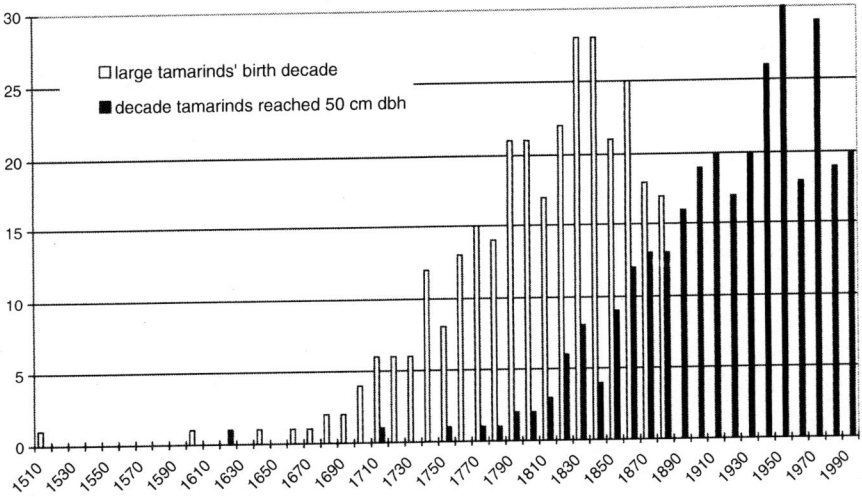

FIGURE 7.11. Decade of birth and decade of reaching 50 cm DBH for surviving large tamarinds, if the assumptions of growth are correct (n = 311).

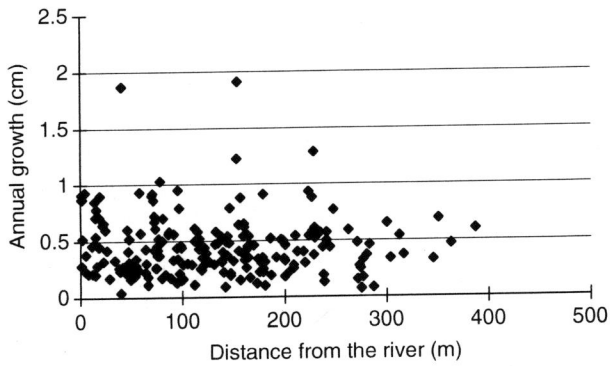

FIGURE 7.12. Distance from the river and annual increase of the size of DBH (n = 205).

7.3.3. Population Dynamics of Kily Trees Within the Main Study Area of 14.2 Hectares

7.3.3.1. Kily Trees Grown in the Ranges of Ringtailed Lemur Troops

We estimated that, in 1982, there were 170 kily trees within the main study area of 14.2 ha (12.0 trees/ha). In 1989, when we started to observe ringtailed lemurs, there might have been 181 kily trees (12.7 trees/ha). Of these, 37 trees (20.4%) died by 2000, and 13 kilys newly joined the population after 1990; the number of surviving trees was 157 in 2000. Thus, the density decreased to 11.1 trees/ha (Table 7.3). We do not know the exact year of death for each kily tree, but most deaths occurred after 1990.

In contrast, the number of ringtailed lemurs increased from 63 in 1989 to 89 in 2000. The number of kily trees per animal decreased from 2.8 in 1989 to 1.8 in 2000 (35.7% per year). Note that there was a great variation in the number of kily trees per animal among troops. Using the number of kily trees in each home range as a indicator of availability of foods, troop CX occupied the richest area (4.7 kily trees per animal), followed by troop C2A (1.9), troop C2B (1.7), troop T1 (1.6), troop C1 (1.3), and troop T2 ranged the poorest area (0.6 kily trees per animal). Thus, availability of kily trees differed among troops.

7.3.3.2. The Harvest of Kily Fruits

The harvest of kily fruits was very poor in 2000. We estimated the availability of fruits and leaves of 157 surviving kily trees by using the abundance scores for each troop (Table 7.4). The mean abundance score of kily fruits was 1.1 for troop CX, which was significantly lower than the average score (1.8) of the main study area over all years.

On the other hand, the trees grown near Mandrare River did not always produce abundant fruits. There was no significant correlation between the mean distance from the river and the mean abundance score of kily leaves ($H = 7.9358$, $DF = 5$, $p = 0.160$) and that of fruits ($H = 3.430$, $DF = 5$, $p = 0.634$).

TABLE 7.3. Number of kily trees within the ranges of each troop in 1989 and 2000.

In 1989	Troop T		Troop B		Troop C		Total
No. of lemurs	22		18		23		63
No. of kily trees	42		51		88		181
No. of trees per animal	1.9		2.8		3.8		2.8
Dead trees	3		11		23		37
Newcomers	4		1		8		13
In 2000	Troop T1	Troop T2	Troop C2A	Troop C2B	Troop C1	Troop CX	
No. of lemurs	20	17	16	6	20	10	89
No. of kily trees	32	11	31	10	26	47	157
No. of kily trees per animal	1.6	0.6	1.9	1.7	1.3	4.7	1.8

TABLE 7.4. Number of kily trees, availability of foods, and the distance from the river.

Troop	n	Abundance scores				Mean distance from the river (m)
		Fruits	Mean	Leaves	Mean	
T1	32	65	2.0	82	2.6	78.7
T2	11	26	2.4	29	2.6	197.7
C2A	31	73	2.4	73	2.4	81.6
C2B	10	18	1.8	25	2.5	105.1
C1	26	45	1.7	65	2.5	113.7
CX	47	53	1.1	109	2.3	87.7
Total	157	280	1.8	383	2.4	101.0

Abundance scores were calculated scoring 0, 1, 2, and 3 to each kily tree. 0, none; 1, 1–33%; 2, 34–66%; 3, 67–100%.

7.4. Discussion

7.4.1. Population Dynamics of Kily Trees

In recent years, several socioecological models have been proposed based on the studies of wild primates (Wrangham, 1980; van Schaik, 1989; Sterck et al., 1997, etc.). However, few long-term studies have been carried out both on wild primate and plant populations. For ringtailed lemurs, several authors have argued about the importance of kily trees (*Tamarindus indica*) as staple foods (Sauther, 1992). In Berenty Reserve, Madagascar, ringtailed lemurs eat kily as a staple food throughout the year (Rasamimanana and Rafidinarivo, 1993), and this species alone accounted for 34.9% of all feeding time (Soma, 2003).

Our long-term data on kily trees suggests that over time, this species is decreasing in number in the gallery forest of Berenty Reserve. In our main study area, its density decreased from 12.7 per ha 1989 to 11.1 per ha in 2000. This decrease was due to the fact that the number of dead kilys exceeded the number of newcomers. The total number of dead kily trees within the broader study area of 30.4 ha was 64 (2.1 per ha) and that within the main study area of 14.2 ha was 37 (2.6 per ha). This scanty repopulation by newcomers may suggest that the natural regeneration of kily trees does not function in this area.

Because the reason of decline of kily trees is uncertain, there may be several possibilities: (1) the climatic effects, in particular those of the droughts that occurred in 1983 and 1992 (on the other hand, Figure 7.1 shows that the amount of annual rainfall showed no consistent correlation with time); (2) lowering of the underground water level; (3) artificial effects from the increase of tourists or illegal felling of trees; and (4) overbrowsing by lemurs. Note that there was no significant correlation between survival rate and the distance from the river.

Based on the DBH, we calculated that mean increase of DBH was 0.45 cm per year, and we estimated that it required 112 years for a kily tree to reach the size of 50 cm in DBH. Chambers et al. (1998) noted that average growth rates varied from 1.0 to 6.4 mm per year among twenty ^{14}C-dated trees in the central Amazon. Our figure of 0.45 cm (or 4.5 mm) is well within these values. Of course, there may be some overestimate in this figure, because we have measured only large trees of more than 50 cm in DBH. Turner (2001) mentioned that large trees, in general, grew faster than small trees. He also mentioned that growth rate differed significantly by the tree species in tropical rainforest, and mean annual increase of DBH was within 0.5–6 mm. Data for small trees under 50 cm in DBH are needed for future studies.

At any rate, the large trees exceeding 130 cm in DBH would be expected to have been 200 or more years old. A tamarind bears fruit at around the age of 6–8 years, and its longevity is 80–200 years (Anonymous, 1999). Therefore, it is likely that longevity of kily trees may be one of the probable causes of death.

7.4.2. Kily Trees Grown in the Ranges of Ringtailed Lemurs

Jolly et al. (2002) pointed out that ringtailed lemurs depend on fruits of trees, and that the chief limiting factor of their population is fruit supply, not water per se. Total population fruit yield is a function of both tree density and individual fecundity (Peters, 1996). Because the production of fruits is not a linear function of size (e.g., the case of *Shorea mombin* belonging to the family Dipterocarpaceae), annual fruit production may increase exponentially up to a diameter of about 50 cm DBH, and then stabilize (Peters, 1996).

In contrast with the decrease of large kily trees, the number of ringtailed lemurs in the main study area increased from 63 in 1989 to 89 in 2000. As a consequence, the home range size of each troop decreased (Koyama et al., 2002). Thus, the number of kily trees per animal decreased from 2.8 in 1989 to 1.8 in 2000. Furthermore, there was a great variation in the number of kily trees per animal among troops. Troop CX occupied the richest area (4.7 kily trees per animal), and troop T2 ranged the poorest area (0.6 kily trees per animal). These conditions may have intensified both between-group resource competition (BGC) and within-group competition (WGC) (Wrangham, 1980; van Schaik, 1983).

On the other hand, the harvest of kily fruits varies from year to year, just as reported for other plant species (Koenig, et al., 1994; Suzuki et al., 1998). In 2000, the productivity of kily fruits was poor, and its abundance score was 1.1 within the home range of troop CX, which was much lower than within the ranges of other troops. So, the number of kily trees per animal did not always represent the richness of kily fruits. Furthermore, in Berenty, there were competitors such as sifakas (*Propithecus verreauxi*), brown lemurs (*Eulemur fulvus*), and fruit-bats (*Pteropus rufus*). In particular, the population of brown lemurs was increasing. Thus, the ecological conditions also intensified interspecies competition over food resources.

7.5 Conclusion

At Berenty Reserve, Madagascar, we have set up a main study area of 14.2 ha to study the demographic changes of ring-tailed lemurs for 11 years from 1989 to 2000. We also have set up a broader study area of 30.4 ha to study the environmental changes, including the main study area. In the main study area, due to the social changes such as troop divisions and evictions, the number of troops increased from three to six. Consequently, home range size of a troop decreased. In the broader study area, there were 475 large trees belonging to 14 species and 9 families. The most abundant species was kily (*Tamarindus indica*) (n=289), the second benono (*Acacia rovumae*) (n=74), the third voleli (*Neotina isoneura*) (n=66). These three species accounted for 90.3% of all large trees. Population density of large trees was 15.6 per ha, and that of kily trees was 10.3 per ha. In 1989, 12.7 kily trees existed per ha within the main study area of 14.2 ha.

In 2000, we re-measured the size of kily trees. Within the main study area of 14.2 ha, the density of kily trees decreased to 11.2 per ha. This decrease was due to the number of dead kily trees exceeded the number of newcomers. In contrast, the number of ring-tailed lemurs over one-year old increased from 63 in 1989 to 89 in 2000. As a result, the number of kily trees per animal was decreased from 2.8 to 1.8. Among troops, there was a great variation in the number of kily trees per animal. Troop CX occupied the richest area (4.7 kily trees/animal), and Troop T2 ranged the poorest area (0.6 kily trees/animal). On the other hand, the amount of kily fruits may fluctuate by each individual tree, area, and year. For example, the harvest of kily fruits was very poor in 2000, and the abundance score of kily fruits for Troop CX was lower than average score of the main study area. The number of brown lemurs (*Eulemur fulvus*) was also increasing. Since brown lemurs' feeding habits were very similar to ring-tailed lemurs, it is likely that both the within species and inter-species competitions are becoming intense for ring-tailed lemurs.

Acknowledgments. We thank A. Randrianjafy, former director of the Botanical and Zoological Park of Tsimbazaza, for his kind permission to perform this research in Madagascar. We also thank the de Heaulme family for providing the opportunity to measure trees and observe lemurs at Berenty Reserve. Mr. C. Rakotomalala provided the data on precipitation at Berenty. Supplemental data were obtained from Masayuki Nakamichi, Megumi Okamoto, Lys Rakototiana, and Naomi Miyamoto. We are grateful to Alison Jolly and Sara S. Berry for their comments on the manuscript. This work was supported by Grants-in-Aid for Scientific Research from the Ministry of Education, Science and Culture of Japan to S. Yamagishi (no. 01041079) and N. Koyama (no. 06610072 and no. 05041088).

References

Anonymous, 1999. Tamarind Factsheet: Fruits for the Future Tamarind. Available at http://www.civil.soton.ac.uk/icuc.

Budnitz, N., and Dainis, K. (1975). *Lemur catta*:ecology and behavior. In: Tattersall, L., and Sussman, R. W. (eds.), *Lemur Biology*. Plenum, New York, pp. 219–235.

Chambers, J. Q., Noguchi, N., and Schimel, J. P. (1998). Ancient trees in Amazonia. *Nature* 391(8):135–136.

Hotta, M., et al. (eds.). 1989. *Dictionary of Useful Plants in the World*. Heibonsya, Tokyo. (in Japanese)

Jolly, A., Dobson, A., Rasamimanana, H. M., Walker, J., O'Connor, S., Solberg, M., and Perel, V. (2002). Demography of *Lemur catta* at Berenty Reserve, Madagascar: Effects of troop size, habitat and rainfall. *Int. J. Primatol.* 23:327–353.

Koenig, W. D., Mumme, R. L., Carmen, W. J., and Stanback, M. T. (1994). Acornproduction by oaks in central coastal California: variation within and among years. *Ecology* 75:99–109.

Koyama, N., Nakamichi, M., Oda, R., Miyamoto, N., Ichino, S., and Takahata, Y. (2001). A ten-year summary of reproductive parameters for ring-tailed lemurs at Berenty, Madagascar. *Primates* 42:1–14.

Koyama, N., Nakamichi, M., Ichino, S., and Takahata, Y. (2002). Population and social dynamics changes in ring–tailed lemurs at Berenty, Madagascar between 1989–1999. *Primates* 43:291–314.

Nakagawa, N. (1998). Ecological determinants of the behavior and social structure of Japanese monkeys: A synthesis. *Primates* 39:375–383.

Peters, C. M. (1996). *The Ecology and Management of Non-Timber Forest Resources.* World Bank technical paper, no. 322. The World Bank, Washington, D. C.

Rasamimanana, H. R., and Rafidinarivo, E. (1993). Feeding behavior of *Lemur catta* females in relation to their physiological state. In: Kappeler, P. M., and Ganzhorn, J. U. (eds.), *Lemur Social Systems and Their Ecological Basis.* Plenum, New York, pp. 123–133.

Sauther, M. L. (1992). *The Effect of Reproductive State, Social Rank and Group Size on Resource Use Among Free-Ranging Ringtailed Lemurs (Lemur catta) of Madagascar.* Ph.D. thesis, Washington University, St. Louis.

van Schaik, C. P. (1989). The ecology of social relationships amongst female primates. In: Standen, V., and Foley, R. A. (eds.), *Comparative Socioecology: The Behavioural Ecology of Humans and Other Mammals.* Blackwell Scientific, Oxford, pp. 195–218.

Soma, T. (2003). *Feeding Ecology of Ring-tailed Lemurs (Lemur catta) at Berenty Reserve, Madagascar.* Master's thesis, Graduate School of Asian and African Area Studies, Kyoto University. (in Japanese)

Sterck, E. H. M. (1999). Variation in langur social organization in relation to the socioecological model, human habitat alteration and phylogenetic constraints. *Primates* 40:199–213.

Sterck, E. H. M., Watts, D. F., and van Schaik, C. P. (1997). The evolution of female social relationships in nonhuman primates. *Behav. Ecol. Sociobiol.* 41:291–309.

Suzuki, S., Noma, N., and Izawa, K. (1998). Inter–annual variation of reproductive parameters and fruit availability in two populations of Japanese macaques. *Primates* 29:313–324.

Turner, I. M. (2001). *The Ecology of Trees in the Tropical Rain Forest.* Cambridge University Press, Cambridge.

Wrangham, R. W. (1980). An ecological model of female-bonded primate groups. *Behaviour* 75:262–300.

8
The Influence of Tamarind Tree Quality and Quantity on *Lemur catta* Behavior

ANNE S. MERTL-MILLHOLLEN, HAJARIMANITRA RAMBELOARIVONY, WENDY MILES, VERONICA A. KAISER, LISA GRAY, LORETTA T. DORN, GEORGE WILLIAMS, AND HANTANIRINA RASAMIMANANA

8.1. Introduction

Ever since the earliest studies of ringtailed lemur behavior at Berenty Reserve, Madagascar (Jolly, 1966; Klopfer and Jolly, 1970; Sussman, 1974; Budnitz and Dainis, 1975), researchers have tried to relate lemur behavior to the environment. Why are troop core areas and borders placed where they are? Why do they choose to feed on some trees and not others? Some troops show great consistency over the 40 years they have been studied (Jolly and Pride, 1999; Mertl-Millhollen, 2000). We hypothesize that there is a pattern to the forest resources that is conducive to this kind of ranging behavior and that water determines much of this pattern.

This research addresses that hypothesis for ringtailed lemur troop D1A. It inhabits the closed canopy gallery forest next to the Mandrare River, the kind of riverine forest that is among the richest of ringtailed lemur habitats remaining (Sussman et al., 2003 and this volume; Goodman et al., this volume). The troop also ranges into open forest farther from the river. By including both a moist and a dry forest environment, there should be a maximum diversity of food sources (Gosling and Petrie, 1981). We examine the forest richness in detail and then correlate troop feeding and ranging behavior with the distribution of that richness.

We focus on the quality and quantity of food provided by the tamarind (*Tamarindus indica*), a leguminous tree that is a keystone food source for the lemurs (Budnitz and Dainis, 1975; Rasamimanana and Rafidinarivo, 1993; Sauther, 1998; Yamashita, 2002) and the dominant species of the riverine forests in the ringtailed lemur range (Sussman and Rakotozafy, 1994). We include a map of all large tamarinds in the D1A range and replicate and expand upon an earlier study (Mertl-Millhollen et al., 2003). It demonstrated that lemur ranging to feed on leaves correlates with tamarind leaf water and protein content. Because only 10 tamarinds were sampled in that study and because it was only done during the birth season, in September/October, we are expanding upon those findings here. We collected leaf samples from 18 tamarinds in the 2002 birth season in order to verify the results with an increased sample size. We then did repeated sampling in other seasons in order to see whether the results generalize.

The earlier study also found that lemurs fed on fruit wherever they could find it (Figure 8.1). Proximity to the river did not correlate with fruit abundance. In fact, in that particular year, there was more fruit farther from the river, and the lemurs traveled to it. This study again compares lemur ranging to feed on fruit with fruit abundance in order to evaluate the consistency of the pattern.

We focus on the months of September and early October because this is the end of the dry season, the harshest time of the year in having little fruit or flowers available at Berenty (O'Connor, 1987; Rasamimanana, 1999). It has the lowest rainfall of the year (Jolly et al., 2002) and includes the end of gestation, the birth season, and the beginning of lactation, a time of great physiological stress for the female lemurs (Sauther, 1994). Lemurs have evolved to adapt to the extreme variable conditions typical of an island environment by optimizing their intake of high-quality food during the rich wet season, the time of lactation and weaning (Wright, 1999; Ganzhorn, 2002). Dietary flexibility is selected for in the tropical dry deciduous forest zone because of the great fluctuations in food resources (Oates, 1987; Hladik, 1988). *Lemur catta* diet varies between seasons, years, and habitats (Budnitz and Dainis, 1975; Rasamimana and Rafidinarivo, 1993; Sauther, 1998; Sauther et al., this volume). This study examines how one Berenty Reserve troop copes with these seasonal challenges.

FIGURE 8.1. Ringtailed lemur eating tamarind fruit in Berenty Reserve. (Photo by Anne Mertl-Millhollen).

8.2. Methods

8.2.1. Site and Subjects

Troop D1A lives adjacent to the river in the closed canopy and open forests in the 1 km^2 Malaza portion of the reserve. In addition to living near to the river, the troop is provisioned with two water troughs. Troop D1A derived from D troop, which has been studied intermittently ever since Jolly's original 1962–1963 study (Jolly, 1966; Mertl-Millhollen, 1988; Jolly and Pride, 1999; Mertl-Millhollen, 2000; Pride, this volume). Troop D fissioned into D1 and D2 in 1993 (Koyama et al., 2002), and D1 fissioned into D1A and D1B in 2000 (E. Pride, pers. comm.).

Not counting infants, troop D1A was composed of 9 individuals in 2002 (4 adult females, 2 adult males, 1 subadult, and 2 juveniles), 12 in 2003 (4 adult females, 3 adult males, 2 subadults, and 3 juveniles), and 8 in 2004 (3 females, 3 males, 1 subadult and 1 juvenile).

8.2.2. Behavioral Observations

Except for April 2003, which had only 7 days of data, we collected data using scan sampling (Altmann, 1974) every 5 minutes during 10 full day follows, 0600–1800 h, for a total of 684 hours of observation divided into six blocks of time (Table 8.1). Following the protocol created by Jolly (Mertl-Millhollen et al., 2003), we mapped the behavior and position of the troop at each scan and noted all intertroop encounters on a 25 m × 25 m quadrat map of the reserve.

Because no distinction was made between foraging and feeding, the word feeding in this study refers to both. We mapped and labeled a tree as a feeding patch whenever the lemurs fed in it for 20 or more minutes (four consecutive scans). When feeding was observed, we recorded the nature or species of the food source (plant, insect, bark, soil, or water) and the plant part eaten (flower, fruit, mature leaves, or new leaves).

Every time two troops came within 20 m of each other, it was recorded as an encounter. The first troop to move away from the other for more than 20 m was

TABLE 8.1. Data collection.

	Behavioral sampling	Leaf sampling	Fruit abundance sampling
Mertl-Millhollen, Miles, Rasamimanana, Jolly	Sept. 18–Oct. 10, 2002	Sept. 18–23, 2002	Yes
Rambeloarivony, Rasamimanana	Oct. 28–Nov. 16, 2002, Apr. 17–25, 2003, Aug. 9–Sept. 6, 2003, Sept. 8–23, 2003	Oct. 25–26, 2002, Nov. 18–19, 2002, Apr. 16–21, 2003, Aug. 27–28, 2003	
Kaiser, Gray, Mertl-Millhollen	Sept. 2–Oct. 3, 2004		Yes

considered to have lost the encounter. We used the positions of agonistic intertroop encounters to determine a defended territorial border within each home range. The border was placed where the troop won the majority of the conflicts. Because there were no encounters in the southern part of the range during this study, we used the range defended in 2000 (Mertl-Millhollen et al., 2003) for Figure 8.2.

8.2.3. Tree Mapping, Plant Sampling, and Laboratory Analyses

In 2002 and 2004, Williams found Geographical Positioning System coordinates to map all the tamarind trees of >25 cm diameter at breast height (DBH) in the troop D1A range onto the 25 m × 25 m quadrat map of the reserve that he had created (Figure 8.2). An average of four measurements was taken around each

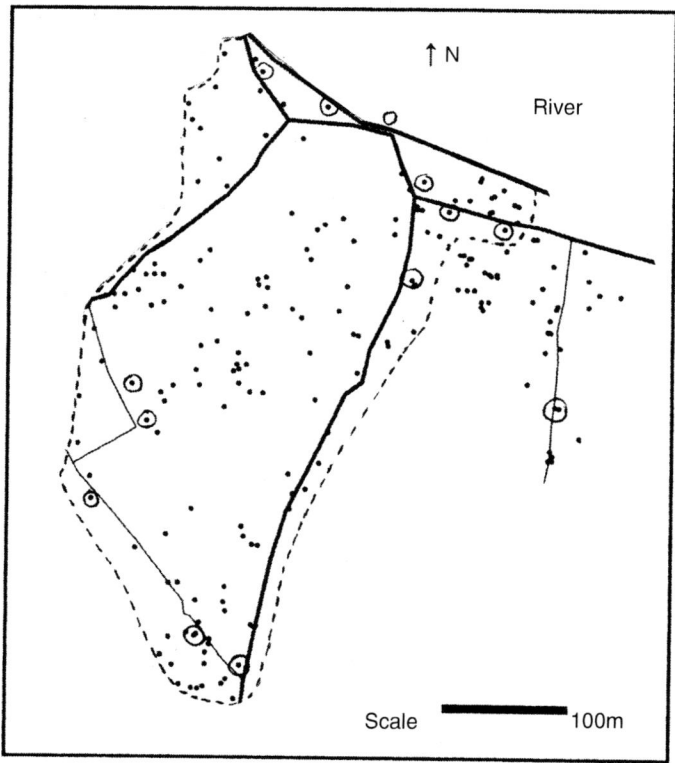

FIGURE 8.2. Tamarind trees of DBH >25 cm in the home range for troop D1A. Dots represent tamarind trees. Circled dots are 14 of the 18 sample trees. Lines are trails and riverbank. Dashed line is the defended troop border in year 2000 (Mertl-Millhollen et al., 2003).

tree to position it on the map. We were then able to compare lemur ranging and feeding with tamarind availability.

To understand how the lemur ranging corresponded with leaf quality and fruit abundance, we resampled the tamarind trees studied in 2000 (Mertl-Millhollen et al., 2003) and increased the sample size to 18 trees. Fourteen of these are shown in Figure 8.2. The rest were south of the home range and off the map. Sample trees were consistently chosen along trails as, perchance, there could be an edge effect because of increased light (Ganzhorn, 1995; Blumenfeld-Jones, this volume) that might result in edge trees being preferred feeding sites (Mittermeier and van Roosmalen, 1981). We measured canopy diameter and tree DBH. For trees with multiple stems, we took the sum of their DBHs. To look at fruit abundance, we used a visual measure of abundance (Chapman et al., 1992). We scanned four 1 m^3 areas on different sides of each tree's canopy to calculate the average fruit abundance as a percentage of that shown for the most abundant tree found in the forest in 2002.

We used the year 2000 methods (Mertl-Millhollen et al., 2003; Ganzhorn, 1988) for collecting, drying, and weighing the leaves in order to study leaf water and protein content. In general, leaf quality is at its highest at the end of the day (Janson and Chapman, 1999), and choosing samples from a consistent height and multiple sides of the tree is important because of the within-tree variability (Denno and McClure, 1983; Perica, 2001). For example, when we looked at three samples collected from near the trunk and two sides of one tree, the water content was 62%, 51%, and 53% and the crude protein content was 5.2%, 4%, and 4.6%. We collected all samples at a uniform height of 5 m from three sides of the tree between 1600 and 1700 h. We pooled the leaflets from each individual tree. We weighed them 2 hours later followed by air drying and weighing again to determine water content. Dorn subsequently analyzed the leaflets for % crude protein (N content × 6.25 expressed as % of dry sample) using a modified Kjeldahl digestion.

8.2.4. Hydrologic Environment

Because we were interested in tamarind leaf water content, we also examined the sources of water to the tamarind trees. Rainfall data were provided to us by M. Rakotomalala, Berenty. Based on 19 years of rainfall data (Jolly et al., 2002; Jolly, this volume), the average annual lemur year rainfall is 524.7 mm ± 175.4 mm, with a range from 265 mm to 894 mm.

Because the original study demonstrated that the leaves from trees near the river had higher water content, we assumed that there was a shallow water table near the river that was important to the trees. We measured the depth to the static water level of a 6.25-m-deep hand-dug well located in the Ankoba part of the reserve at a distance of 65 m from the riverbank. This well is located in what appears to be a former river channel and had been deepened in 1993 because of a falling water table. We also did preliminary leaf and soil sampling from a tamarind located 1.1 km south of the current river but located in the river's former channel.

We measured shallow soil moisture next to 16 of the leaf sample trees in 2002. We collected two soil cores, at a 5 m distance to either side of each tree, to a depth of 25 cm. We noted soil texture and nature, depth of leaf litter, and core weights. We then combined the cores from around each tree, air dried them in the sun for 5 days, and weighed them again to determine soil moisture.

8.2.5. Statistical Analyses

We analyzed the nature of range use by comparing the quadrats containing encounters and feeding patches with those that did not. We correlated that with the number of tamarinds available, the quality of the tamarind leaves, and the quantity of fruit. Significance of differences in range use was determined using the χ^2, standard deviations of the sample were given with means, fruit abundance in different years was compared with the Wilcoxon matched-pairs, signed-ranks test, and correlations were determined using the Spearman rank correlation coefficient r_s adjusted for ties, with $\alpha \leq 0.05$ (Siegel, 1956).

8.3. Results

8.3.1. Hydrologic Environment

The Mandrare River did not dry up during the years of this study. The static water level in the water well was 5 m 87 cm below ground level in September 2002 and was 5 m 36 cm below ground level in September 2004, after an 11 mm rain. The water table appeared to recover quickly after pumping and returned to the static water level in 6 minutes after pumping out 3000 L.

We consider annual rainfall less than 450 mm to be a drought, and rainfall data indicate that only lemur year October 2001–September 2002, with 526 mm of rainfall, was not a drought year during the period of this study. The other two years had 423 mm and 424 mm, respectively.

The shallow soil water content was directly related to distance from the river (N = 16, r_s = 0.96, p $\leq$ 0.01) and ranged from 3% water content near the river to 12% at 810 m from the river. This correlated with soil texture that was sandy in the closed canopy forest samples near the river but had a high clay content in the open forest samples distant from the river.

We also collected samples from a point on the former river channel, south of the part shown in Figure 8.3. The leaf sample had a water content of 61.3%. This is as high as for tamarinds near the river (Table 8.2). The soil sample had a high clay content and contained 16.5% water, higher than any of the other soil samples.

8.3.2. Tamarind Leaf Quality

The water content of mature tamarind leaves collected during the birth season in September, 2002 significantly correlated with proximity to the river (Table 8.2).

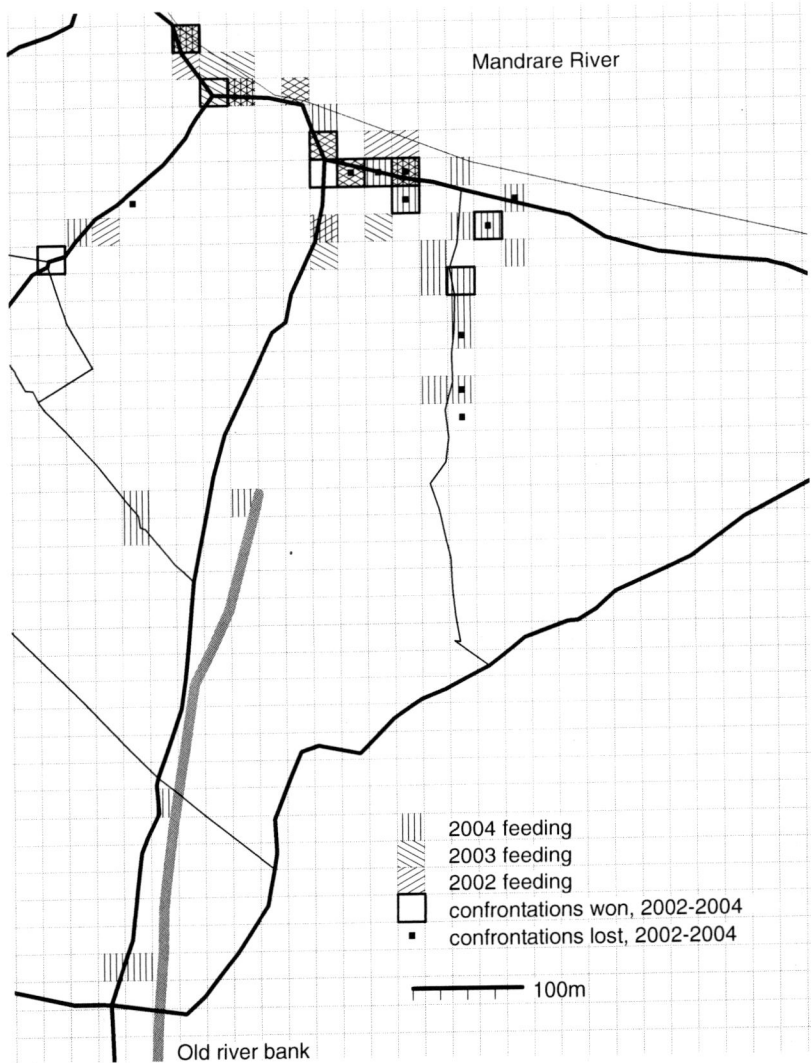

FIGURE 8.3. Tamarind feeding patches and confrontations during September/October birth seasons in years 2002–2004. Lines are trails and riverbank. Wider gray line is former river channel. Quadrats are 25 m × 25 m.

The crude nitrogen content of the leaves was also significantly higher closer to the river. However, water content in samples collected in late October and early November 2002 during lactation did not correlate significantly (Table 8.3). Because this was the beginning of the rainy season, the late October samples were a mix of leaf buds and mature leaves, and the November samples were entirely new leaves. Water content of mature leaf samples collected in April

TABLE 8.2. Percent water by weight and crude nitrogen content of selected tamarind tree leaves during the lemur birth season 2002.

Distance from river (m)	% Leaf water content*	% Leaf crude protein*
Closed canopy forest		
0 (F)		
9 (F)	68	8.71
12 (F)	64	7.41
36 (F)	65	8.21
45 (F)	62	6.22
50	63	
Open canopy forest		
107 (F)	52	4.87
150	54	3.97
279	52	3.21
286	61	5.53
362	56	3.47
430	57	4.83
438	58	4.72
448	51	3.36
510	50	3.75
621	54	5.50
626	48	1.75
810	56	4.6

(F), feeding patch for leaves or for both leaves and fruit; 438 m marks the end of their defended range; 810 m is the farthest outside of their range that they fed, only feeding on fruit, not leaves.
Significant correlations: *Spearman $r_s = 0.66$, $p < 0.01$.

2003, the end of the rainy season and the time of copulation, and in August 2003, the dry season and time of gestation, again correlated positively with proximity to the river.

8.3.3. Tamarind Fruit Abundance

We measured tamarind fruit abundance in the canopy of the 18 sample trees in 2002 and 2004 (Table 8.4). There was no correlation between fruit abundance and either distance from the river, DBH, canopy diameter, or the same trees in different years. There was markedly more fruit available in 2002. The sample trees averaged 43% abundance in 2002 but only 7% abundance in 2004. This is a significant difference in fruit abundance between the two years (Wilcoxon matched-pairs, signed-ranks test, N = 17, T = 0, $p < 0.01$.).

8.3.4. Tamarind Tree Abundance and Behavioral Observations

During the birth season in the years 2002–2004, troop D1A spent an average of 47% of their time feeding, and they fed on tamarinds an average of 59% of that feeding time (Table 8.5). Their feeding range, an area of 11.4 ha, is shown in

TABLE 8.3. Percent water by weight of selected tamarind tree leaves during different seasons.

Distance from river (m)	Late Oct. 2002 (lactation)	Nov. 2002 (lactation)	Apr. 2003 (mating)	Aug. 2003 (gestation)
Closed canopy forest				
0	(F)	81 (F)	76	70 (F)
9	69	66	76	72
12	66	78	66	65 (F)
36	64 (F)	66 (F)	72	62
45	62	80	69	56
Open canopy forest				
107	61	61	67	54
150	60	76	60	51
279	78	68	58	51
430	59	78	60	62
438	53	70	60	57
510	80	72	61	56
621	63	76		42
810	65			
Spearman r_s	0.11, NS	0.06, NS	0.82, $p < 0.01$	0.70, $p < 0.05$

(F), feeding patch for leaves or for both leaves and fruit.

Figure 8.2 and is based on the combination of the data from this study with year 2000 ranging and intergroup defense (Mertl-Millhollen et al., 2003). More than 200 large tamarind trees with a DBH >25 cm are contained in this range. However, the lemurs did not even visit many of the available trees during this study. This area contains 182 of the 25 m × 25 m quadrats, but the maximum number of quadrats that were visited for feeding occurred in 2004 when they fed in 68 quadrats.

The farthest they fed from the river in any observation period in 2002–2003 was 210 m, and the farthest from the river that they defended their range during an encounter was 235 m (Table 8.5; Figure 8.3). If we take this shorter distance as the southern edge of their feeding range, the feeding range includes 144 quadrats, an area of 9 ha. It contains 129 tamarind trees, 14.3 trees/ha. Considering troop D1A to average 10 animals during this period, that feeding range contains 13 tamarind trees/lemur.

Tamarind quantity alone does not explain their ranging behavior. The highest number of tamarinds/quadrat was 6, contained in one quadrat within their range. A feeding patch was identified in this quadrat only once, in the 2003 birth season. There were 4 quadrats that each contained 4 tamarinds. One of these was never used, and one contained feeding patches and was the site of intergroup encounter all three birth seasons (Figure 8.3) and is the highly contested area of overlap to the east with troop D1B (Mertl-Millhollen et al., 2003). However, quadrats to the west in the other highly contested area of overlap with troop

TABLE 8.4. Tamarind fruit abundance.

Distance from river (m)	Total DBH (cm)	Canopy diameter (m)	% Fruit/m³ 2002*	% Fruit/m³ 2004*
Closed canopy forest				
0	30	6	0	0
9	125	21	55	10
12	110	24	50	25
36	165	26	70	5
45	85	18	60	10
50	147	20	25	10
Open canopy forest				
107	160	19	85	5
150	147	30	25	5
279	123	16	75	10
286	70	12	25	5
362	77	19	35	0
430	82	19	65	5
438	98	21	65	8
448	27	9	38	0
510	147	27	2	0
621	91	22	10	5
626	150	20	40	10
810	150	24	50	8

*Significant difference in fruit abundance between years (Wilcoxon matched-pairs signed ranks test, N = 17, T = 0, p < 0.01).

CX that also contained multiple feeding patches each of the three years, averaged only one tamarind/quadrat.

The lemurs fed on leaves an average of 31% of their time feeding on tamarinds during the birth seasons (Table 8.5) and did this feeding close to the river where leaf quality was the highest. They fed on fruit an average of 69% of their tamarind feeding time. That feeding was also close to the river in the 2002 and 2003 birth seasons and during the intervening months. The 2002 measure of fruit abundance showed it to be abundant throughout the range (Table 8.4). However, in 2004, when there was little fruit on any but a few trees positioned seemingly randomly, troop D1A increased their feeding/foraging time for fruit, doubled the number of feeding patches and quadrats they utilized within their range, and traveled not only to the farthest point from the river within their range (roughly 438 m) but also to trees at 650 m and 810 m from the river in quadrats adjacent to a former path of the river (Table 8.5; Figure 8.3). These are the same trees visited in 2000 when there was not only a shortage of tamarind fruit (Mertl-Millhollen et al., 2003) but also loss of food sources in general after a cyclone (Pride, this volume).

Unlike 2000, the extra ranging in 2004 did not result in increased confrontations with other troops (Table 8.5). In 2000, the troop had 43 encounters, and troops seemed to be wandering everywhere in the reserve, often seeking out water troughs. Even though troop D1A traveled as far out of their range in 2004 as in 2000, they only had 14 encounters. In 2004, it was sometimes hard to find other

TABLE 8.5. Lemur ranging and foraging/feeding during the birth season.

	2002	2003	2004
% time feeding	49	38	54
Feeding range: no. quadrats	36	20	68
% of total feeding time on tamarinds	60	48	68
No. of quadrats with feeding patches	11	11	26
% of tamarind feeding time on leaves	35	37	20
Farthest from river feed on leaves (m)	193	127	205
% of tamarind feeding time on fruit	65	63	80
Farthest from river feed on fruit (m)	107	129	810
No. of intergroup encounters	17	15	14
No. of quadrats with encounters	4	8	8
% encounters won	53	53	50
% encounter quadrats with or adjacent to feeding patches	75	88	100

troops in order to do our census. Some scrub troops moved into the spiny forest to feed and two troops traveled nearly 2 km out of the reserve to feed on sisal flowers (Jolly, pers. comm.).

The lemurs often fed in the areas that were also zones of confrontation with other troops. The majority of the quadrats where encounters occurred either contained or were adjacent to feeding patches and to trails (Table 8.5; Figure 8.3). A disproportionate amount of time was spent in the areas of overlap with other troops. These included areas rich in high-quality trees along the path paralleling the river. For instance, in the 2002 birth season, they entered 44 quadrats but spent 44% of their time in the 4 quadrats where there were also intertroop encounters. Those 4 quadrats accounted for only 11% of their range that year. In 2004, they spent 15% of their time in the east quadrat where they won three out of four confrontations (Figure 8.3). They fed there on a tamarind tree that had abundant fruit but was within D1B's range. This would appear to be an example of the successful raiding described by Pride (this volume) in a year of fruiting failure.

8.4. Discussion

The 11.4-ha home range of troop D1A in gallery forest next to the Mandrare River contains more than 200 large tamarind trees, their keystone food species. Only some of these available trees were used during this study. The troop's range includes closed canopy forest next to the river that in general contains larger and more numerous tamarinds and a higher density of not only ringtailed lemurs (Budnitz and Dainis, 1975; Jolly et al., 2002; Koyama et al., this volume) but also *Propithecus verreauxi* (Jolly et al., 1982) and *Lepilemur mustelinus leucopus* (Charles-Dominique and Hladik, 1971) than does the adjacent open forest and brush and scrub. Lemur biomass also correlates with richness of forest at Beza Mahafaly Special Reserve (Sussman, 1991; Sussman and Rakotozafy, 1994; Sauther et al., 1999; Whitelaw and Sauther, 2003). The extent of these gallery forests is probably limited by the water table (O'Connor, 1987), which was found in this study to be only about 5.5 m below ground level when measured in a well near the Mandrare River. Despite the fact that there was less than average rainfall during the years of this study, the Mandrare River never dried out.

In our shallow soil samples, we found significantly higher water content in the clay soils distant from the river than in the sandy soils adjacent to the river. Deeper core samples (20–40 cm deep) analyzed by O'Connor (1987) also showed higher clay content in the brush and scrub area far from the river than in the closed canopy forest close to the river. Because tamarind leaf water content was higher near to the river, we are assuming that the water retained by the shallow clay soil is not contributing greatly to the adult tamarinds farther from the river but may be significant to seedling survival. Based on current data, we hypothesize that the shallow water table is responsible for the higher leaf water content near the river. As pointed out by Blumenfeld-Jones (this volume), vegetation gradients at Berenty are more complex than simply distance from the river. The high leaf and soil water content we found in our preliminary sampling near the former channel of the river is an example of that complexity. One of our future projects will be to expand our sampling away from the river but near former riverbeds in order to see how that affects tamarind leaf water content and lemur feeding.

For troop D1A, ranging and tamarind leaf feeding behavior correlate with leaf nutrient content, verifying the results of the year 2000 study (Mertl-Millhollen et al., 2003). During the time of lemur births in the late dry season during this study, the mature tamarind leaves had significantly higher leaf water and protein content when taken from the forest near the river than from the open forest farther from the river. The new leaves that appeared with the onset of spring rains did not show this correlation with proximity to the river; however, mature leaves gathered at the end of the wet season and in the dry season again showed the correlation. We interpret these results to mean that when new leaves are produced, leaf water content is more dependent on the onset of rains than on the water table, but that mature leaf water content is strongly influenced by proximity to the river and shallow water table. Troop D1A ate leaves only in the part of their range near the river where the leaves had the higher water and protein content.

When there was a fruit failure in D1A's range in the 2004 birth season, rather than increasing their leaf consumption, they actually ate a lower proportion of leaves than in other years and traveled out of their range to locate fruit. Optimal diet theory has typically used energy as the currency to explain foraging (Belovsky, 1990), and the complexity of leaf chemistry causes nutrient and digestive constraints (Sih and Christensen, 2001). Both fiber and tannin content can vary between individual trees and negatively impact protein availability, leaf palatability, and primate biomass (Ganzhorn, 1992; Mowry, 1998; Simmen et al., 1999; Chapman et al., 2003, 2004; Simmen et al., this volume) and may partly explain why the lemurs in this study expended energy to find energy-rich tamarind fruit during a fruit shortage rather than simply eating more mature leaves. They do eat soil (Rasamimanana and Rafidinarivo, 1993; Sauther et al., 1999; Mertl-Millhollen et al., 2003), which may counter the negative effects of tannins (Krishnamani and Mahaney, 2000). However, as with the primate population at Cocha Cashu, Peru, increased consumption of mature foliage does not seem to be an option (Terborgh and van Schaik, 1987). Dorn is continuing the chemical analyses of the leaf samples to determine fiber and tannin influence on leaf nutrient content.

Although ringtailed lemurs are a frugivorous/folivorous species (Simmen et al., this volume), they have also been called opportunistic frugivores (Ganzhorn, 1986; Rasamimanana and Rafidinarivo, 1993), and the fruiting pattern shown by the tamarind trees during this study correlates with the feeding and opportunistic ranging patterns by troop D1A. Fruit abundance did not correlate with distance from the river as was also found by Koyama et al. (this volume). When fruit was abundant in the closed canopy forest near the river, such as in 2002 and 2003, they fed no farther than 193 m from the river. In those years, the combination of fruit yield and tree density determined where the lemurs fed as was also seen for golden lion tamarins (Miller and Dietz, 2004). However, in the years when fruit production was very low, such as in 2004 and 2000 (Mertl-Millhollen et al., 2003), they traveled out of their range to a distance of 810 m from the river to get to trees that did have fruit. As seen in other primates (Garber, 1989), troop D1A demonstrated a spatial memory for fruiting trees located out of their range by traveling directly to them in both years. This flexible ranging response to locating tamarind fruit has also been seen in other studies (Sauther, 1998; Yamashita, 2002) and can positively impact survival (Berenstain, 1986; Johns and Skorupa, 1987).

Fruiting failure is common in tropical forests because of between-year climatic variation (Janson and Chapman, 1999), and fruit abundance may be the chief limiting factor in *Lemur catta* survival (Jolly et al., 2002). Tamarind fruiting patterns show a relationship with rainfall (Jolly et al. 2002; Jolly, this volume), and changes in rainfall can lead to higher lemur mortality rates (Sauther, 1998; Gould et al., 1999). Lemurs in general cope with dry season food shortages by decreasing their metabolism (Pereira, 1993; Pereira et al., 1999), traveling out of their normal range to find food or water (Sauther, 1998; Jolly and Pride, 1999; Scholz and Kappeler, 2004; Pride, this volume), and conserving their energy and optimizing their food intake in the rainy season (Jolly, 1966; Tilden, 1997; Wright, 1999; Ganzhorn et al., 2002). *Lemur catta* troops defend their ranges, thus ensuring a heritable food supply (Jolly, this volume), and, when conditions are good again after a drought or disaster, their population rebounds quickly (Gould et al., 1999).

Lemur ranging appears to be also influenced by social factors, and there may even be a historical component to their intergroup conflicts (Jolly et al., 1993). The areas of group overlap in forest adjacent to the river are rich in resources, but not so obviously richer than other areas to explain the disproportionate amount of time spent feeding in them. It would appear that they are showing exploitation (scramble) competition by feeding there so long. A similar combination of scramble and contest competition has been seen before by lemur troops (Mertl-Millhollen, 1988) and by platyrrhines (Peres, 2000).

Gallery forests near rivers in Madagascar are a more productive habitat than areas farther from rivers, contain a higher number of lemur groups than do dry areas, and are endangered by deforestation (Budnitz and Dainis, 1975; Whitelaw and Sauther, 2003; Sussman et al., 2003). Although lemurs are adapted to living in arid, patchy environments and are not limited by access to water (Sussman, 1974 and this volume; Goodman et al., this volume), the loss of gallery forests would greatly impact the number of surviving lemurs in Madagascar (Sussman et al., 2003). This study shows that water plays a crucial role in not only the quantity of the gallery forest trees but also in their quality, which in turn influences lemur ranging. Troop D1A ranging behavior includes defending a small, rich home range, maximizing the water and protein content of the leaves they feed on, feeding on fruit in that range when available, and having the behavioral flexibility to feed outside of their range when needed. That flexibility may be crucial to their survival.

Acknowledgments. Our heartfelt thanks go to Alison Jolly for providing the scientific environment in which this research could flourish, to Richard Jolly for his critical and enthusiastic help and support, to Kathryn Blumenfeld-Jones for her keen insights, to Michael Millhollen for help with the figures, to Jean Benoit Damy for creating the leaf pruner, and to the Berenty Reserve staff who took care of us in so many ways. We also thank Fort Hays State University students Christopher Schneider, Kaysha Schwartz, and Laci Franklin Schrant for their assistance with the leaf nitrogen analyses and Sigma Xi for funding a portion of Rambeloarivony's research. Most of all, we thank the de Heaulme family for their hospitality all these years and for conserving Berenty Reserve for the future.

References

Altmann, J. (1974). Observational study of behavior: Sampling methods. *Behaviour* 49:227–267.
Belovsky, G. E. (1990). How important are nutrient constraints in optimal foraging models or are spatial/temporal factors more important? In: Hughes, R. N. (ed.), *Behavioral Mechanisms of Food Selection.* Springer-Verlag, Berlin, pp. 255–280.
Berenstain, L. (1986). Responses of long-tailed macaques to drought and fire in eastern Borneo: A preliminary report. *Biotropica* 18(3):257–262.
Budnitz, N., and Dainis, K. (1975). *Lemur catta*: Ecology and behavior. In: Tattersall, I. and Sussman, R. W. (eds.), *Lemur Biology.* Plenum, New York, pp. 219–235.
Chapman, C. A., Chapman, L. J., Wangham, R., Hunt, K., Gebo, D., and Gardner, L. (1992). Estimators of fruit abundance of tropical trees. *Biotropica* 24(4):527–531.
Chapman, C. A., Chapman, L. J., Rode, K. D., Hauck, E. M., and McDowell, L. R. (2003). Variation in the nutritional value of primate foods: Among trees, time periods, and areas. *Int J Primatol* 24(2):317–333.
Chapman, C. A., Chapman, L. J., Naughton–Treves, L, Lawes, M. J., and McDowell, L. R. (2004). Predicting folivorous primate abundance: Validation of a nutritional model. *Am J Primatol* 62:55–69.
Charles-Dominique, P., and Hladik, C. M. (1971). Le *Lepilemur* du sud de Madagascar: Ecologie, alimentation et vie sociale. *Extrait de la Terre et la Vie* 1:3–66.

Denno, R. R., and McClure, M. S. (eds.). (1983). *Variable Plants and Herbivores in Natural and Managed Systems*. Academic Press, New York.

Ganzhorn, J. U. (1986). Feeding behavior of *Lemur catta* and *Lemur fulvus*. *Int J Primatol* 7:17–30.

Ganzhorn, J. U. (1988). Food partitioning among Malagasy primates. *Oecologia* 75:436–450.

Ganzhorn, J. U. (1992). Leaf chemistry and the biomass of folivorous primates in tropical forests: Test of a hypothesis. *Oecologia* 91:540–547.

Ganzhorn, J. U. (1995). Low-level forest disturbance effects on primary production, leaf chemistry, and lemur populations. *Ecology* 76(7):2084–2096.

Ganzhorn, J. U. (2002). Distribution of a folivorous lemur in relation to seasonally varying food resources: Integrating quantitative and qualitative aspects of food characteristics. *Oecologia* 131:427–435.

Ganzhorn, J. U., Klaus, S., Ortmann, S., and Schmid, J. (2002). Adaptations to seasonality: Some primate and nonprimate examples. In: Kappeler, P. M., and Pereira, M. E. (eds.), *Primate Life Histories and Socioecology*. University of Chicago Press, Chicago, pp. 132–144.

Garber, P. A. (1989). Role of spatial memory in primate foraging patterns: *Saguinus mystax* and *Saguinus fuscicollis*. *Am J Primatol* 19:203–216.

Gould, L., Sussman, R. W., and Sauther, M. L. (1999). Natural disasters and primate populations: The effects of a 2-year drought on a naturally occurring population of ring-tailed lemurs (*Lemur catta*) in southwestern Madagascar. *Int. J. Primatol.* 20:69–84.

Gosling, L. M., and Petrie, M. (1981). The economics of social organization. In: Townsend, C. R., and Callow, P. (eds.), *Physiological Ecology: An Evolutionary Approach to Resource Use*. Blackwell Scientific Publications, Oxford, pp. 315–345.

Hladik, C. M. (1988). Seasonal variations in food supply for wild primates. In: DeGarine, I., and Harrison, G. A. (eds.), *Coping with Uncertainty in Food Supply*. Clarendon Press, Oxford, pp. 1–25.

Janson, C. H., and Chapman, C. A. (1999). Resources and primate community structure. In: Fleagle, J. G., Janson, C. H., and Reed, K. E. (eds.), *Primate Communities*. University Press, Cambridge, pp. 237–267.

Johns, A. D., and Skorupa, J. P. (1987). Responses of rain-forest primates to habitat disturbance: A review. *Int. J. Primatol.* 8(2):157–191.

Jolly, A. (1966). *Lemur Behavior*. University of Chicago Press, Chicago.

Jolly, A., and Pride, E. (1999). Troop histories and range inertia of *Lemur catta* at Berenty, Madagascar: A 33-year perspective. *Int. J. Primatol.* 20:359–373.

Jolly, A., Gustafson, H., Oliver, W. L. R., and O'Connor, S. M. (1982). *Propithecus verreauxi* population and ranging at Berenty, Madagascar, 1975 and 1980. *Folia Primatol.* 39:124–144.

Jolly, A., Rasamimanana, H. R., Kinnaird, M. F., O'Brien, T. G., Crowley, H. M., Harcourt, C. S., Gardner, S., and Davidson, J. M. (1993). Territoriality in *Lemur catta* groups during the birth season at Berenty, Madagascar. In: Kappeler, P. M., and Ganzhorn, J. U. (eds.), *Lemur Social Systems and Their Ecological Basis*. Plenum, New York, pp. 85–109.

Jolly, A., Dobson, A., Rasamimanana, H. M., Walker, J., O'Connor, S., Solberg, M., and Perel, V. (2002). Demography of *Lemur catta* at Berenty Reserve, Madagascar: Effects of troop size, habitat and rainfall. *Int. J. Primatol.* 23(2):327–353.

Klopfer, P. H., and Jolly, A. (1970). The stability of territorial boundaries in a lemur troop. *Folia Primatol.* 12:199–208.

Koyama, N., Nakamichi, M., Ichino, S., and Takahata, Y. (2002). Population and social dynamics changes in ring–tailed lemur troops at Berenty, Madagascar between 1989–1999. *Primates* 43:291–314.

Mertl-Millhollen, A. S. (1988). Olfactory demarcation of territorial but not home range boundaries by *Lemur catta*. *Folia Primatol.* 50:175–187.

Mertl-Millhollen, A. S. (2000). Tradition in *Lemur catta* behavior at Berenty Reserve, Madagascar. *Int. J. Primatol.* 21:287–297.

Mertl-Millhollen, A. S., Moret, E. S., Felantsoa, D., Rasamimanana, H., Blumenfeld-Jones, K. C., and Jolly, A. (2003). Ring-tailed lemur home ranges correlate with food abundance and nutritional content at a time of environmental stress. *Int. J. Primatol.* 24(5):969–985.

Miller, K. E., and Dietz, J. M. (2004). Fruit yield, not DBH or fruit crown volume, correlates with time spent feeding on fruits by wild *Leontopithecus rosalia*. *Int. J. Primatol.* 25(1):27–39.

Mittermeier, R. A., and van Roosmalen, M. G. M. (1981). Preliminary observations on habitat utilization and diet in eight Surinam monkeys. *Folia Primatol.* 36:1–39.

Mowry, C. B. (1998). Feeding ecology of *Lemur catta*: Implications for captive populations. International Primatological Society, XVIIth Congress, University of Antananarivo, Madagascar, Abstract 265.

Oates, J. F. (1987). Food distribution and foraging behavior. In: Smuts, B. B., Cheney, D. L., Seyfarth, R. M., Wrangham, R. W., and Struhsaker, T. T. (eds.), *Primate Societies*. University of Chicago Press, Chicago, pp. 197–209.

O'Connor, S. M. (1987). *The Effect of Human Impact on Vegetation and the Consequences to Primates in Two Riverine Forests, Southern Madagascar*. PhD diss., Cambridge University. Cambridge.

Pereira, M. E. (1993). Seasonal adjustment of growth rate and adult body weight in ring-tailed lemurs. In: Kappler, P. M., and Ganzhorn, J. U. (eds.), *Lemur Social Systems and Their Ecological Basis*. Plenum Press, New York, pp. 205–221.

Pereira, M. E., Strohecker, R. A., Cavigelli, S. A., Hughes, C. L., and Pearson, D. D. (1999). Metabolic strategy and social behavior in Lemuridae. In: Rakotosamimanana, B., Rasamimanana, H., Ganzhorn, J., and Goodman, S. M. (eds.), *New Directions in Lemur Studies*. Kluwer Academic/Plenum Publishers, New York, pp. 93–118.

Peres, C. A. (2000). Territorial defense and the ecology of group movements in small–bodied neotropical primates. In: Boinski, S., and Garber, P. A. (eds.), *How and Why Animals Travel in Groups*. University of Chicago Press, Chicago, pp. 100–123.

Perica, S. (2001). Seasonal fluctuation and intracanopy variation in leaf nitrogen level in olive. *J. Plant Nutr.* 24(4&5):779–787.

Rasamimanana, H. (1999). Influence of social organization patterns on food intake of *Lemur catta* in the Berenty Reserve. In: Rakotosamimanana, B., Rasamimanana, H., Ganzhorn, J., and Goodman, S. M. (eds.), *New Directions in Lemur Studies*. Kluwer Academic/Plenum Publishers, New York, pp. 173–188.

Rasamimanana, H., and Rafidinarivo, E. (1993). Feeding behavior of *Lemur catta* females in relation to their physiological state. In: Kappeler, P. M., and Ganzhorn, J. U. (eds.), *Lemur Social Systems and Their Ecological Basis*. Plenum Press, New York, pp. 123–133.

Sauther, M. L. (1994). Wild plant use by pregnant and lactating ringtailed lemurs, with implications for early hominid foraging. In: Etkin, N. L. (ed.), *Eating on the Wild Side: The Pharmacologic, Ecologic and Social Implications of Using Non-cultigens*. University of Arizona Press, Tucson, pp. 240–256.

Sauther, M. L. (1998). Interplay of phenology and reproduction in ring-tailed lemurs: Implications for ring-tailed lemur conservation. *Folia Primatol.* 69(Suppl.1): 309–320.

Sauther, M. L., Sussman, R. W., and Gould, L. (1999). The socioecology of the ringtailed lemur: Thirty-five years of research. *Evol. Anthropol.* 8(4):120–132.

Scholz, F., and Kappeler, P. M. (2004). Effects of seasonal water scarcity on the ranging behavior of *Eulemur fulvus rufus*. *Int. J. Primatol.* 25(3):599–613.

Siegel, S. (1956). *Nonparametric Statistics for the Behavioral Sciences.* McGraw-Hill, New York.

Sih, A., and Christensen, B. (2001). Optimal diet theory: When does it work, and when and why does it fail? *Anim. Behav.* 61:379–390.

Simmen, B., Hladik, A., Ramasiarisoa, P. L., Iaconelli, S., and Hladik, C. M. (1999). Taste discrimination in lemurs and other primates, and the relationships to distribution of plant allelochemicals in different habitats of Madagascar. In: Rakotosamimanana, B., Rasamimanana, H., Ganzhorn, J., and Goodman, S. M. (eds.), *New Directions in Lemur Studies.* Kluwer Academic/Plenum Publishers, New York, pp. 201–219.

Sussman, R. W. (1974). Ecological distinctions in sympatric species of *Lemur*. In: Martin, R. D., Doyle, G. W., and Walker, A. C. (eds.), *Prosimian Biology.* University of Pittsburgh Press, Pittsburgh, pp. 75–108.

Sussman, R. W. (1991). Demography and social organization of free-ranging *Lemur catta* in the Beza Mahafaly Reserve, Madagascar. *Am. J. Phys. Anthropol.* 84:43–58.

Sussman, R. W., and Rakotozafy, A. (1994). Plant diversity and structural analysis of a tropical dry forest in southwestern Madagascar. *Biotropica* 26(3):241–254.

Sussman, R. W., Green, G. M., Porton, I., Andrianasolondraibe, O. L., and Ratsirarson, J. (2003). A survey of the habitat of *Lemur catta* in southwestern and southern Madagascar. *Primate Conserv.* 19:32–57.

Terborgh, J., and van Schaik, C. P. (1987). Convergence vs. nonconvergence in primate communities. In: Gee, J. H. R. and Giller, P. S. (eds.), *Organization of Communities Past and Present.* Blackwell Scientific Publications, Oxford, pp. 205–226.

Tilden, C. D. (1997). Low rates of maternal reproductive investment characterize lemuriform primates. *Am. J. Phys. Anthropol. Suppl.* 24:228.

Whitelaw, D. C., and Sauther, M. L. (2003). A preliminary survey and GIS analysis of ring-tailed lemur habitat use in and around Beza-Mahafaly Reserve, Madagascar. *Am. J. Phys. Anthropol. Suppl.* 36:224.

Wright, P. C. (1999). Lemur traits and Madagascar ecology: Coping with an island environment. *Yrbk. Phys. Anthropol.* 42:31–72.

Yamashita, N. (2002). Diets of two lemur species in different microhabitats in Beza Mahafaly Special Reserve, Madagascar. *Int. J. Primatol.* 23(5):1025–1051.

9
Feeding Competition Between Introduced *Eulemur fulvus* and Native *Lemur catta* During the Birth Season at Berenty Reserve, Southern Madagascar

SUSAN PINKUS, JAMES N.M. SMITH, AND ALISON JOLLY

Dedicated to the memory of Jamie Smith, 1944–2005

9.1. Introduction

We examined resource competition between the introduced population of brown lemurs (*Eulemur fulvus rufus*) and the native population of ringtailed lemurs (*Lemur catta*) at Berenty Reserve, Madagascar. Conditions facing ringtailed lemurs at Berenty may place them at high risk from introduced competitors. Berenty is part of an isolated 400-ha fragment of gallery forest and xeric scrub forest. It has been altered extensively, including clearing of edge habitat and provision of drinking water for lemurs. Ringtailed lemurs are well adapted to surviving in xeric scrubby or edge habitats (Gould et al., 1999; Godfrey et al., 2004; Goodman et al., this volume; Sussman et al., this volume). They also use adjacent gallery forest (Sussman 1972; Sussman et al., 2003), but in this habitat they are poor competitors with other similar-sized lemurs. At Berenty, ringtailed lemurs are mainly found in gallery forest habitat.

The IUCN currently lists ringtailed lemurs as *Vulnerable* (Ganzhorn et al., 2000), but recent habitat surveys suggest this listing should be upgraded to *Endangered* (Sussman et al., 2003). Of the six protected areas containing wild ringtailed lemurs, Berenty holds one of the largest populations, about 500 individuals. There is therefore concern that the presence of introduced brown lemurs is threatening the survival of the ringtailed lemurs at Berenty. In this chapter, we examine one aspect of this threat: competition for food. The population dynamics of Berenty's brown lemurs and their demographic impact on Berenty's ringtailed lemurs are described elsewhere (Pinkus 2004; Pinkus et al., in preparation).

Ringtailed lemurs and brown lemurs share similar life histories, morphology, seasonal growth patterns, reproductive biology and seasonality, maternal investment, and juvenile development (Sussman, 1972; Pereira, 1993). In gallery forests, the diets of brown lemurs and ringtailed lemurs are similar. Both species primarily eat fruit and leaves, supplemented with occasional animal prey. Fruit and leaves from the tree *Tamarindus indica*, a keystone dry season food species for ringtailed lemurs (e.g., Sauther 1991), make up much of the diet of both species. Brown lemurs deplete patches of mutually preferred foods to lower levels than do ringtailed lemurs (Ganzhorn, 1986). Work by Sussman (1972) suggested that in gallery forests, brown lemurs reach much higher population densities than do ringtailed lemurs.

Despite first appearances, the introduction of brown lemurs may not place Berenty's ringtailed lemur population at substantial risk. Brown lemurs and ringtailed lemurs are naturally sympatric in part of their natural ranges, about 300 km northwest of Berenty (Sussman, 1972). The forest at Berenty is ecologically similar to forests in which the species are naturally sympatric, both in degree of seasonality and in the composition of food, predator, and other primate species (Sussman, 1972; Sussman and Rakotozafy, 1994; Pinkus, unpublished data). *T. indica* is abundant at Berenty.

In his study of ringtailed lemurs and brown lemurs in natural sympatry at Anserananomby, Sussman (1972) found little interspecific overlap in habitat use or diet during the late dry season and rarely observed interspecific interactions of any sort. He described brown lemurs as arboreal dietary specialists. They ate primarily mature leaves and fruit from *T. indica*, seldom ventured outside of closed canopy forest, and spent less than 3% of their time on the ground. Ringtailed lemurs, in contrast, were semiterrestrial habitat and dietary generalists. They ate a much greater variety of plant species than brown lemurs and were found at all levels of the forest canopy (Sussman, 1972). Like brown lemurs, ringtailed lemurs inhabited closed canopy forest, but most of their territories included areas of transition forest and scrub at the forest edge (Sussman, 1972). Ringtailed lemurs outside of sympatry with brown lemurs act much as Sussman (1972) observed at Anserananomby (Sussman, 1972; Sauther, 1991; Yamashita 2002). Where they are allopatric with ringtailed lemurs, brown lemurs live in both dry forest and rainforest and are much less specialized in diet and habitat use than they are at Anserananomby (e.g., Overdorff, 1991; Scholz and Kappeler, 2004).

Although Sussman (1972) did the only study of ringtailed lemurs and brown lemurs in natural sympatry, Ganzhorn (1985, 1986) studied ringtailed lemurs and brown lemurs housed together in large enclosures at Duke Primate Center. Simmen et al. (2003) studied foraging by sympatric brown lemurs and ringtailed lemurs in the "tourist front" and "Ankoba" areas of Berenty. We did not use the results of Simmen et al. for comparison here because the ecology and population dynamics of ringtailed lemurs differs in these highly modified areas of Berenty (Jolly et al., this volume).

Our goal in this study was to test the hypothesis that the brown lemur population at Berenty is competing directly with ringtailed lemurs for limiting resources during the birth (late dry) season. Long-term interspecific diet overlap will result in exploitation competition, and a resulting decline in one or both competing populations, if the shared resources are limiting and can be depleted (e.g., Caughley and Sinclair, 1994). Ringtailed lemurs' birth season food sources meet these criteria; their availability critically affects ringtailed lemurs' survival and reproduction (Sauther, 1991; Gould et al., 1999). Food is scarce in the birth season and female lemurs are energetically stressed, even in years with normal *T. indica* fruit crops (Sauther, 1991; Jolly et al., 2002). Our study took place during the birth season of a year in which the *T. indica* fruit crop failed. Usually, primate diets diverge when food is most scarce, minimizing interspecific exploitation competition (Gautier-Hion, 1980; Terborgh, 1983; Overdorff, 1991; Richard and Dewar, 1991; Tutin et al., 1997). We thus expected interspecific diet overlap at Berenty to be at its lowest during our study, allowing us the best chance of disproving our hypothesis.

We show here that the large interspecific differences in diet and habitat use seen in the dry season at Anserananomby are not evident at Berenty during the birth (late dry) season. At Berenty, both ringtailed and brown lemurs ate similar plant species and parts, fed at similar heights, and used similar patch sizes. In fact, the differences in diet and habitat use between troops of the same lemur species were greater than the differences between paired troops of different lemur species. Ringtailed and brown lemurs at Berenty compete directly for limiting resources during the birth season.

These results suggest that brown lemurs have the potential to exert a strong demographic impact on ringtailed lemurs at Berenty. Surprisingly, as of 2000, there was not yet evidence of such an impact even though the brown lemur population had increased dramatically from the 8 animals introduced in 1975 to 310 animals and colonized all parts of Berenty Reserve, even the edge and scrub habitat normally monopolized by ringtailed lemurs (Pinkus, 2004). It has, however, been shown previously that a population's size may remain stable, or even increase, as environmental conditions worsen, and then decline abruptly (Abrams, 2002). Indeed, our most recent data (2004, 2005) show abrupt and marked changes in ringtailed lemur demographics at Berenty and suggest that the introduced brown lemurs may in fact threaten the long-term stability of Berenty's ringtailed lemur population (Jolly, unpublished data).

9.2. Methods

9.2.1. Study Site

The data used in this study were collected in Malaza Forest, a 97-ha parcel of closed-canopy gallery forest, open-canopy transition forest, and xeric scrub habitat located at the center of Berenty Reserve. Ringtailed lemurs and brown lemurs in Malaza are habituated to human presence and can be approached within 2–5 m without noticeably altering their behavior. We conducted this study in Malaza's Gallery and Scrub habitat types, where there is extensive range overlap between the two lemur species. For descriptions of Gallery and Scrub habitats in Malaza, see Jolly et al., this volume.

Malaza abuts a tourist development to the west, the Mandrare River to the east, 40 ha of closed canopy forest to the north, and 210 ha of degraded open-canopy forest and subdesert spiny forest to the south. Ringtailed lemurs and brown lemurs are present in all forest parcels. Ringtailed, but not brown, lemurs range throughout the tourist development and subdesert forest.

On January 25, 1975, eight juvenile red-fronted brown lemurs (*Eulemur fulvus rufus*) were introduced to Berenty. The animals were being kept as pets but escaped from cages into the forest during a cyclone (Jolly, 2004). Their provenance was Analabe Reserve, a deciduous dry forest 300 km northwest of Berenty (M. Jean de Heaulme, interview, September 1998; see also Jolly et al., 1982). In addition, up to nine collared lemurs (*Eulemur collaris*), originating from the Anhohahela region east of Berenty, were introduced between 1975 and 1983 (O'Connor 1987; M. Jean de Heaulme, pers. comm.). Genetic analysis showed that a third of 88 animals sampled in 1996 were hybrids of *E. collaris* and *E. fulvus*, and pelage color surveys suggested that about half of all "browns" were hybrids (Jekielek, 2002). Here, we will use the term "brown lemur" to mean *Eulemur fulvus rufus, E. collaris,* or *E. f. rufus X E. collaris* hybrids.

During the past 20 years, Berenty has become one of the most visited tourist destinations in Madagascar. Most disturbances in Malaza are the direct or indirect result of tourism, although the main wide trails have existed at least since the 1950s. Water troughs, filled sporadically during the dry season, are present throughout the forest. The distances between Malaza's nine water troughs range from approximately 150 m in Gallery forest to 500–1000 m in Scrub forest. Food provisioning with bananas occurred in the tourist area outside Malaza between 1985 and 1999 but does not affect troops inside the forest (S.P., pers. obs.).

9.2.2. *Diet Overlap and Activity Patterns*

We compared ringtailed lemur and brown lemur diets, activity patterns, and microhabitat use during the birth/early lactation season of 1998. We observed seven troop-pairs, each composed of one troop of brown lemurs and one troop of ringtailed lemurs. All troops were of similar size (±2 adult animals) and had nearly or completely overlapping home ranges and hence access to similar resources. Troop sizes ranged from 8 to 13 animals. Four troop-pairs in Gallery forest and three troop-pairs in Scrub forest were each sampled for one 12-hour "follow" per troop.

Ringtailed lemurs generally have consistent ranging patterns on successive days, with most changes in ranging patterns corresponding with changes in phenology of food plants. Occasional long "excursions" of 500 to 1500 m occur. No day ranges containing such "excursions" were included in this study. We assume that one 11- to 12-hour sample reflects a given troop's foraging and ranging patterns at a given stage of the availability of certain plant foods during the southern spring. To control for changing resource availability, paired troops were followed within 4 days of each other. Most follows lasted from 0600 to 1800 h; a few began later when particular troops were hard to find, but always before 0715 h. Data were collected between September 5 and 20, 1998.

We used teams of 5–8 observers to collect intensive ecological data on each troop. Follows were done by teams of at least two observers for every three lemurs. This allowed us to account for all feeding by all members of each troop. Susan Pinkus or Alison Jolly supervised all data collection to standardize data categories. R. Ranaivojaona from the Park Botanique et Zoologique de Tsimbazaza in Antananarivo identified all plant species.

During follows, 5-minute scan samples (Martin and Bateson, 1993) were used to record activity patterns and resource use. Data collection for one scan took from 30–90 seconds. At each scan, we recorded the major activity of the troop (feed, travel between patches, move within a patch, rest, sun, interspecific encounter) and the 25 m × 25 m quadrat location of the majority of the troop. "Majority" was defined as the greatest number of animals doing the same activity or in the same location. For each food patch being fed in by a troop member, we recorded species and size of the patch, plant part being fed on (young leaf, mature leaf, old leaf, unripe fruit, ripe fruit, flower, insect, drink, other), majority height of troop members in the patch, number of animals visible, number feeding, and total number known to be in the patch.

In this study, we use the word "patch" to refer to a discrete food source fed in by a lemur. A patch could be a tree, a bush, a swarm of caterpillars on a dead log, a clump of herbs on the ground, and so forth. A patch is equivalent to one plant of a given species if it is a tree or bush. The diameter of the patch is then its canopy diameter, estimated by eye. For ground vegetation, a patch is a group of plants, dead leaves, insects, and so forth, in which a lemur can feed continuously. For these patches and for patches of lianas, patch diameter is the greatest width of the patch from edge to edge, estimated by eye.

9.2.3. *Analysis of Data*

For each follow, we summed the number of scans in which each activity was recorded and expressed them as a proportion of the total scans recorded that day. To compare the proportion that each diet component made up of a troop's total feeding time, we summed the number of animals recorded feeding on a particular resource type, resource part, or height and expressed it as a proportion of the total number of animals recorded feeding on any patch that day. "Animal minute" refers to one animal feeding in one patch during one scan. Feeding data are presented as proportions of total animal minutes spent feeding.

Most resource types were trees, shrubs, lianas, or forbs identified to species level. Two types represent more than one food species: "Litter" refers to dry plant matter found in the leaf litter, and "Insect" refers to any invertebrate eaten before it could be identified to species. Four other resource types were not plants: "Phromnia" is glucose-rich secretions of the flower-mimic insect *Phromnia rosea*; "Egg" is bird eggs, "Acacia Worm" is swarms of the caterpillars of a recognized (though unidentified) butterfly species; and "Dirt" is soil without a visible plant or any animal material in it.

We calculated Horn's Index of Overlap (Krebs, 1999) to estimate diet overlap between all pairwise combinations of lemur troops. We compared estimates of overlap between members of the same troop-pair with the mean of the estimated overlap between each member of the troop pair and all other troops in that habitat type. These data should be considered minimum estimates of diet overlap. Simulations have shown that Horn's Index underestimates overlap, particularly when sample sizes are small or uneven, as they are in these diet data (Krebs, 1999).

Diet breadth, or degree of diet specialization, can be thought of as the number of resource types making up some minimum proportion of the diet (Krebs, 1999). We characterized diet breadth as the number of resource types making up at least 4% of total diet for each species in each habitat. To estimate diet breadth for each habitat type, we summed the number of animal minutes each lemur species spent feeding on a given resource type and expressed it as a proportion of the total number of animal minutes that species spent feeding. Our diet data may underestimate brown lemurs' dietary breadth because brown lemurs feed both by day and by night, and we only collected feeding data during the day.

We used a significance level of 5% and two-tailed parametric tests. In some cases, however, we comment on trends where sample sizes were low and alpha was > 0.05. Variability in data is displayed using 95% confidence intervals (C.I.).

9.3. Results

9.3.1. Overall Diet Breadth and Overlap

Forty-four foods were used in all, but diets overall were narrow for both species. Most food types made up less than 4% of total feeding time in both Gallery and Scrub habitats. Brown lemurs ate few foods not eaten by ringtailed lemurs (five spp. in Gallery, four spp. in Scrub).

Sussman's previous characterization of brown lemurs as specialists, and ringtailed lemurs as generalists (1972), did not apply at Berenty. The number of food types making up the threshold criterion for contributing significantly to the diet (≥4%; Krebs, 1999) did not differ significantly between the two species. Diet breadth in Scrub was four species for ringtailed lemurs and two species for brown lemurs. In Gallery, diet breadth was three species for ringtailed lemurs and four species for brown lemurs. Four plant species accounted for 84% of the diet of ringtailed lemurs in Scrub, 87% for ringtailed lemurs in Gallery, 90% of the diet of brown lemurs in Scrub, and 94% for brown lemurs in Gallery. Both ringtailed lemurs and brown lemurs ate all of these species.

Two tree species, *Tamarindus indica* and *Celtis philippensis*, dominated diets in both habitats, together making up more than 74% of feeding time by both brown and ringtailed lemurs (Figures 9.1A and 9.1B; Table 9.1). Two additional species, *Quivisianthe papinae* and *Azima tetracantha*, made up more than 4% of intake for ringtailed lemurs in scrub habitat (Figure 9.1B). *A. tetracantha* also

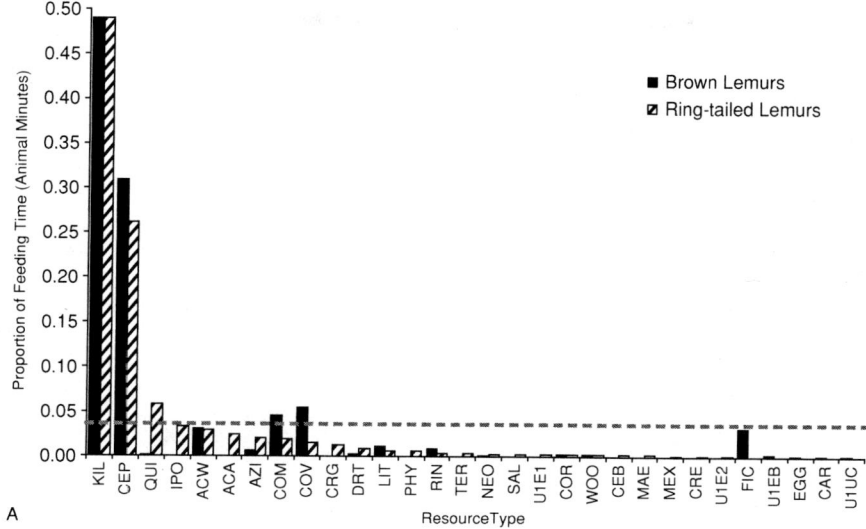

FIGURE 9.1. (A) Diet overlap between ringtailed and brown lemurs in Gallery habitat. Bars above the dashed line are resource types that make up more than 4% of the species' diet; these resources were used to calculate diet breadth. Resource types are species of plant or insect fed on by either lemur species. See Table 9.1 for a key to the resource type codes used in this figure.

made up more than 4% of intake for ringtailed lemurs in Gallery habitat. *Combretum albiflorum* and *Cordia varo* accounted for more than 4% of the diet of brown lemurs in Gallery habitat (Figure 9.1A).

In both Scrub and Gallery habitats, ringtailed lemurs ate more different food species (Gallery and Scrub = 27 spp.) than brown lemurs (Scrub = 18 spp.; Gallery = 15 spp.). However, ringtailed lemurs ate only small amounts of the species not eaten by brown lemurs. No food plant species eaten by only one lemur species represented more than 2.5% of the diet of that lemur species. As a result, diet breadth was similar between ringtailed lemurs and brown lemurs despite the much larger number of foods sampled by ringtailed lemurs. To put these data in perspective, given the sample sizes in this study, a single food type will represent 0.5–1% of a lemur species' diet in one habitat type if half of a troop of lemurs was recorded feeding on this food type during one 5-minute scan.

9.3.2. Diet Overlap Across Troop-pairs and Habitats

There was high diet overlap within each troop-pair. The mean value of Horn's Index of Overlap (I_o) was 0.84 (95% C.I. 0.74–0.94, n = 4 troop-pairs) for troop-pairs in Gallery, and 0.79 (95% C.I. 0.75–0.83, n = 3 troop-pairs) in Scrub. Troop-pair diet overlap was similar to overlap of either member of the pair with other troops of the opposite species (95% C.I. Gallery 0.73–0.85, n = 4

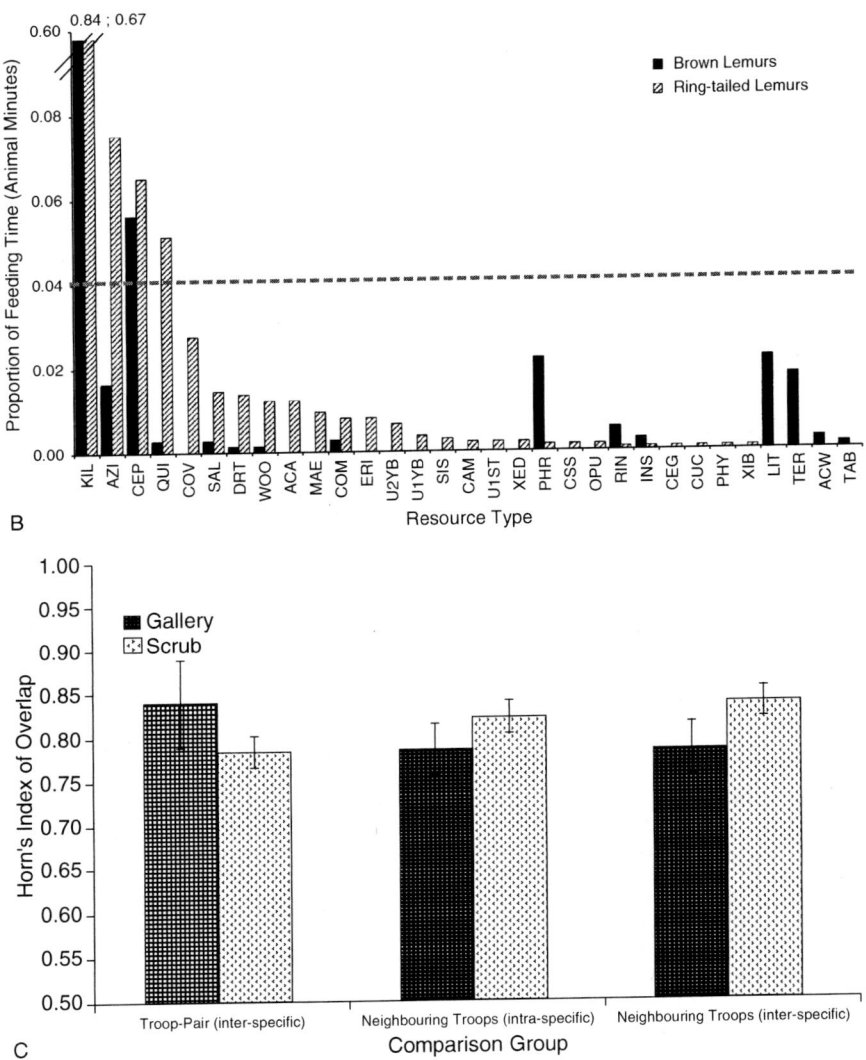

FIGURE 9.1. (*Continued*) (B) Diet overlap between ringtailed and brown lemurs in Scrub habitat. Scale expanded to show less used species. Bars above the dashed line are resources making up more than 4% of the species' diet; these resources were used to calculate diet breadth. Resource types are species of plant or insect fed on by either lemur species. See Table 9.1 for a key to the resource type codes used in this figure. (C) Inter- and intraspecific overlap of food types eaten by troop pairs and neighboring troops in Gallery and Scrub habitat. Each bar represents the mean value of Horn's Index of Overlap between pairs of troops. Comparison groups are paired troops of (i) one ringtailed lemur and one brown lemur troop in overlapping ranges, (ii) two troops of the same species in adjacent ranges, and (iii) one ringtailed and one brown lemur troop in adjacent ranges. Error bars are ± 1 SE for the mean of Horn's Index of Overlap for paired troops.

TABLE 9.1 Resource types fed on by ringtailed and brown lemurs during this study. Abbreviations appear in Figures 9.1A and 9.1B.

Abbreviation	Species/description
ACA	*Acacia rovumae*
ACW	Caterpillar (unknown sp.)
AZI	*Azima tetracantha*
CAM	Unknown plant sp.
CAR	Unknown plant sp.
CEB	*Celtis bifida*
CEG	*Celtis gomphophylla*
CEP	*Celtis philippensis*
COM	*Combretum albiflorum*
COR	*Cordia rothii*
COV	*Cordia varo*
CRE	*Crateva excelsa*
CRG	*Crateva greveana*
CSS	*Cissus quadrangularis*
CUC	Unknown plant sp.
DRT	Soil/Clay
EGG	Bird egg (unknown sp.)
ERI	*Erigeron sp.*
FIC	*Ficus spp.*
INS	Insect (unknown sp.)
IPO	*Ipomoea cairica*
KIL	*Tamarindus indica*
LIT	Leaf litter (unknown plant sp.)
MAE	*Maerua filiformis*
MEX	*Argemone mexicana*
NEO	*Neotina isoneura*
OPU	*Opuntia vulgaris*
PHR	Secretions from the homopteran *Phromnia rosea*
PHY	*Phyllanthus casticum*
QUI	*Quivisianthe papinae*
RIN	*Rinorea greveana*
SAL	*Salvadora angustifolia*
SIS	*Agave sisalana*
TAB	*Tabernaemontana sp.*
TER	*Terminalia mantaly*
U1E1	Unknown plant sp.
U1E2	Unknown plant sp.
U1EB	Unknown plant sp.
U1ST	Unknown plant sp.
U1YB	Unknown plant sp.
U2YB	Unknown plant sp.
WOO	Rotting wood (unknown sp.)
XED	*Xerosicyos perrieri*
XIB	Unknown plant sp.

troop-pairs; Scrub 0.80–0.88, n = 3 troop-pairs). Within each habitat type, intraspecific diet overlap was almost identical to interspecific diet overlap and was not significantly different from overlap within troop-pairs (Figure 9.1C). Seven of eight troops in Gallery, and four of six troops in Scrub, had greater diet overlap with their pair-troop than with other troops of the same species. Comparing across habitats, inter- and intraspecific diet overlap was slightly higher within than between habitats (interspecific I_O between habitats: brown lemurs 0.75; ringtailed lemurs 0.71; intraspecific I_O between habitats: brown lemurs 0.76; ringtailed lemurs 0.72).

What stands out from these results is that diet overlap is greatest between different species of lemur using the same home range. The diets of brown and ringtailed lemurs are so similar that, even within a habitat type, different species using the same home ranges are more similar than troops of the same species using different home ranges. Species-specific differences in diet were minor compared with the effect of small-scale habitat variation.

9.3.3. Feeding Height

Even with nearly complete diet overlap, competition could be reduced if brown lemurs and ringtailed lemurs fed at different heights. Naturally sympatric populations of brown lemurs and ringtailed lemurs do forage at different average heights, with ringtailed lemurs spending more time near the ground and brown lemurs using the highest canopy levels (Sussman, 1972; Ganzhorn, 1985), thus depleting different "slices" of shared resources. In contrast, at Berenty there was almost complete overlap in the feeding heights in both Scrub and Gallery (Figures 9.2A and 9.2B). The amounts of time spent feeding on the ground, close to the ground, and in the upper canopy were all similar. Though the ringtailed lemurs spent 4–6% of their time feeding 2–5 m above the ground, a height class that was barely used by brown lemurs, this result stemmed from the behavior of only two out of seven ringtailed lemurs troops—one in Gallery and one in Scrub. The remaining five ringtailed lemur troops used the 2–5 m layer similarly to brown lemurs.

9.3.4. Use of Tamarindus indica, a Keystone Resource

Another potential mechanism for reducing competition, given a high degree of diet overlap, is for the competing consumer species to feed on different individuals of a given plant species, or on different parts of the same individual. Because *T. indica* is a keystone resource for ringtailed lemurs during the dry season, we compared brown lemurs' and ringtailed lemurs' use of different plant parts (fruit, leaves, buds, etc.) and patch sizes of *T. indica*. We present here data for Scrub habitat, in which *T. indica* comprises a much larger portion of both species' diets; the pattern, however, is similar for Gallery.

Although *T. indica* trees have very deep canopies with apparently abundant feeding sites at all strata above 5 m in height, there was little difference in the

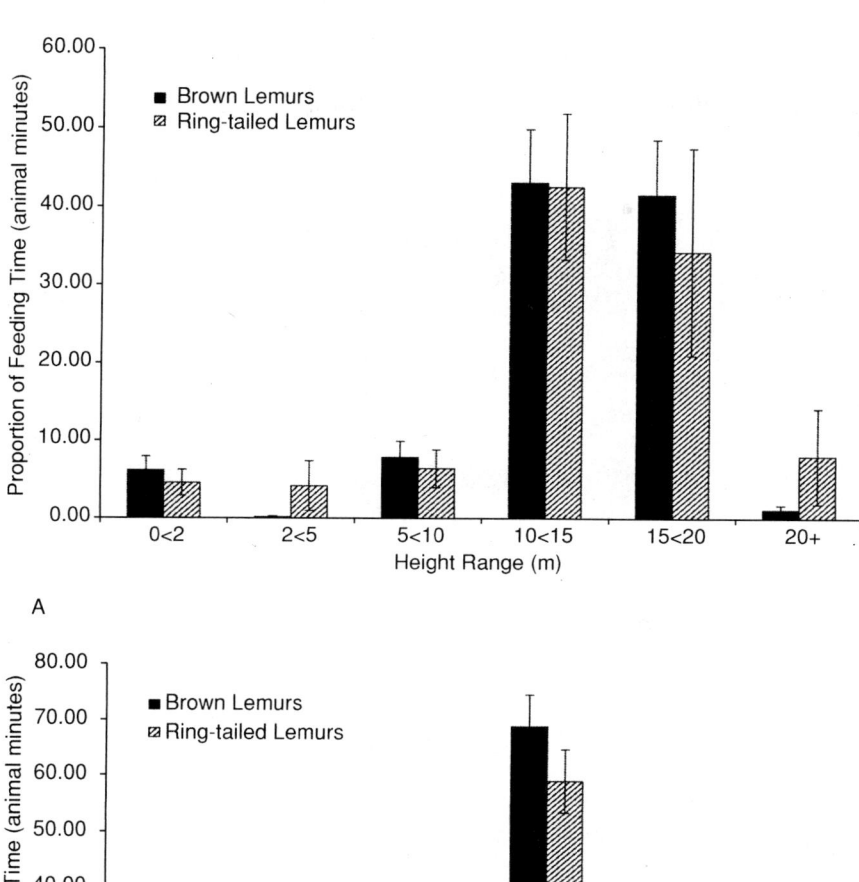

FIGURE 9.2. (A) Mean feeding heights of ringtailed and brown lemurs in Gallery habitat. Error bars are ± 1 SE. (B) Mean feeding heights of ringtailed and brown lemurs in Scrub habitat. Error bars are ± 1 SE.

feeding heights chosen by brown lemurs and ringtailed lemurs (Figures 9.2A and 9.2B). There was also no significant difference in use of the most frequently chosen patch diameters, 10–15 m and 15–20 m (Figure 9.2C). There was a statistically insignificant trend for brown lemurs to use patches of 5–10 m, and >20 m in diameter to a greater extent than ringtailed lemurs (Figure 9.2C).

There was high overlap in plant parts used (Figure 9.2C), with no significant difference except for leaf buds, which were fed on only by ringtailed lemurs. Because leaf buds are an ephemeral resource, opening into new leaves within 2 days, and brown lemurs and ringtailed lemurs feed on new leaves to a similar extent, it is likely that this diet difference reflects the day that troops were followed rather than interspecific diet preferences. In choice of patch diameter and plant part, as with diet as a whole, intraspecific variation equalled or exceeded interspecific variation. Ripe and unripe *T. indica* fruit are the most frequently eaten items. The relative frequency with which brown lemurs and ringtailed lemurs eat each part varies more within than between species (Figure 9.2D).

9.3.5. Diurnal Activity Budgets

Feeding at different times of day could reduce contest competition for resources, though the slow rate at which plant parts regenerate makes it unlikely that this

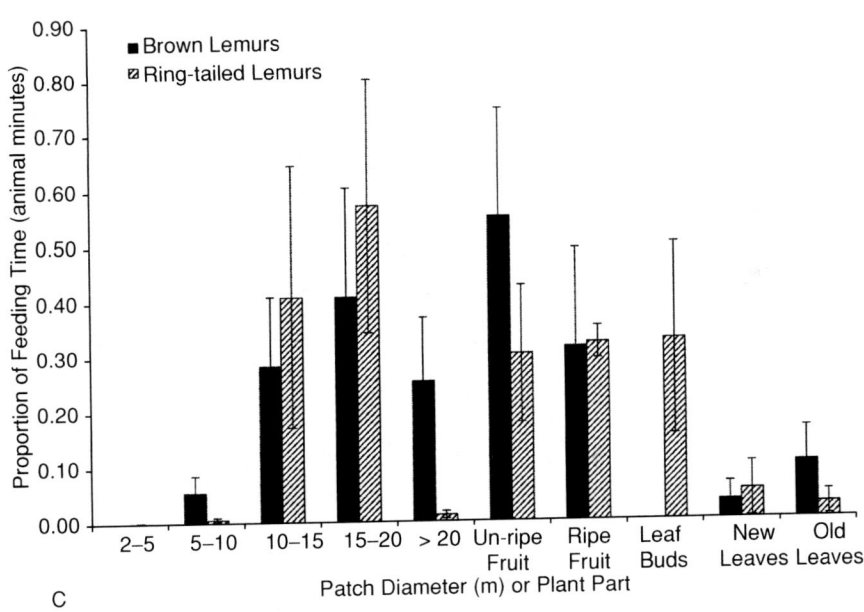

FIGURE 9.2. (*Continued*) (C) Patch size and plant part of *Tamarindus indica* used by ringtailed and brown lemurs in Scrub habitat. Error bars are ± 1 SE.

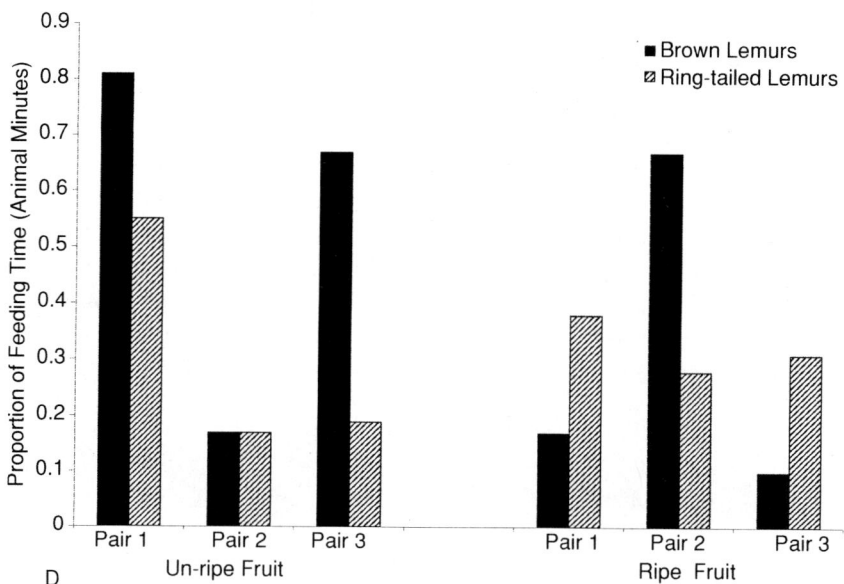

FIGURE 9.2. (*Continued*) (D) *Tamarindus indica* fruit consumption by three troop-pairs in Scrub habitat.

mechanism would reduce scramble competition markedly (Ganzhorn and Kappeler, 1996). In both Gallery and Scrub habitat, brown lemurs at Berenty spend a smaller proportion of daylight time feeding than ringtailed lemurs (two-sample t-tests, n = 14 troops, p < 0.05), and a greater proportion of time sleeping or resting (two-sample t-tests, n = 14 troops, p < 0.05; Table 9.2). Brown lemurs at Berenty are frequently observed feeding and moving after dark, so their daylight time spent feeding is likely only a part of their total feeding time. We did not make quantitative observations at night.

TABLE 9.2. Daytime activity of ringtailed and brown lemurs at Berenty.

Lemur species	% Time feeding (95% C.I.)		% Time resting (95% C.I.)	
	Gallery	Scrub	Gallery	Scrub
Ringtailed	0.27–0.45	0.28–0.40	0.44–0.64	0.37–0.69
Brown	0.16–0.26	0.10–0.22	0.62–0.72	0.67–0.85

Percent of 5-minute scan samples between 6 a.m. and 6 p.m. spent feeding, and resting or sleeping, in Gallery and Scrub habitats (for each species, n = 4 troops in Gallery; n = 3 troops in Scrub). Data are presented as 95% confidence intervals (C.I.). In both habitats, brown lemurs fed significantly less (two-sample t-tests, one-tailed, p < 0.01) and slept or rested significantly more (two-sample t-tests, one-tailed, p < 0.05) than ringtailed lemurs.

9.4. Discussion

9.4.1. Interspecific Diet Overlap

Twenty-three years after their introduction to Berenty, brown lemurs and ringtailed lemurs overlapped almost completely in diet and habitat use during the birth season, and to a much greater extent than reported for other sympatric populations of the two species (Figures 9.1C, 9.2A, 9.2B; Sussman, 1972; Ganzhorn, 1985, 1986). Overlap was high at Berenty because the ringtailed lemurs there behaved as they do in natural sympatry, but the brown lemurs did not. Ringtailed lemurs at Berenty, as in all known populations, had generalist habitat use and a diet dominated by a few plant species, particularly *Tamarindus indica*, and rounded out with many lightly used food species (e.g., Sussman, 1972; Ganzhorn, 1985, 1986; Sauther, 1991; Rasamimanana and Rafidinarivo, 1993; Yamashita, 2002; Simmen et al., 2003). At Berenty, brown lemurs did not behave as the specialized arboreal folivores described in other populations sympatric with ringtailed lemurs (Sussman, 1972; Ganzhorn, 1985). Instead, both lemur species at Berenty competed directly for food during the birth season, a time of resource scarcity that profoundly impacts ringtailed lemurs' reproductive success (Jolly et al., 2002).

High overlap during periods of food scarcity occurs occasionally in sympatric primate populations (e.g., Gautier-Hion, 1980; Waser, 1980; Richard and Dewar, 1991; Vasey, 2000), but it is more common for diets to diverge most when food is most scarce (Gautier-Hion, 1980; Terborgh, 1983; Overdorff, 1991; Richard and Dewar, 1991; Tutin et al., 1997). Numerous studies have shown that when food is scarce and primates of similar body size eat the same diet, divergence in microhabitat or feeding height tends to increase (e.g., Gautier-Hion, 1980; Gautier-Hion, 1988; Vasey, 2000; see also Waser, 1980; Terborgh, 1983; Ganzhorn, 1988; Ganzhorn and Kappeler, 1996). In our study, however, high diet overlap was not mediated by directional differences in feeding height, patch size, or forest sites chosen. Brown lemurs fed for fewer daylight hours than ringtailed lemurs (Table 9.2) and may have fed more at night, yet these differences would not have reduced scramble competition for slowly renewed food types (Ganzhorn and Kappeler, 1996). Two other dimensions along which sympatric primates' diets can diverge, feeding substrate and leaf chemistry, were not measured in our study (e.g., Emmons et al., 1983; Glander and Rabin, 1983; Harcourt and Nash, 1986; Ganzhorn, 1986, 1988, 1989; Tomlin and Cranford, 1994; Vasey, 2000, Simmen et al., this volume). Because brown and ringtailed lemurs have the same mass and locomotory habits, they are unlikely to diverge greatly in feeding substrate. The possibility of interspecific differences in preferred leaf chemistry is more promising (e.g., Simmen et al., this volume).

9.4.2. Interspecific Differences in Dietary Flexibility

On average, the foraging behavior of the seven brown lemur troops in this study was similar to that of the ringtailed lemur troops. They foraged at similar heights, ranged in overlapping habitat, and had similar diets. The only diet component that differed consistently between species was new leaves; as found in other studies, brown lemurs ate fewer new leaves than ringtailed lemurs (Berenty: Jolly et al., unpublished; Simmen et al., 2003; captivity: Ganzhorn, 1986). However, although brown and ringtailed lemurs' foraging behavior was similar on average, we observed far more variation in diet and foraging heights among individual brown lemur troops than among ringtailed lemur troops (Figure 9.2D). For example, certain brown lemur troops in this study, and in another study done at the same time (Simmen et al., 2003), exhibited very different foraging strategies from each other. *T. indica* accounted for from 25% to 94% of the brown lemur troops' diets, while all ringtailed lemur troops in overlapping ranges ate a similar proportion (60%) of *T. indica*. Following many troop-pairs allowed us to establish that brown lemurs at Berenty have greater intertroop variability in diet than ringtailed lemurs. This observation could easily have been missed if we had done multiple follows on each of a smaller number of troop-pairs.

The flexible behavior of Berenty's brown lemurs is not surprising. Brown lemurs exhibit much variation among habitat types and populations (e.g., Sussman, 1972; Ganzhorn, 1985, 1986; Overdorff, 1991, 1993, 1996; Simmen et al., 2003; Scholz and Kappeler, 2004). They adjust their foraging strategy in response to small-scale changes in food availability over time and space. Unlike ringtailed lemurs, they can exist at high density while feeding primarily on mature leaves (Sussman, 1972). Yet, they are highly frugivorous where fruit is plentiful (Overdorff, 1991). At Berenty, they opportunistically feed on insects, lizards, bird eggs, and even baby ringtailed lemurs (Jolly et al., 2000; Walker pers. comm.; Jolly and Pinkus, unpublished data). Thus, brown lemurs act as serial specialists (see also Ganzhorn, 1986) at the population and troop levels.

Ringtailed lemurs, in contrast, vary their foraging strategy much less in response to changing resource availability, and are more rigidly frugivorous. This foraging strategy works well when exploiting varied edge habitats or a habitat gradient that includes both closed canopy forest and edge (see Gould et al., 1999; Wright, 1999; Godfrey et al., 2003; discussed further below). It may, however, place ringtailed lemurs at a competitive disadvantage in habitats where they are sympatric with brown lemurs, because of the brown lemurs' capacity for serial specialization (see also Ganzhorn, 1986).

9.4.3. Interspecific Differences in Competitive Ability

Serial specialization gives brown lemurs greater resilience than ringtailed lemurs to temporal variation in food availability. Strikingly, brown lemur numbers at Berenty continued to grow in years of fruit scarcity when ringtailed lemur numbers fell (Pinkus et al., in preparation). Though they sometimes eat mainly fruit, brown lemurs can become almost complete folivores during periods of fruit scarcity. So, like white sifakas (Richard et al., 2000, 2002; Godfrey et al., 2004), brown lemurs may be limited by leaf rather than fruit availability (Pinkus et al., in preparation). The availability of high-quality leaves is more predictable than the availability of fruit and is not as closely correlated with total rainfall (Janson and Chapman, 1999). As a result, periods of reduced leaf availability are less severe than those of reduced fruit availability. Generalizing until resources become scarce, and then specializing, can allow populations of facultative specialists to out-compete obligate generalists (Robinson and Wilson, 1998). This result lends support to the idea that serial specialization by brown lemurs may give them a competitive advantage over ringtailed lemurs.

9.4.4. Behavioral Dominance of Brown Lemurs Over Ringtailed Lemurs

Though brown lemurs and ringtailed lemurs seldom interact in natural sympatry (Sussman, 1972), *Eulemur fulvus* subspecies show behavioral dominance over other lemur species both in rainforest and in captivity at Duke Primate Center (Overdorff, 1991; K. Glander, pers. comm.). At Berenty, brown lemurs dominate ringtailed lemurs in contest competition at food and water sources (pers. obs.; N. Koyama, J. Walker, pers. comm.). In areas of high population density at Berenty, each ringtailed lemur troop's range is overlapped with several brown lemur troops, and some ringtailed lemurs troops spend 25% of their feeding time in patches occupied by brown lemurs (Jolly and Pinkus, unpublished data). During this study, brown lemur and ringtailed lemur troops regularly entered food patches occupied by the other species. These intrusions were twice as likely to lead to aggressive conflict when brown lemurs invaded patches occupied by ringtailed lemurs than vice versa. It seems likely that, if the density of brown lemurs continues to increase, contest competition with ringtailed lemurs will intensify.

9.4.5. Seasonal Food Scarcity and Tamarindus indica as a Keystone Species

Tamarindus indica fruit is a keystone dry season resource for ringtailed lemurs and is often the dominant component of their diet (Jolly, 1966, Sussman, 1972; Sauther, 1991; Rasamimanana and Rafidinarivo, 1993; Yamashita, 2002; Koyama, this volume, Mertl-Millhollen et al., 2003; Simmen et al., this volume). At Berenty, *T. indica* trees bear heavy fruit crops during droughts that occur about every 7 years (Jolly et al., 2002). Fruit production then decreases dramatically over the next 1 to 2 years (Jolly et al., 2002; Simmen et al., 2003). Throughout ringtailed lemurs' range, areas with higher densities of *T. indica* trees support higher densities of ringtailed lemurs (Gould et al., 1999; Hawkins, 1999; Sussman et al., 2003). However, ringtailed lemurs' reliance on *T. indica* fruit means that droughts affect them more severely than sympatric species with more folivorous diets (Gould et al., 1999; Richard et al., 2000; Jolly et al., 2002).

Food availability in southern Malagasy dry forests is low during the late dry season, even in years of average food abundance (Ganzhorn et al., 2003).

At Berenty, female ringtailed lemurs end gestation and begin lactation during the late dry season (Koyama et al., 2001; Jolly et al., 2002). Energy appears to be particularly limiting for female ringtailed lemurs during gestation and lactation (Young et al., 1990; Pereira, 1993; Pereira et al., 1999; Godfrey et al., 2004). A female's ability to rear surviving offspring in consecutive years may depend on weight gain during this time (Pereira, 1993). Thus, ringtailed lemur populations may be particularly vulnerable to food shortages during the late dry season.

Ringtailed lemurs have a suite of physiological adaptations to seasonal resource scarcity during the dry season, including reduced activity, food intake and body mass, decreased metabolic rate, and semitorpor (Pereira, 1993; Ganzhorn et al., 2003). In years when keystone dry season foods, particularly *T. indica*, are scarce, these strategies are inadequate to prevent high mortality and reproductive failure (Gould et al., 1999; Jolly et al., 2002; Godfrey et al., 2004). In years of average resource availability, reproductive success hinges on the degree and outcome of resource competition during the late dry season.

T. indica fruit also dominates brown lemurs' diet during the late dry season at Berenty, but mortality of brown lemurs does not increase in postdrought years when fruit is scarce (Pinkus et al., in preparation). In fact, recruitment of juvenile brown lemurs in Scrub habitat, where *T. indica* is least dense, appeared to increase in postdrought years, when juvenile recruitment by ringtailed lemurs in this habitat decreased (Jolly et al., 2002; Pinkus et al., in preparation). Brown lemurs in some other populations seldom eat *T. indica* (e.g., Overdorff 1991; Ganzhorn and Kappeler, 1996); they appear to favor *T. indica* fruit when it is available but thrive on alternate foods when it is not. In contrast, as a keystone food during the late dry season, *T. indica* is an irreplaceable resource for ringtailed lemurs.

9.4.6. *The Role of Edge Habitat in Ringtailed Lemurs' Resilience to Food Scarcity*

Ringtailed lemurs are able to weather periods of food scarcity and rebound rapidly from population declines (e.g., Ross, 1992; Gould et al., 1999; Godfrey et al., 2004). This allows their numbers to persist despite frequent *T. indica* crop failures. Their use of edge habitat may play a key role in this demographic resilience. Successional vegetation in edge habitats provides a more consistent (Opler et al., 1980) and diverse (Ganzhorn, 1995; Strier, 1999) supply of higher quality fruit and foliage (Brugiere et al., 2002) than vegetation in the forest interior. Plant species in edge habitats also tend to have higher water content than those in canopy forest. Thus, frugivorous primate populations without access to edge or other early successional habitat may be at a high risk of starvation when crop failures occur in keystone fruit resources (Strier, 1999; Brugiere et al., 2002).

Surveys of naturally sympatric brown and ringtailed lemur populations suggest that ringtailed lemurs tend to use both closed canopy forest and scrubby edge habitat, whereas brown lemurs remain in the closed canopy forest (Sussman, 1972; Hawkins, 1999; Sussman et al., 2003). Sussman (1972) surveyed habitat fragments occupied by ringtailed lemurs only. All populations he found had access to either scrubby habitat or river edge habitat. He found no populations in closed canopy forest without edge, though he surveyed this habitat type, and in some cases found that it was occupied by brown lemurs. The apparent equivalence of large areas of scrub habitat and small areas of river edge is interesting. Edge habitat may be important to ringtailed lemurs as a source of water, whether it comes directly from a river or from high-moisture fruit and foliage.

The manner in which brown lemurs use closed canopy forests at sites other than Berenty (Sussman, 1972; Hawkins, 1999) suggests that they are poorly adapted to edge habitat. They have less flexible thermoregulatory behavior than ringtailed lemurs (Ganzhorn, 1985) and travel shorter distances than ringtailed lemurs in dry forest (Sussman, 1972; Ganzhorn, 1985; Jolly and Pinkus, unpublished). Brown lemurs conserve energy and water by ranging in shady habitat, moving relatively little, and subsisting on food types that are abundant nearby. Conserving water may be especially important for dry forest populations of brown lemurs because, compared with ringtailed lemurs, their diets contain fewer young leaves, leaves from succulent plants, and fleshy fruit (Figure 9.2C; Sussman, 1972). It is thus surprising that brown lemurs at Berenty have colonized edge habitat as effectively as ringtailed lemurs. We suspect that this is made possible by water provisioning in Scrub habitat at Berenty (Pinkus et al., in preparation).

9.5. Conclusions

We found high interspecific diet overlap between brown and ringtailed lemurs in both Gallery and edge/Scrub habitats at Berenty. Interspecific competition, particularly for *T. indica* fruit, is likely to have greater consequences for ringtailed lemurs than for brown lemurs because brown lemurs are behaviorally dominant and resilient to fruit shortages. In natural sympatry, ringtailed lemurs foraging in edge habitat face little or no competition from similar-sized frugivores; edge habitat likely functions for ringtailed lemurs as a refuge from resource competition. Edge habitat at Berenty has been cleared extensively and, since 1995, ringtailed lemurs have been forced to share with brown lemurs almost all the edge habitat that remains.

Brown lemur numbers at Berenty have continued to increase since 1998, when this resource use study was completed (Pinkus et al., in preparation). Ringtailed lemur numbers have remained relatively stable at the highest density of any known wild ringtailed lemur population. Brown lemurs at Berenty have colonized edge/Scrub habitat, reaching three times the density of ringtails (Pinkus et al., in preparation). High brown lemur density in edge habitat at Berenty is recent and has not yet led to a decline in numbers of ringtailed lemurs (Pinkus et al., in preparation). Though the number of ringtailed lemurs has not declined, we have recent evidence of marked seasonal home range shifts from gallery and edge habitat to marginal scrub and agricultural habitat (Jolly, unpublished data). These changes may presage a population decline of ringtailed lemurs. Continued monitoring and management of the brown lemur population at Berenty, as well as active management and restoration of edge habitat, may be crucial to the long-term sustainability of Berenty's ring-tailed lemur population.

Acknowledgments. We thank the de Heaulme family, Earthwatch, and the Natural Sciences and Engineering Research Council of Canada for financial and in-kind support. We thank Bob Sussman and Sarah Townsend whose suggestions greatly improved this paper. We thank John Walker for invaluable help with every aspect of this project. Armand Ball, Beverly Ball, Sabine Day, Alison Flood, Roland Ranaivojoana, Voajanahary Ranaivosoa, Hantanirina Ravololomamahary, John Walker, and George Williams helped collect these data. *Misoatra betsaka* to you all!

References

Abrams, P. A. (2002). Will small populations warn us of impending extinctions? *Am. Nat.* 160:293–305.

Brugiere D., Gautier, J. P., Moungazi, A., and Gautier–Hion, A. (2002). Primate diet and biomass in relation to vegetation composition and fruiting phenology in a rain forest in Gabon. *Int. J. Primatol.* 23:999–1024.

Caughley, G., and Sinclair, A. R. E. (1994). *Wildlife Ecology and Management.* Blackwell Science, Don Mills, Ontario.

Emmons, L. H., Gautier-Hion, A., and Dubost, G. (1983). Community structure of the frugivorous–folivorous forest mammals of Gabon. *J. Zool. Lond.* 199:209–222.

Ganzhorn, J. U. (1985). Habitat separation of semifree-ranging *Lemur catta* and *Lemur fulvus*. *Folia Primatol.* 45:76–88.

Ganzhorn, J. U. (1986). Feeding behavior of *Lemur catta* and *Lemur fulvus*. *Int. J. Primatol.* 7:17–30.

Ganzhorn, J. U. (1988). Food partitioning among Malagasy primates. *Oecologia* 75: 436–450.

Ganzhorn, J. U. (1989). Niche separation of seven lemur species in the eastern rainforest of Madagascar. *Oecologia* 79:279–286.

Ganzhorn, J. U. (1995). Low-level forest disturbance effects on primary production, leaf chemistry, and lemur populations. *Ecology* 76:2084–2096.

Ganzhorn, J. U., and Kappeler, P. M. (1996). Lemurs of the Kirindy Forest. *Primate Report* 46:257–274.

Ganzhorn, J. U., Klaus, S., Ortmann, S., and Schmid, J. (2003). Adaptations to seasonality some primate and nonprimate examples. In: Kappeler, P. M., and Pereira, M. E. (eds.), *Primate Life Histories and Socioecology*. University of Chicago Press, Chicago, pp. 132–144.

Ganzhorn, J. and Members of the Primate Specialist Group. (2000). *Lemur catta*. In: IUCN 2003. 2003 IUCN Red List of Threatened Species. Available at www.redlist.org.

Gautier-Hion, A. (1980). Seasonal variations of diet related to species and sex in a community of Cercopithecus monkeys. *J. Anim. Ecol.* 49:237–269.

Gautier-Hion, A. (1988). The diet and dietary habits of forest guenons. In: Gautier-Hion, A., Bourlière, F., and Gautier, J. P. (eds.), *A Primate Radiation: Evolutionary Biology of the African Guenons*. Cambridge University Press, Cambridge, pp. 257–283.

Glander, K. E., and Rabin, D. P. (1983). Food choice from endemic North Carolina tree species by captive prosimians (*Lemur fulvus*). *Am. J. Primatol.* 5:221–229.

Godfrey, L. R., Samonds, K. E., Jungers, W. L., and Sutherland, M. R. (2003). Dental development and Primate life histories. In: Kappeler, P. M., and Pereira, M. E. (eds.), *Primate Life Histories and Socioecology*. University of Chicago Press, Chicago, pp. 177–203.

Godfrey, L. R., Samonds, K. E., Jungers, W. L., Sutherland, M. R., and Irwin, M. T. (2004). Ontogenetic correlates of diet in Malagasy lemurs. *Am. J. Phys. Anth.* 123:250–276.

Gould, L., Sussman, R. W., and Sauther, M. L. (1999). Natural disasters and primate populations: The effects of a 2-year drought on a naturally occurring population of ring-tailed lemurs (*Lemur catta*) in southwestern Madagascar. *Int. J. Primatol.* 20:60–84.

Gould, L., Sussman, R. W., and Sauther, M. L. (2003). Demographic and life-history patterns in a population of ring–tailed lemurs (*Lemur catta*) at Beza Mahafaly Reserve, Madagasar: A 15-year perspective. *Am. J. Phys. Anth.* 120:182–194.

Harcourt, C. S., and Nash, L. T. (1986). Species differences in substrate use and diet between sympatric galagos in two Kenyan coastal forests. *Primates* 27:41–52.

Hawkins, A. F. (1999). The primates of Isalo National Park, Madagascar. *Lemur News* 4:10–14.

Janson, C., and. Chapman, C. A. (1999). Resources and primate community structure. In: Fleagle, J. G., Janson, C., and Reed, K. E. (eds.), *Primate Communities*. Cambridge University Press, New York, pp. 237–267.

Jekielek, J. (2002). *Hybridization of Brown Lemurs at Berenty Reserve, Madagascar*. MSc thesis, Department of Biological Sciences, University of Alberta, Calgary.

Jolly, A. (1966). *Lemur Behavior*. University of Chicago Press, Chicago.

Jolly, A. (2004). *Lords and Lemurs: Mad Scientists, Kings with Spears, and the Survival of Diversity in Madagascar*. Houghton Mifflin, Boston.

Jolly, A., Gustafson, H., Oliver, W. L. R., and O'Connor, S. M. (1982). Population and troop ranges of *Lemur catta* and *Lemur fulvus* at Berenty, Madagascar: 1980 census. *Folia Primatol.* 39:115–123.

Jolly, A., Caless, S., Cavigelli, S., Gould, L., Pereira, M. E., Pitts, A., Pride, R. E., Rabenandrasana, H. D., Walker, J. D., and Zafison, T. (2000). Infant killing, wounding, and predation in *Eulemur* and *Lemur*. *Int. J. Primatol.* 21:21–40.

Jolly, A., Dobson, A., Rasamimanana, H. M., Walker, J., O'Connor, S., Solberg, M. E., and Perel, V. (2002). Demography of *Lemur catta* at Berenty Reserve, Madagascar: Effects of troop size, habitat and rainfall. *Int. J. Primatol.* 23:327–353.

Koyama, N., Nakamichi, M., Oda, R., Miyamoto, N., and Takahata, Y. (2001). A ten-year summary of reproductive parameters for ring-tailed lemurs at Berenty, Madagascar. *Primates* 42:1–14.

Krebs, C. J. (1999). *Ecological Methodology*, 2nd ed. Addison-Wesley, Don Mills, Ontario.

Martin, P., and P. Bateson. (1993). *Measuring Behaviour*, 1st ed. Cambridge University Press, New York.

Mertl-Millhollen, A. S., Moret, E. S., Felantsoa, D., Rasamimanana, H., Blumenfeld-Jones, K. C., and Jolly, A. (2003). Ring-tailed lemur home ranges correlate with food abundance and nutritional content at a time of environmental stress. *Int. J. Primatol.* 24:969–985.

O'Connor, S. M. (1987). *The Effect of Human Impact on Vegetation and the Consequences to Primates in Two Riverine Forests, Southern Madagascar*. PhD thesis, Department of Applied Biology, Cambridge University, Cambridge.

Opler, P. A., Baker, H. G., and Frankie, G. W. (1980). Plant reproductive characteristics during secondary succession in Neotropical lowland forest ecosystems. *Biotropica* 12(Suppl. 1):24–39.

Overdorff, D. J. (1991). *Ecological Correlates to Social Structure in Two Prosimian Primates: Eulemur fulvus rufus and Eulemur rubriventer in Madagascar*. PhD thesis, Department of Biological Anthropology and Anatomy, Duke University, Durham.

Overdorff, D. J. (1993). Ecological and reproductive correlates to range use in red-bellied lemurs (*Eulemur rubriventer*) and rufous lemurs (*Eulemur fulvus rufus*). In: Kappeler, P. M., and Ganzhorn, J. U. (eds.), *Lemur Social Systems and Their Ecological Basis*. Plenum, New York, pp. 167–178.

Overdorff, D. J. (1996). Ecological correlates to activity and habitat use of two prosimian primates: *Eulemur fulvus rufus* and *Eulemur rubriventer* in Madagascar. *Am. J. Primatol.* 40:327–342.

Pereira, M. E. (1993). Seasonal adjustments of growth rate and adult body weight in ringtailed lemurs. In: Kappeler, P. M., and Ganzhorn, J. U. (eds.), *Lemur Social Systems and Their Ecological Basis*. Plenum, New York.

Pereira, M. E., Strohecker, R. A., Cavigelli, S. A., Hughes, C. L., and Pearson, D. D. (1999). Metabolic strategy and social behaviour in Lemuridae. In: Rakotosamimanana, B., Rasamimanana, H., Ganzhorn, J. U., and Goodman, S. M. (eds.), *New Directions in Lemur Studies*. Plenum, New York, pp.98–119.

Pinkus, S. E. (2004). *Impact of an Introduced Population of Eulemur fulvus on a Native Population of Lemur catta at Berenty Reserve, Madagascar*. MSc thesis, Department of Zoology, University of British Columbia, Vancouver.

Pinkus, S., Jolly, A., and Walker, J. (in preparation). Demography and distribution of introduced *Eulemur fulms* and native *Lemur catta* at Berenty Reserve, Southern Madagascar.

Rasamimanana, H. R., and Rafidinarivo, E. (1993). Feeding behavior of *Lemur catta* females in relation to their physiological state. In: Kappeler, P. M., and Ganzhorn, J. U. (eds.), *Lemur Social Systems and Their Ecological Basis*. Plenum, New York, pp. 123–134.

Richard, A. F., and R. E. Dewar. (1991). Lemur ecology. *Annu. Rev. Ecol. Syst.* 22:145–175.

Richard, A. F., Dewar, R. E., Schwartz, M., and Ratsirarson, J. (2000). Mass change, environmental variability and female fertility in wild *Propithecus verreauxi*. *J. Human Evol.* 39:381–391.

Richard, A. F., Dewar, R. E., Schwartz, M., and Ratsirarson, J. (2002). Life in the slow lane? Demography and life histories of male and female sifakas (*Propithecus verreauxi verreauxi*). *J. Zool. Lond.* 256:421–436.

Robinson, B. W., and Wilson, D. S. (1998). Optimal foraging, specialization, and a solution to Liem's Paradox. *Am. Nat.* 151:223–235.

Ross, C. (1992). Environmental correlates of the intrinsic rate of natural increase in primates. *Oecologia* 90:383–390.

Sauther, M. L. (1991). *The Effect of Reproductive State, Social Rank and Group Size on Resource Use Among Free-ranging Ringtailed Lemurs (Lemur catta) of Madagascar*. PhD thesis, Washington University, St. Louis.

Scholz, F., and Kappeler, P. M. (2004). Effects of seasonal water scarcity on the ranging behavior of *Eulemur fulvus rufus*. *Int. J. Primatol.* 25:599–613.

Simmen, B., Hladik, A., and Ramasiarisoa, P. (2003). Food intake and dietary overlap in native *Lemur catta* and *Propithecus verreauxi* and introduced *Eulemur fulvus* at Berenty, Southern Madagascar. *Int. J. Primatol.* 24:949–968.

Strier, K. B. (1999). Population viabilities and conservation implications for Muriquis (*Brachyteles arachnoides*) in Brazil's Atlantic forest. *Biotropica* 32:903–913.

Sussman, R. W. (1972). *An ecological study of two Madagascar Primates: Lemur fulvus rufus (Audebert) and Lemur catta (Linnaeus)*. PhD thesis, Department of Anthropology, Duke University, Durham.

Sussman, R. W., and Rakotozafy, A. (1994). Plant diversity and structural analysis of a tropical dry forest in southwestern Madagascar. *Biotropica* 26:241–254.

Sussman, R. W., Green, G. M., Porton, I., Andrianasolondraibe, O. L., and Ratsirarson, J. (2003). A survey of the habitat of *Lemur catta* in southwestern and southern Madagascar. *Primate Conservation* 19:32–57.

Terborgh, J. (1983). *Five New World Primates: A Study in Comparative Ecology*. Princeton University Press, Princeton, N.J.

Tomlin, D. C., and Cranford, J. A. (1994). Ecological niche differences between *Alouatta palliata* and *Cebus capucinus* comparing feeding modes, branch use, and diet. *Primates* 35:265–274.

Tutin, C. E. G., Ham, R. M., White, L. J. T., and Harrison, M. J. S. (1997). The primate community of the Lopé Reserve, Gabon: Diets, responses to fruit scarcity, and effects on biomass. *Am. J. Primatol.* 42:1–24.

Vasey, N. (2000). Niche separation in *Varecia variegata rubra* and *Eulemur fulvus albifrons*: I. Interspecific patterns. *Am. J. Phys. Anthropol.* 112:411–431.

Waser, P. M. (1980). Polyspecific associations of *Cercocebus albinigena*: geographical variation and ecological correlates. *Folia Primatol.* 33:57–76.

Wright, P. C. (1999). Lemur traits and Madagasar ecology: Coping with an island environment. *Yrbk. Phys. Anth.* 42:31–72.

Yamashita, N. (2002). Diets of two lemur species in different microhabitats in Beza Mahafaly Special Reserve, Madagascar. *Int. J. Primatol.* 23:1025–1051.

Young, A. L., Richard, A. F., and Aiello, L. C. (1990). Female dominance and maternal investment in strepsirhine primates. *Am. Nat.* 135:473–488.

10
Tradition and Novelty: *Lemur catta* Feeding Strategy on Introduced Tree Species at Berenty Reserve

TAKAYO SOMA

10.1. Introduction

Berenty Reserve is located in southeastern Madagascar. It is one of the most famous places of the country for tourists and also for researchers. This area is constituted by a patch of gallery forest of approximately 240 ha surrounded by the Mandrare River at the north and sisal plantation on the other sides. It was established in 1936 as a private reserve. Since the 1980s, tourism has been developed. The forest is divided into a strict gallery forest with closed canopy and a transitional forest in which canopy is more open. Both are dominated by *Tamarindus indica* (O'Connor, 1987), which is considered to be a keystone species for *Lemur catta* here (Jolly, 1966; Rasamimanana and Rafidinarivo, 1993; Mertl-Millhollen et al., 2003) as also at Beza Mahafaly Reserve (Sauther, 1998). In the tourist area, exotic species have been planted as ornamental trees, mainly before 1960' (e.g., *Azadirachta indica*, *Cordia sinensis*, and *Eucalyptus* sp.; Simmen et al., 2003) or for livestock fodder, planted in early 1990' (*Leucaena leucocephala*, J. de Heaulme, pers. comm). Besides *L. catta*, five other species of lemur inhabit the reserve: Verreaux' sifaka (*Propithecus verreauxi verreauxi*), reddish-gray mouse lemur (*Microcebus griseorufus*), Gray mouse lemur (*M. murinus*), white-footed sportive lemur (*Lepilemur leucopus*), and an introduced hybrid population of brown lemurs (*Eulemur fulvus rufus x E. f. collaris*) (Pitts, 1995; Jolly et al., 2002).

L. catta is described as an opportunistic frugivorous–folivorous species that includes flowers and invertebrates in its diet (Ganzhorn, 1986; Rasamimanana and Rafidinarivo, 1993; Sauther, 1993; Sauther et al., 1999). Numerous studies are focused on its feeding ecology at Berenty (Sussman, 1977; Budnitz, 1978; Rasamimanana and Rafidiarivo, 1993; Rasamimanana, 1999; Simmen et al., 2003) and at Beza Mahafaly (Sauther, 1994, 1998; Yamashita, 1996, 2002), but none of these have described the impact of introduced plant species on *L. catta*'s feeding ecology.

Southern Madagascar is characterized by alternate dry and wet seasons during which rainfall can vary tremendously. Because of this unpredictable climate, the dry forests of Madagascar are considered as low plant productivity areas with poor soil and strict fruit seasonality, hence Malagasy lemurs have evolved their own special strategies. In order to deal with these severe environmental conditions including important variations in food availability, lemurs use many different feeding strategies (Wright, 1999).

However, there is little knowledge about the relation between longer term environmental changes and lemur ecology. How do lemurs react to these changes? Although *L. catta* has been studied for a long time, there are not many studies from this point of view. To investigate this question, Berenty Reserve is a suitable site, lying in a region of severe natural environmental challenge in southern Madagascar. It is a forest fragmented by human activity (O'Connor, 1987), with a complex of natural and introduced plant vegetation, and the special characteristics of high lemur population density and stability of territory (Jolly, 1966; Klopfer and Jolly, 1970; Sussman, 1977; Budnitz, 1978; Mertl-Millhollen et al., 1979; Mertl-Millhollen, 2000).

The following report announces the preliminary results of a long-term study on the change in the feeding habits of *L. catta* related to the food availability of introduced tree species *versus* endemic tree species. I classified endemic tree species both as true endemic tree species (e.g., *Rinorea greveana*) and tree species that have grown in the study area at least for some centuries (e.g., *T. indica*). Introduced tree species are trees planted these past 70 years for ornamental considerations or for livestock by Berenty Reserve's owner since 1936. This study is undertaken not only with the goal of increasing knowledge of the ecology of lemurs but also with the aim of better managing of fragmented forests.

10.2. Methods

During the period from February 2001 to January 2002, I observed four adult females and three adult males living in one troop of *L. catta* known as troop CX. At the beginning of the study, the troop was composed of 11 individuals (Table 10.1). Troop structure has changed as follows during the study. After their accept-

TABLE 10.1. Changes in age–sex composition of troop CX during the study period from February 2001 to January 2002.

Age class	Feb.–Mar.		Apr.–Aug.		Sept.–Feb.	
	Female	Male	Female	Male	Female	Male
Adult	4	1	4	3	3	3
Sub-adult	1	1	1	1	1	1
Juvenile	0	0	0	0	1	0
Infant	2	0	2	0	1	1
Total	7	2	7	4	6	5

(1) Two males were accepted by females in the estrus season in April.
(2) Subordinate female died by injury from conflict with invading troop in September.
(3) Female juvenile of subordinate had depression, ended up dying 2 weeks after mother's death.
(4) Male infant died in conflict with adjacent troop in February.

ance by residents, two males entered into the troop at the mating season (in April 2001). One subordinate female died from severe injuries during a conflict with a neighboring troop during the birth season (in September 2001). After this death, her juvenile fell into "depression" and disappeared 2 weeks later (in early October 2001). All of the infants born in September 2001 died, except the one of the dominant females. At the end of the study, the troop CX was composed of 10 individuals (in January 2002).

In this study, each year was divided into two seasons from the ombrothermic graph (Figure 10.1) on a rainfall data set of 4 consecutive years with temperatures recorded during the field study. Rainfall data were recorded by C. Rakotomalala, plantation manager, in a local meteorological station. Thus, the dry season corresponds with the period May–October and the wet season with the period November–April.

10.2.1. Activity Budget and Diet

To estimate the activity budget, I recorded in each 5-minute interval the focal animal activity (resting, feeding, actual ingestion, sitting, moving—including foraging, grooming, social behavior such as playing, marking and agonistic behaviors). To estimate diet, I recorded all the occurrences of feeding behavior and timed each feeding bout. Plant categories are fruits, mature leaves, new leaves, flowers, bud, stem, insects, and others.

Each focal animal was followed (Altmann, 1974) from dawn to dusk (wet season: 0500–1900 h; dry season: 0600–1800 h). All the animals were evenly

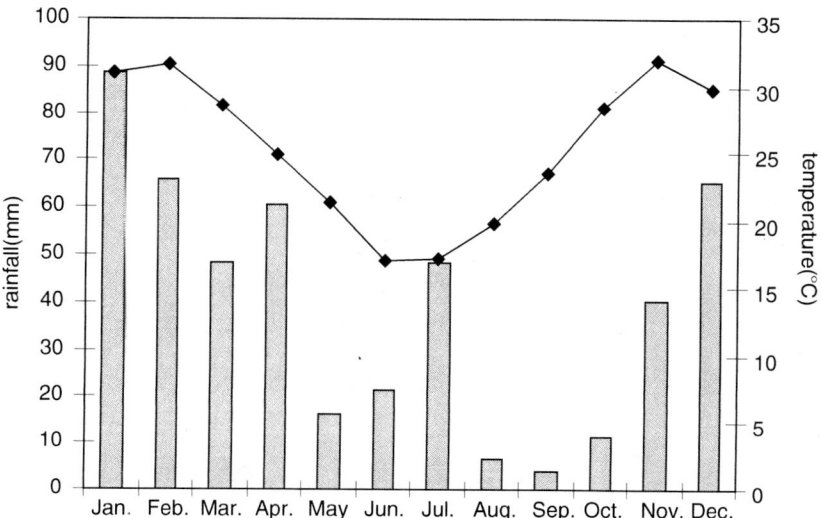

FIGURE 10.1. Ombrothermic graph for Sept. 2000–2002.

observed, allowing for the fact that I started to collect male data from April 2001 when the two peripheral males had been accepted by all the female residents and I had to stop collecting data for the dead subordinate female in October 2001. This study corresponds with 190 days of data, which represent 2280 hours of observations spread over the whole year.

10.2.2. Home Range and Territory

I located the troop's home range and troop's territory (territory is defined as area defended against neighboring troops in encounters) by recording the location of the focal animal for each scan and each feeding occurrence. Mapping is based on the study of Koyama et al. (2002) who divided the home range of the seven different troops of *L. catta* studied by Koyama and his colleagues into squares of 25 m × 25 m. It appeared that the territory and home range of troop CX lies in gallery forest and was limited on its northern boundary by the Mandrare river (Figure 10.2). Gallery forest is mainly composed of endemic tree species including many individual trees of *T. indica* (O'Connor, 1987). All the seven lemur troops studied by our team inhabit home ranges including many trees of this species. These trees are used as sleeping trees for troop CX.

10.2.3. Plant Phenology

I estimated plant phenology of the 20 main *L. catta* food plant species from 300 marked plant individuals. To estimate plant part availability, I used a scale graded from 0 (none) to 4 (abundant, meaning maximum). I recorded plant phenology every 15 days, all through the year.

10.3. Results

10.3.1. Diet of L. catta at Berenty

L. catta in troop CX ate 144 different food items (53 plant species, 27 plant families) including plant parts from 12 introduced tree species. In detail, 35 species of trees, 4 of lianas, 11 of herbs, and 2 of succulents, 3 species of caterpillars, some species of cicadas, and soil were consumed (Table 10.2). Licking behaviors on dead wood, stone, and tourist bungalow walls were also observed. In the birth season, females ate their own placentas after giving birth. Animals drank water in artificially provisioned water pools. In spite of the numerous plant parts eaten by *L. catta*, their diet was quantitatively composed of limited number of plant species. It was mainly composed of 10 plant species that totaled more than 85% of their feeding time (Table 10.3). *T. indica* is the most important species, of which various parts are consumed by *L. catta* spending 34.8% of their feeding time on it.

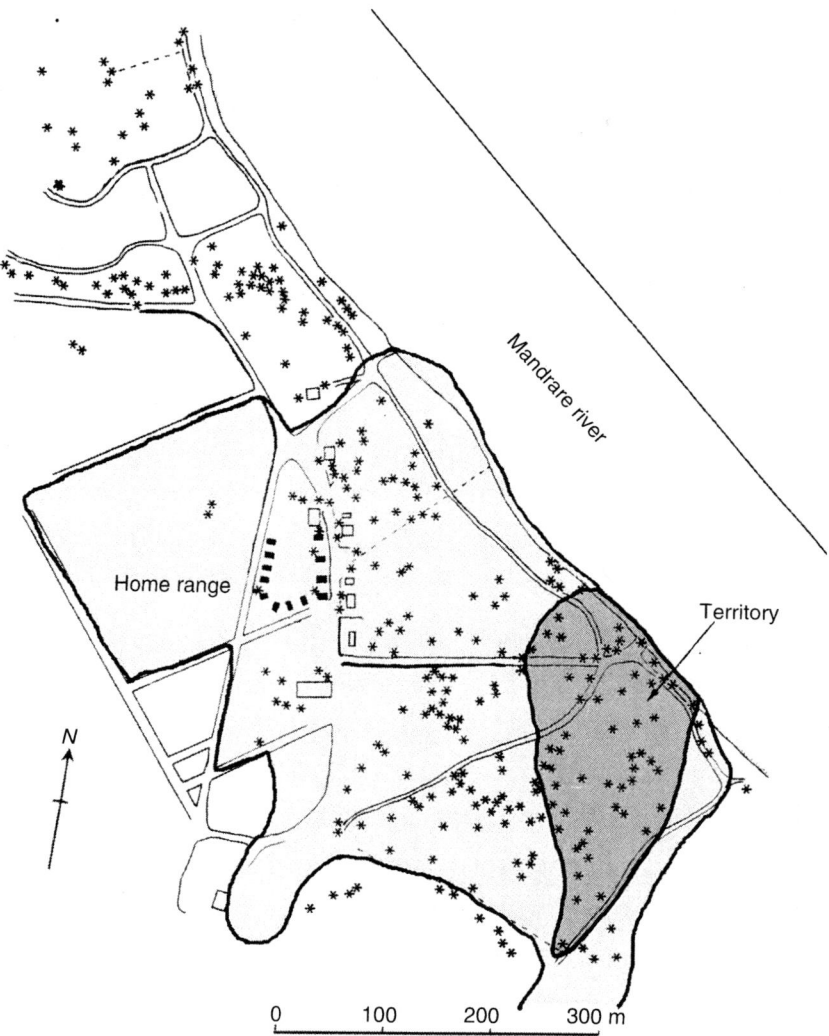

FIGURE 10.2. Territory and home range of troop CX from February 2001 to January 2002. Asterisks indicate the emplacement of *Tamarindus indica* trees with DBH above 50 cm (from Koyama et al., 2002).

L. catta spent almost an equal amount of time consuming fruits and mature leaves (40.3% and 40.2%, respectively, Figure 10.3a). But time allocated to such resources varied according to monthly availability (Figure 10.4). In the wet season, *L. catta* spend 51.5% of their feeding time on fruits (with a peak in February 61.2%, Figure 10.3b). In the dry season, they replaced fruits by leaves, allocating 51.4% of their feeding time to the latter (with a peak in June 80.1%,

TABLE 10.2. Food items ingested by *Lemur catta* in troop CX.

Plant family	Species	Vernacular name	Part eaten	Int.
Anacardiaceae	*Mangifera indica*	mangy	fF,skn	x
	Sclerocarya birrea	sakoa	F	x
Apocynaceae	*Hazunta modesta*	feka	yl,ml	
Bignoniaceae	*Fernandoa madagascariensis*	somotsoy	ml	
Boraginaceae	*Cordia caffra*	varo	F,yl,fl,fb,b	
	Cordia sinensis	varombazaha	F,f,ml,yl,fl,fb	x
Cactaceae	*Opuntia vulgaris*	raketa	stm, ystm,ex	x
Caesalpiniaceae	*Tamarindus indica*	kily	F,fF,f,ff,ml,yl,fl,fb,b	
	Senna siamena		fl,fb	x
	Delonix regia	flamboyant	ml	x
Capparidaceae	*Crateva* sp.	keleon	F,fF,f,ff,fl,fb,p,b	
	Capparis sepiaria	ropiteko	F,f,fF,ff,ml,yl,fl	
	Maerua filiformis	solety	ml	
Combretaceae	*Combretum albiflorum*	tamenaka	ml,fl	
Commelinaceae	*Commelina bengalensis*	andranahake	ml	
Convolvulaceae	*Ipomoea* sp.	tsimatavendrano	ml	
Gramineae	*Panicum maximum*	ahebe	ml	
	Cynodon sp.	ahepoly	ml	
Euphorbiaceae	*Antidesma madagascariensis*	voafona	F,f	
	Phyllanthus seyrigii	sangira	F,f,ml,yl,fl,fb	
Liliaceae	*Aloe vahombe*	vahombe	ml	
Lythraceae	*Lawsonia inermis*	kotica	f,ml	
Meliaceae	*Azadirachta indica*	leranomby	F,f,fF,ff,ml,yl	x
	Melia azedarach	voandereka	ml	x
	Quivisianthe papinae	variandro	fl,f?,ml	
Mimosaceae	*Albizia polyphylla*	halomblo	ml	
	Acacia rovumae	benono	yl	
	Leucaena leucocephala	kantsakantsa	ml	x
Moraceae	*Ficus sycomorus*	adabo	fF, ml	
	Ficus pachyclada	mange	F,fF,dl	
Musaceae	*Musa* sp.	akondro	fF,skn	x
Myrtaceae	*Eucalyptus* sp.	kininiy	fl,fb	x
Nyctaginaceae	*Boerhavia diffusa*	bendrahy	ml	
	Bougainvillea spectabilis		yl,fl	x
Polygonaceae	*Polygonum senegalense*	leranomby	ml	
Rubiaceae	*Enterospermum* sp.	mantsaka	F,f,ml,yl,fl,fb,	
	Tricalysia sp.	hazombalala	ml,yl,yf	
Salvadoraceae	*Azima tetracantha*	filofilo	f,ml,yl,fl,fb	
	Salvadora angustifolia	sasavy	ml	
Sapindaceae	*Neotina isoneura*	volely	yl,b	
Tiliaceae	*Grewia saligna*	taorankafitra	fl,fb	
Ulmaceae	*Celtis bifida*	bemavo	ml,yl,fl,fb,b	
	Celtis philippensis	tsilikatrifaka	F,f,fl,fb	
Violaceae	*Rinorea greveana*	tsatsake	F,f,ml	
Tree 1	Unidentified		ml	
Liana1	Unidentified		ml	
Liana2	Unidentified		ml	
Herb 1	Unidentified	eboebo	ml,fl	
Herb 2	Unidentified	beamena	ml	

Herb 3	Unidentified	ml
Herb 4	Unidentified	ml
Herb 5	Unidentified	ml
Herb 6	Unidentified	ml
Cicada 1	Unidentified	
Cicada 2	Unidentified	
Cicada 3	Unidentified	
Caterpillar 1	Unidentified	
Caterpillar 2	Unidentified	
Caterpillar 3	Unidentified	
Insect	Unidentified	
Placenta	*Lemur catta*	
Soil		
Licking behavior	Wood	
	Dead wood	
	Dead leaf	
	Stone	
	Concrete	
	Bungalow wall	

Ripe fruit (F), unripe fruit (f), ripe fallen fruit (fF), unripe fallen fruit (ff), mature leaf (ml), young leaf (yl), flower (fl), stem (stm), flower bud (fb), ripe seed (Sd), unripe seed (sd); exudate (ex), introduced species (Int.).

TABLE 10. 3. Proportion of feeding time on main plant species.

	Plant species	Rate (%)	Bud	Young leaf	Mature leaf	Flower	Fruit	Int.
1	*Tamarindus indica*	34.8	○	○	○	○	○	
2	*Leucaena leucocephala*	16.3			○			*
3	*Azadirachta indica*	8.8			○	△	○	*
4	*Rinorea greveana*	6.7		△	△		○	
5	*Tricalysia* sp.	5.0			○		△	
6	*Celtis philippensis*	3.6		△			○	
7	*Cordia sinensis*	3.4			△	△	○	*
8	*Boerhavia diffusa*	2.5			○			
9	*Celtis bifida*	2.0			△	△	△	
10	*Cordia caffra*	1.9			○		△	

○: Species eaten representing more than 10% of total time per month.
△: Species eaten representing less than 10% of total time per month
Int.: introduced plant species.

Figure 10.3c). In November and December, they spent 15.7 % and 14.1% of their feeding time foraging on insects such as cicadas and caterpillars. They increased their time spent drinking in the wet season probably according to the high temperatures.

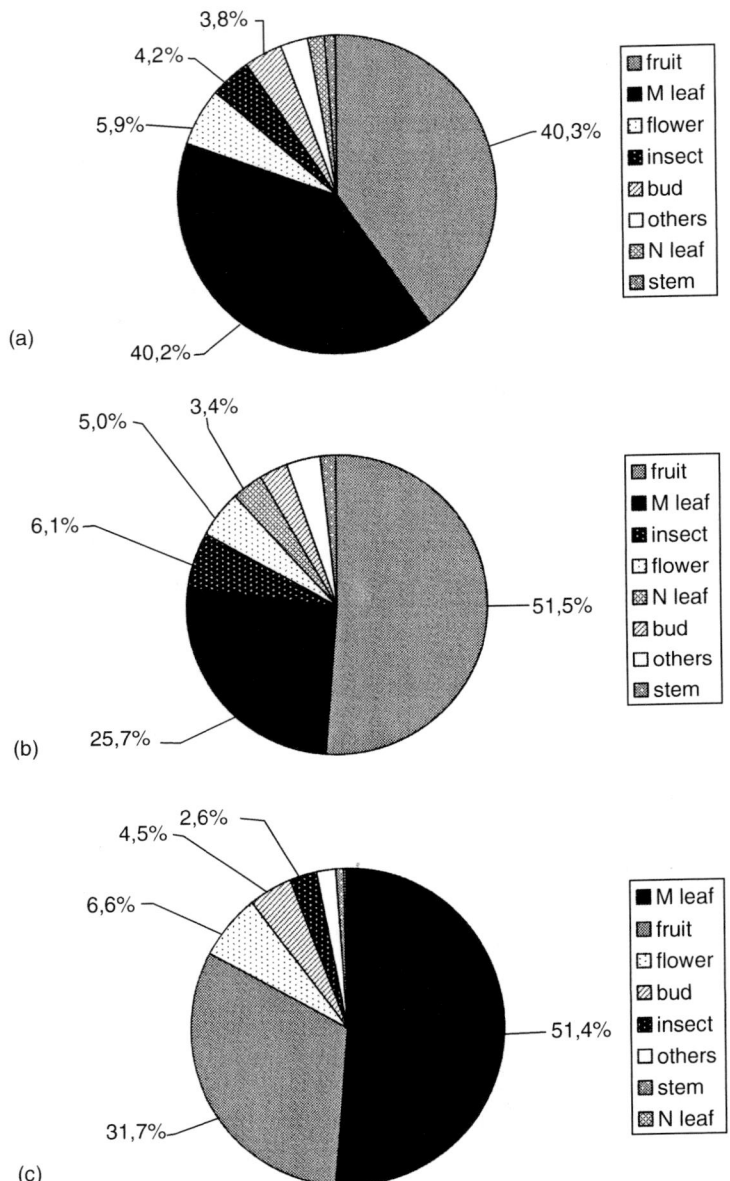

FIGURE 10.3. (a) Proportion of plant parts eaten all over the year. (b) Proportion of plant parts eaten during the dry season. (c) Proportion of plant parts eaten during the wet season.

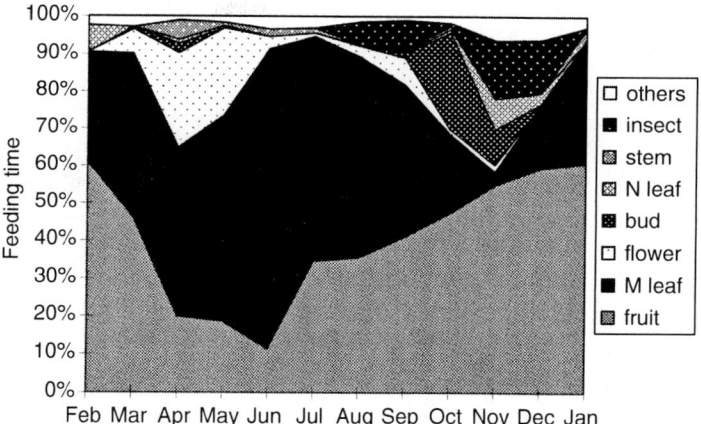

FIGURE 10.4. Variation of proportion of plant parts eaten by month.

10.3.2. Food Availability and Consumption: The Importance of Introduced Tree Species

Many of the endemic tree species produce fruits, flowers, new leaves and buds during the wet season (Figure 10.5). In my study period, different tree species produced its fruits in the wet season. For example, *R. greveana* bore fruits early in the wet season and then followed *Neotina isoneura* and *Crateva* sp. However, the fruiting tree species showed two different patterns. One type of tree species bore fruit throughout the year but considerable individual variation and without synchronization; the other species bore fruits during the dry season. Some introduced plant species also bore fruits during the wet season (e.g., *A. indica*), but

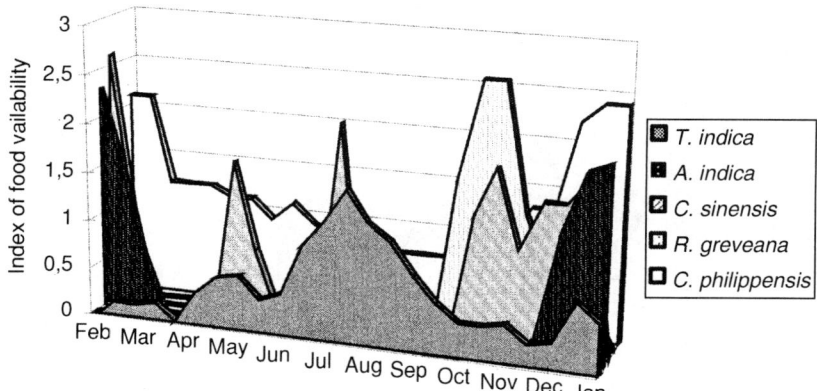

FIGURE 10.5. Index of fruit availability for 5 important plant species related to *Lemur catta*'s diet.

some species such as *C. sinensis* and *L. leucocephala* had fruits several times a year, including the dry season when the fruit availability of endemic tree species was scarce.

L. catta started to feed on endemic *T. indica* when its fruits increased in availability in the late dry season. Moreover, they could feed on fallen pods of this species over the wet season (Figure 10.6). When the wet season started, they switched to the fruits of endemic species such as *R. greveana* and after these of *A. indica* and *C. sinensis* (second part of the wet season). In the early dry season (a period without fruits), they intensively fed on *L. leucocephala* leaves, a species that continued to be consumed throughout all the dry season. Later, they again ate fruits of *C. sinensis*. For *C. sinensis*, they would organize excursions into neighbouring territories. In the dry season, the proportion of feeding time allocated by *L. catta* to feeding on introduced plant species was higher than that in the wet season (42.9% versus 21.5%). For example, in the early dry season (May), they spend 52.8% of their feeding time on introduced tree species, and in the mid-dry season (August), 51.0%.

On the contrary, in the early wet season (November and December), they did not feed on introduced tree species. In those months, they concentrated feeding on the endemic plant species such as *R. greveana* as described above (Figure 10.7).

10.3.3. *Territory, Home Range, and Distribution of Introduced Tree Species*

Introduced tree species were found mainly in the open area of the gallery forest where were the touristic complex in. For example, *A. indica* and *C. sinensis* are planted in the tourist area and along the road, whereas *L. leucocephala*

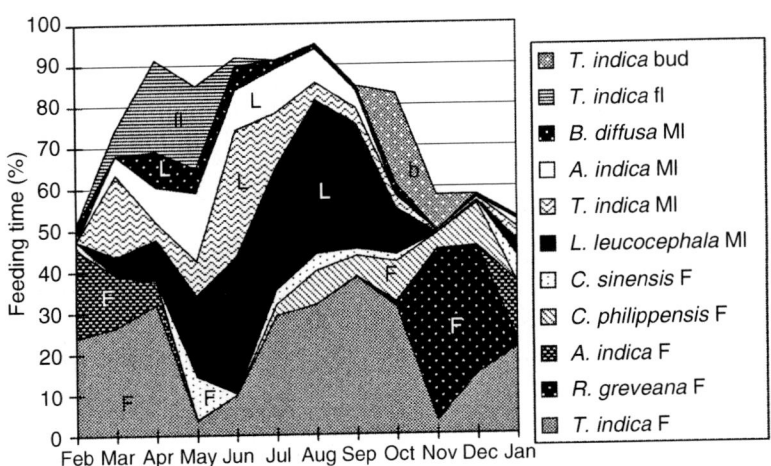

FIGURE 10.6. Proportion of time spend feeding on the 11 main plant species eaten by *Lemur catta* all over the study year.

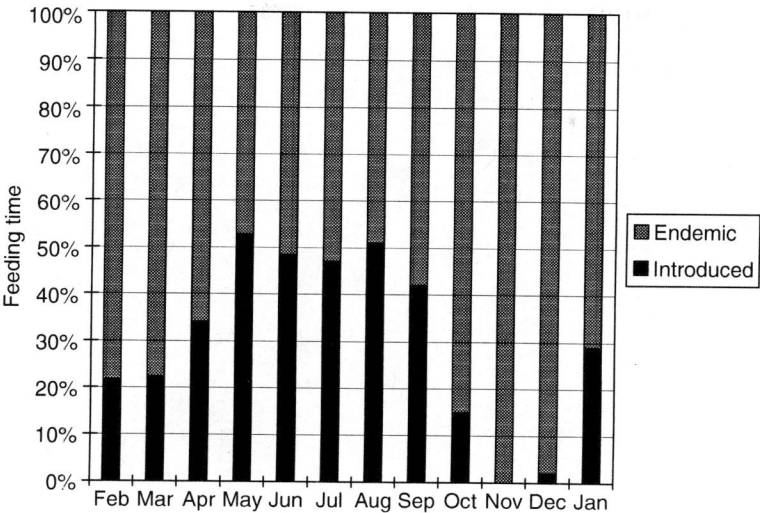

FIGURE 10.7. Monthly Variation of proportion of introduced and endemic tree species eaten by *Lemur catta* all over the study year (Endemic: *Tamarindus indica*, *Rinorea greveana* and *Celtis philippensis*, Introduced: *Azadirachta indica*, *Cordia sinensis*).

was planted in the open space of the river bank, spreading by self-seeding (Figure 10.8).

In the early dry season (May–June), when feeding time on introduced plant species was maximal, troop CX made excursions outside their territory and fed on introduced plant species (Figure 10.9a): *C. sinensis* (fruits), *A. indica* (mature leaves), and *L. leucocephala* (mature leaves). In the late dry season (July–August), they also spent 89.2% of their feeding time out of their territory, feeding on same species as in the early dry season (Figure 10.9b). On the contrary, in the early wet season (November and December), they stayed for 88.1% of feeding time within their territory and fed on endemic species such as *R. greveana* fruit (Figure 10.9c). This implies that they went out from their territory in the dry harsh season specifically in order to feed on the introduced species.

10.4. Discussion

10.4.1. Is There a Change of Feeding Tradition in L. catta?

In this study, the most important endemic plant species for *L. catta* of troop CX was *T. indica*, as reported in previous studies (Rasamimanana and Rafidinarivo, 1993), but they also allocated a considerable proportion of time feeding on several introduced tree species, especially when "endemic" food was scarce in the dry season. Troop CX is a division of the original troop C, which inhabited the

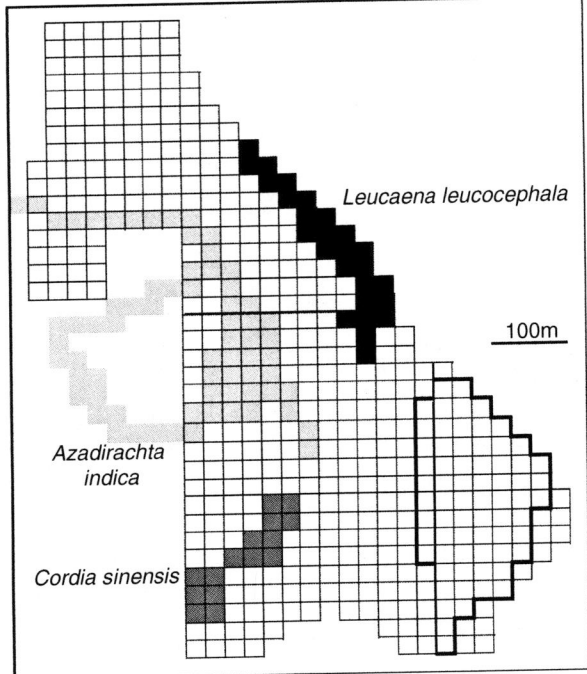

FIGURE 10.8. Distribution of introduced tree species in the home range of troop "CX". The grids scale of maps are 25m X 25m.

gallery forest. Thirty years ago, *T. indica* leaves were already the most important food for troop C, especially during the dry season (Budnitz, 1978; Mertl-Millhollen et al., 1979; Koyama 1991; Jolly and Pride, 1999). Troop CX has maintained troop C's original core area (Jolly et al., 1993; Mertl-Millehollen, 2000; Koyama et al., 2002). It appears that this troop made excursions into territories of other troops to feed on introduced species such as *L. leucocephala* leaves and *C. sinensis* and *A. indica* fruits, even though there was relatively high availability of *T. indica* mature leaves or young pods in their own territory in the dry season.

10.4.2. *Are Introduced Species Efficient Food Items?*

Why did a *L. catta* troop in its "natural" habitat with abundant keystone species in its territory go in the dry season to eat plant parts of *L. leucocephala* (especially leaves)? The main explanation is that leaves of *L. leucocephala* are more nutritious and energetic than those of *T. indica*. *L. leucocephala* mature leaves contain 36.1 g of crude protein, and 11.4 g of ADF (acid detergent fiber) per 100 g of dry weight. *T. indica* mature leaves contain 17.3 g of crude protein and 26.2 g of ADF

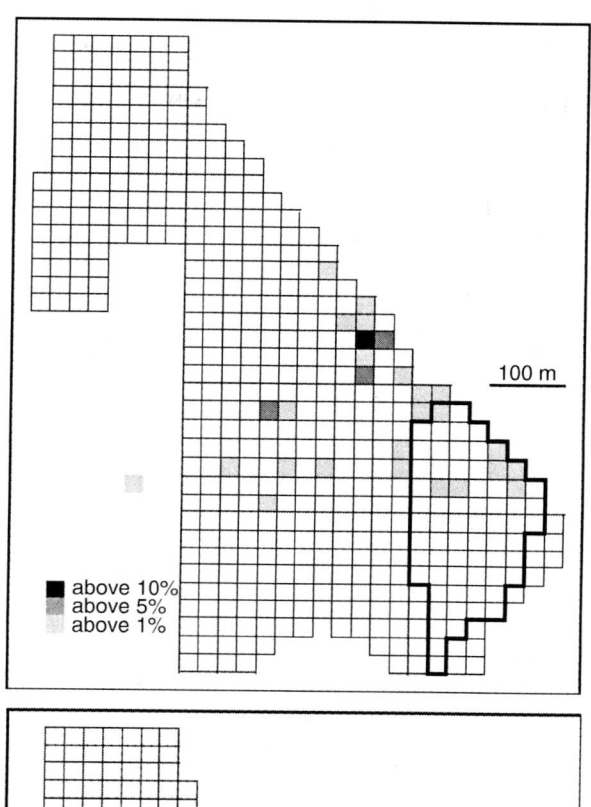

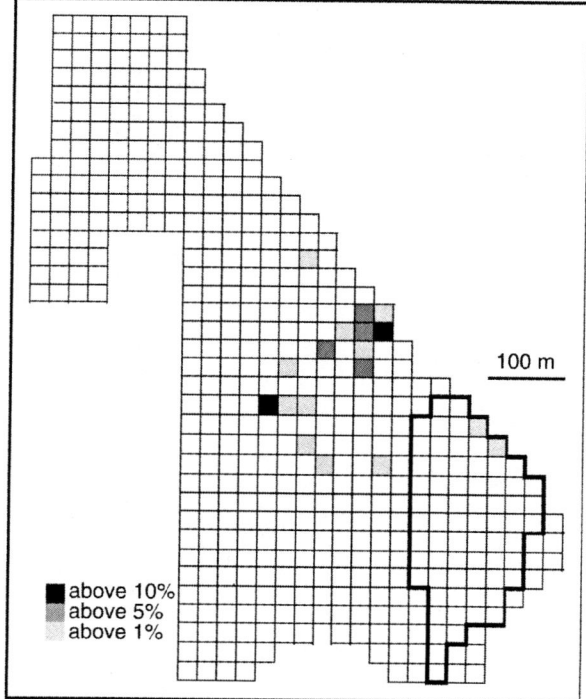

FIGURE 10.9. (A) Proportion of time spent feeding in different place of the home range by the resident troop "CX" in the early dry season (May-June). The proportion of time allocated to feed in their territory is 12.9%. (B) Proportion of time spent feeding in different place of the home range by the resident troop "CX" in the late dry season (July-August). The proportion of time allocated to feed in their territory is 10.8%.

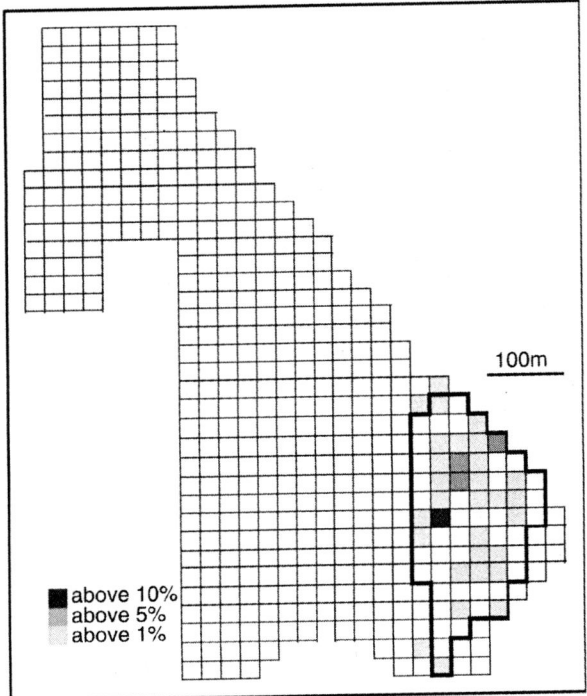

FIGURE 10.9. (C) Proportion of time spent feeding in different place of the home range by the resident troop "CX" in the early wet season (November-December). The proportion of time allocated to feed in their territory is 88.1%.

per 100 g of dry weight in the dry season (Soma and Tarnaud, unpublished data). The larger the amount of fiber, the less edible are the proteins. The ratio of crude protein/ADF is an index of leaf quality. *L. leucocephala* leaves (crude protein/ADF = 3.17) seem to be a much more efficient food than *T. indica* leaves (crude protein/ADF = 0.66). It need quantitative analyse of consumption in *L. catta* among both species in the future study.

10.4.3. Has L. catta Already Adopted a Strategy to Deal with the Harsh Season by Consuming Introduced Tree Species as fall back Food?

Many primate species deal with the period of scarcity of their top ranking diet by adopting different strategies. Yamakoshi (1998) suggested that *Pan troglodytes* in Bossou, living in a fragmented secondary forest, switched their diet to two of the three keystone foods by using tools for cracking nuts and pestle-pounding. Folivorous *Propithecus verreauxi* (Richard, 1977), *Gorilla gorilla* (Yamagiwa et al., 1994), and omnivorous *Macaca fuscata* (Watanuki et al., 1994) have been observed to consume bark during periods of food scarcity. Diet of the folivorous *Indri indri* varies according to the availability of foods (Britt et al., 2002). *Trachypithecus geei* were able to switch over to a fruit and seed diet when foliage was scarce (Gupta and Chivers, 2000). *Eulemur fulvus* in Mayotte focused on just a few species during the dry season (Tarnaud, 2004), and *Eulemur fulvus rufus* has been observed migrating to other areas (Overdorff, 1993).

L. catta switch to new foods according to seasons, and they can make long excursions out of their normal home range to monitor seasonal resources in a food-scarce season, as already described (Budnitz, 1978; Sussman, 1991; Rasamimanana and Rafidinarivo, 1993; Sauther, 1994; Gould et al., 1999; Rasamimanana, 1999; Sauther et al., 1999), or they can expand their home ranges as at Beza Mahafaly (Sauther, 1993).

In the current study, troop CX (like other studied troops of *L. catta* at Berenty) switched to new foods and especially *L. leucocephala* from which they consumed not only leaves but also flowers and unripe seed (Soma, unpublished data; Tarnaud, pers.comm.). In only about 15 years since its introduction, this tree species has become one of the most important food species for troops of *L. catta* inhabiting the tourist area at Berenty. In addition, they also eat large amounts of fruits and leaves of the introduced species *A. indica* and *C. sinensis* (Simmen et al., 2003; Pride, 2003; Soma, unpublished data). Thus, these foods have became fall back (Wrangham et al., 1996) foods during the dry season or when fruit availability of endemic tree species is decreasing (perhaps also when the nutritious quality and calories are lower).

Furthermore, the spatial distribution of introduced tree species could also explain why some of their plant parts are consumed by *L. catta*. *A. indica* trees and *C. sinensis* trees are planted in patches in Berenty Reserve, often in open space outside the forest (Figure 10.8). For *L. catta*, which is an opportunistic forager, the introduced tree species are planted in dense and limited patches in which food is abundant and which can be relatively easily defended.

Thus, *L. catta* have changed their feeding traditions under the conditions of fragmented habitat and high population density, but they still maintain their opportunistic strategy (Sussman, 1977; Rasamimanana and Rafidinarivo, 1993; Rasamimanana, 1999; Sauther, 1998; Sauther et al., 1999; Simmen et al., 2003; Pride, 2003), as they can easily switch to available introduced foods as endemic resources fluctuate in availability.

10.4.4. *Why Didn't They Change Their Territory to Include the Introduced Species Area?*

L. catta in Berenty are territorial. CX has maintained the original troop C core area as their own territory (Mertl-Millhollen et al., 1979; Jolly and Pride, 1999). If the plant parts of the introduced tree species are more efficient and more available during the harsh season than endemic tree species, why did not troop CX not change the boundaries of their territory to include some or all of the introduced tree patches? Although individuals of troop CX allocated a lot of time to consume plant parts of introduced tree species in the dry season, they mainly depended on endemic species like *T. indica* or *R. greveana*. For example, the fruiting of *T. indica* is synchronized with the late gestation period and flush/bud starts at the beginning of lactation. Other endemic tree species are also important: *Crateva sp.*, *Cordia caffra*, and *Enterospermum sp. T. indica* is also important as a sleeping site. Even if *L. catta* fed on *L. leucocephala* or another introduced species until late evening far from their territory, afterward they came back to their preferred individual *T. indica* trees. *L. catta* of troop CX depended on endemic species in the wet season when availability synchronized with lactation and weaning time. Hence, they keep their territory where endemic species are abundant.

10.4.5. What Is the Impact of Introduced Species on the Population and Health of L. catta?

At Berenty, the population of *L. catta* is higher than in those other forests (Jolly and Pride, 1999; Jolly et al., 2002; Koyama et al., 2002). The population of *L. catta* increased threefold between 1985 and 1997 (Jolly and Pride, 1999). In addition, the introduced brown lemur population, whose diet overlaps with that of *L. catta* (Simmen et al., 2003), is also increasing, and both species may compete for food. If it is difficult to predict whether one of the two populations will decrease from food scarcity resulting from overpopulation. Certainly, introduced tree species mitigate the food competition by increasing the availability of food resources (especially in the dry season). *L. leucocephala*, especially, has grown from its introduction in 1990' to widespread thickets of mature trees, thus offering an increasing food supply during the past 15 years. Introduced tree species certainly also buffer the influence of severe seasonal climate effect on these species, and consequently these introduced tree species may allow higher population density of *L. catta* at Berenty.

Nevertheless, eating plant parts of some introduced tree species could be dangerous. Indeed, since 1998, several *L. catta* individuals in different troops, including individuals of troop CX (tourist area), have exhibited symptoms of alopecia (Jolly et al., this volume; Crawford et al., this volume; Tew, 2003; Soma, unpublished data). *L. leucocephala* contains mimosine, a toxic compound whose ingestion provokes decrease in weight, infertility, goitre, paralysis of extremities, and cataract (Jones and Hegarty, 1984; see Crawford et al., this volume). If plant parts of *L. leucocephala* are efficient sources of protein and/or energy, there is a possibility that this the introduced tree species could impose serious adverse side effects on lemur individuals.

10.5. Conclusion

At Berenty reserve, *L. catta* eat introduced tree species during the dry season. Although they switch to efficient introduced species in the harsh season by organizing extraterritorial excursions, they keep their core areas where they depend on the keystone *T. indica* and other endemic tree species during the critical periods of lactation and juvenile development. Plant parts of introduced tree species may also be consumed because of (1) availability of top ranking plant parts like fruit in the harsh season; (2) dense patch distribution; (3) food intake efficiency and, perhaps, high nutrient content. These results suggest that introduced plant species seem to buffer the influence of the severe seasonal climate for *L. catta* at Berenty Reserve and could allow a higher population density of this species than either kind of forest alone, especially in the presence of another lemur species with which *L. catta* is in food competition.

Acknowledgments. I thank my supervisor, Dr. N. Koyama, for giving me the opportunity to conduct this study, Dr. A. Jolly for encouraging me every time, and Mr. Jean de Heaulme for permitting and facilitating my research at Berenty Reserve. Dr. L. Tarnaud and Dr. B. Simmen advised and helped me. Dr. Koyama gave me information on troops and feeding plant species of *L. catta*, Dr. A. Hladick identified plant species, Dr. A. Jolly and Dr. A. Mertl-Millohollen gave me information about the history of feeding tradition of troop C. Dr. S. Ichino helped me to identify lemurs and gave fundamental information about lemurs. Mr. Nzaka, guardian at the researcher house, and his wife Mrs. Genevieve, provided me with comfortable living condition at Berenty.

References

Altmann, J. (1974). Observational study of behavior: Sampling methods. *Behaviour* 49:227–267.
Britt, A., Randriamandratonirina, N. J., Glasscock, K. D., and Iambana, B. R. (2002). Diet and feeding behaviour of *Indri indri* in low-altitude rain forest. *Folia Primatol.* 73:225–239.
Budnitz, N. (1978). Feeding behavior of *Lemur catta* in different habitats. In: Bateson, P. P. G. and P. H. Klopfer (eds.), *Perspectives in Ethology*, Vol. 3. Social Behavior. Plenum Press, New York, pp. 85–108.
Ganzhorn, J. U. (1986). Feeding behavior of *Lemur catta* and *Lemur fulvus*. *Int. J. Primatol.* 7(1):17–30.
Gould L., Sussman R. W., and Sauther, M. L. (1999). Natural disasters and primate populations: The effects of a 2-year drought on a naturally occurring population of ring-tailed lemurs (*Lemur catta*) in southwestern Madagascar. *Int. J. Primatol.* 20:69–84.
Gupta, A. K., and Chivers, D. J. (2000). Feeding ecology and conservation of the golden langur *Trachypithecus geei* Khajuria in Tripura, Northeast India. *Journal of Bombay Natural History Society* 97(3):349–362.
Jolly, A. (1966). *Lemur Behavior*. University of Chicago Press, Chicago.
Jolly A., Rasamimanana H. R., Kinnaird, M. F., O'Brien, T. G., Crowley, H. M., Harcourt, C. S., Gardner, S., Davidson, J. M. (1993). Territoriality in *Lemur catta* groups during the birth season at Berenty, Madagascar. In: Kappeler, P. M., and Ganzhorn, J. U. (eds.), *Lemur Social Systems and Their Ecological Basis*. Plenum Press, New York, pp. 85–109.

Jolly, A., and Pride, E. (1999). Troop histories and range inertia of *Lemur catta* at Berenty, Madagascar: A 33-year perspective. *Int. J. Primatol.* 20(3):359–373.

Jolly, A., Dobson, A., Rasamimanana, H. M., Walker, J., O'Connor, S., Solberg, M., and Perel, V. (2002). Demography of *Lemur catta* at Berenty Reserve, Madagascar: Effects of troop size, habitat and rainfall. *Int. J. Primatol.* 23:327–353.

Jones, R. J., and Hegarty, M. P. (1984). The effect of different proportions of *Leucaena leucocephala* in the diet of cattle on growth, feed intake, thyroid function and urinary excretion of 3–hydroxy 4(IH)–pyridone. *Aust. J. Agric. Res.* 35:317–325.

Klopfer, P. H., and Jolly, A. (1970). The stability of territorial boundaries in a lemur troop. *Folia Primatol.* 12:199–208.

Koyama, N. (1991). Troop division and inter-troop relations of ring-tailed lemurs (*Lemur catta*) at Berenty, Madagascar. In: Ehara, A., Kimura, T., Takenaka, O., and Iwamoto, M. (eds.), *Primatology Today*. Elsevier, Amsterdam, pp. 173–176.

Koyama, N., Nakamichi, M., Ichino, S., and Takahata, Y. (2002). Population and social dynamics changes in ring-tailed lemur troops at Berenty, Madagascar between 1989–1999. *Primates* 43(4):291–314.

Mertl-Millhollen, A. S. (2000). Tradition in *Lemur catta* behavior at Berenty Reserve, Madagascar. *Int. J. Primatol.* 21:287–297.

Mertl-Millhollen, A. S., Gustafson, H. L., Budnitz, N., Dainis, K., and Jolly, A. (1979). Population and territory stability of the *Lemur catta* at Berenty, Madagascar. *Folia Primatol.* 31:106–122.

Mertl-Millhollen, A. S., Moret E. S., Felantsoa D., Rasamimanana H., Blumenfeld-Jones, K. C., and Jolly A. (2003). Ring-tailed lemur home ranges correlate with food abundance and nutritional content at a time of environmental stress. *Int. J. Primatol.* 24:969–985.

O'Connor, S. M. (1987). *The Effect of Human Impact on Vegetation and the Consequences to Primates in Two Riverine Forests, Southern Madagascar*. PhD diss., Department of Applied Biology, Cambridge University, Cambridge.

Overdorff, D. J. (1993). Similarities, differences, and seasonal patterns in the diets of *Eulemur rubriventer* and *Eulemur fulvus rufus* in the Ranomafana National Park, Madagascar. *Int. J. Primatol.* 14:721–753.

Pitts, A. (1995). Predation by *Eulemur fulvus rufus* on an infant *Lemur catta* at Berenty, Madagascar. *Folia Primatol.* 65(3):169–171.

Pride, R. E. (2003). *The Socio-Endocrinology of Group Size in Lemur catta*. PhD diss., Princeton University, Princeton, N.J.

Richard, A. (1977). The feeding behaviour of *Propithecus verreauxi*. In: Clutton-Brock, T. H. (ed.), *Primate Ecology*. Academic Press, New York, pp. 71–96.

Rasamimanana, H. R. (1999). Influence of social organization patterns on food intake of *Lemur catta* in the Berenty Reserve. In: Rakotosamimanana, B., Rasamimanana, H., Ganzhorn, J. U., and Goodman, S. M. (eds.), *New Directions in Lemur Studies*. Kluwer Academic/Plenum, New York, pp. 173–188.

Rasamimanana, H. R. (2004). La dominance des femelles makis (*Lemur catta*): quelles strategies energetiques et quelle qualite de ressources dans la reserve de Berenty au sud de Madagascar? Dissertation de Museum natural díhistoire naturalle ecole doctorale.

Rasamimanana, H. R., and Rafidinarivo, E. (1993). Feeding behavior of *Lemur catta* females in relation to their physiological state. In: Kappeler, P. M., and Ganzhorn, J. U. (eds.), *Lemur Social Systems and Their Ecological Basis*. Plenum Press, New York, pp. 123–133.

Sauther, M. L. (1993). Resource competition in wild populations of ring-tailed lemurs (*Lemur catta*): Implications for female dominance. In: Kappeler, P. M., and Ganzhorn, J. U. (eds.), *Lemur Social Systems and Their Ecological Basis*. Plenum Press, New York. pp. 135–152.

Sauther, M. L. (1994). Wild plant use by pregrant and lactating ring-tailed lemurs, with implications for early hominid foraging. In: Etkin, N. L. (ed.), *Eating on The Wild Side: The Pharmacologic, Ecologic, and Social Implications of Using Noncultigens*. University of Arizona Press, Tucson, pp. 240–256.

Sauther, M. L. (1998). Interplay of phenology and reproduction in ring-tailed lemurs: Implications for ring-tailed lemur conservation. *Folia Primatol.* 69:309–320.

Sauther, M. L., Sussman, R. W., and Gould, L. (1999). The socioecology of the ringtailed lemur: Thirty-five years of research. *Evol. Anthropol.* 8:120–132.

Simmen, B., Hladik, A., and Ramasiarisoa, P. (2003). Food intake and dietary overlap in native *Lemur catta* and *Propithecus verreauxi* and introduced *Eulemur fulvus* at Berenty, Southern Madagascar. *Int. J. Primatol.* 24:949–968.

Sussman, R. W. (1977). Feeding Behaviour of *Lemur catta* and *Lemur fulvus*. In: Clutton-Brock, T.H. (ed.), *Primate Ecology: Studies of Feeding and Ranging Behaviour in Lemurs, Monkeys and Apes*, Academic Press, New York, pp. 1–36.

Sussman, R. W. (1991). Demography and social organization of free-ranging *Lemur catta* in the Beza Mahafaly Reserve, Madagascar. *Am. J. Phys. Anthropol.* 84:43–58.

Tarnaud, L. (2004). Ontogeny of feeding behavior of *Eulemur fulvus* in the dry forest of Mayotte. *Int. J. Primatol.* 25:803–824.

Tew, A. (2003). Social relationships and social correlates of hair loss in a troop of ring-tailed lemurs at Berenty Reserve, Madagascar. Master's thesis, Oxford Brookes University.

Yamagiwa, J., Mwanza, N., Yumoto, T., and Maruhashi, T. (1994). Seasonal change in the composition of the diet of eastern lowland gorillas. *Primates* 35:1–14.

Yamakoshi, G. (1998). Dietary responses to fruit scarcity of wild chimpanzees at Bossou, Guinea: Possible implications for ecological importance of tool use. *Am. J. Phys. Anthropol.* 106:283–295.

Yamashita, N. (1996). Seasonality and site specificity of mechanical dietary patterns in two Malagasy lemur families (Lemuridae and Indriidae). *Int. J. Primatol.* 17:355–387.

Yamashita, N. (2002). Diets of two lemur species in different microhabitats in Beza Mahafaly Special Reserve, Madagascar. *Int. J. Primatol.* 23(5):1025–1051.

Watanuki, Y., Nakayama, Y., Azuma, S., and Ashizawa, S. (1994). Foraging on buds and bark of mulberry trees by Japanese monkeys and their range utilization. *Primates* 35:15–24.

Wrangham, R. W., Chapman C. A., Clack-Arcadi A.P., and Isabirya-Basuta G. (1996). Social ecology of Kanyawara chimpanzees: implications for understanding the costs of great age groups. In: McGrew, W.C. Marchant L.F., and Nishida T. (eds.), *Great Ape Societies*. Cambridge: Cambridge University Press, pp. 45–57.

Wright, P. (1999). Lemur traits and Madagascar ecology: Coping with an island environment. *Yrbk. Phys. Anthropol.* 42:31–72.

11
Diet Quality and Taste Perception of Plant Secondary Metabolites by *Lemur catta*

B. Simmen, S. Peronny, M. Jeanson, A. Hladik, and A. Marez

11.1. Introduction

Before addressing specific issues on the relationships between food choices of free-ranging ringtailed lemurs (*Lemur catta*) and taste perception focused on plant secondary metabolites, it is useful to consider briefly some basic traits of the taste system shared by all primates including humans, from which specific sensory adaptations can be identified. Many behavioral, psychophysical, and neurophysiological studies carried out in primates and other mammals dispute the idea of a taste system organized around few basic, discrete taste qualities like sweet, salty, bitter, and sour (e.g., Schiffman and Erickson, 1971, 1980; Critchley and Rolls, 1996). Whereas taste qualities actually refer to semantic descriptors varying according to human societies (Faurion, 1988), and thus bear poor evolutionary information, primate taste perception seems best described in terms of a gross dichotomous organization that corresponds with perception of attractive versus deterrent substances (Hladik et al., 2002). This functional opposition has been inferred from different sets of psychophysical and electrophysiological data, especially from patterns of activation of neural cells of the chorda tympani proper nerve (one main peripheral taste nerve) in response to a variety of taste stimuli applied on the primate tongue. In marmosets (*Callithrix jacchus*), macaques (*Macaca mulatta*), and chimpanzees (*Pan troglodytes*), there are two main clusters of fibers showing high affinities toward either soluble sugars on the one hand or quinine hydrochloride and various tannins on the other hand—taste cell responses to substances sour or salty to humans are more ambiguous (Hellekant et al., 1997a, 1997b, 1998; Danilova et al., 1998; Hladik et al., 2003). At higher levels of the gustatory pathway, in the primary cortex of the macaque (*Macaca fascicularis*), groups of neurons responsive to either beneficent or toxic stimuli have also been found, even though taste cells are less specific than taste nerve fibers at the peripheral level (Scott et al., 1986). In humans, in which a correlation matrix has been established using taste recognition thresholds for various compounds eliciting bitter, sweet, sour, salty, and tannic tastes, it is remarkable that a similar dichotomous figure has been depicted (Hladik et al., 2002).

Altogether, these data converge to suggest that the taste system of non-human primates and humans has been shaped (1) in relation to nutrients available in the environment in the context of optimizing energy input to meet metabolic requirements; (2) as an immediate control mechanism functioning to avoid toxins and antinutrient substances the consumption of which can have deleterious effects or decrease fitness (Hladik et al., 2003).

The fact that sucrose and other readily digestible carbohydrates have a taste attractive to all primates, as demonstrated experimentally in a range of species (Glaser, 1986; Simmen and Hladik, 1998), including ringtailed lemurs (Simmen, 2004), has been interpreted as a result of co-evolution of primates with angiosperms. Fruits with high energy contents were selected by arboreal and flying vertebrates by virtue of the immediate sensory reward produced by sugars in contact with taste buds together with their nutritional effect, while animals, in return, dispersed more seeds of these plant species (e.g., Herrera, 1985; Hladik, 1993). Contrary to such relatively recent seed-dispersal syndromes, plant–herbivore interactions leading to the synthesis or accumulation of chemical defenses goes far back in the past, before primates appeared in the Paleocene. Secondary metabolites including antinutritional substances, like tannins, and toxins, have evolved in relation to the selective pressures exerted by pathogens and early herbivores. This long-lasting co-evolutionary history in which herbivorous insects presumably played a predominant role given their abundance and high diversity may explain that "poisons" are widespread in many plant taxa and ecosystems (Ehrlich and Raven, 1964; Hladik and Hladik, 1977; Janzen, 1978; Swain, 1979).

Studies in the field of chemical ecology indicate that the composition and richness of forests in secondary metabolites is highly variable, even at a small geographical scale, and that this could determine local primate densities (Oates et al., 1990; Ganzhorn, 1995; Struhsaker et al., 1997; Simmen et al., 2005). Overcoming the noxious effects of plant allelochemicals, either through immediate taste rejection or through detoxification mechanisms, may be especially challenging to ringtailed lemurs considering results of screening tests performed on a range of plant species in some of its habitats. In the southern gallery forest and Didiereaceae spiny forest of Madagascar, ca. 60% of plant species are likely to contain alkaloids in their leaves, and relatively high proportions of plants rich in phenolic compounds, including tannins, are found (Simmen et al., 2003a). In this frugivorous–folivorous primate species, the timing and synchronization of major reproductive events across individuals, either across lactating females, juveniles at weaning, or active males during the mating season, appear to be closely tied to plant phenological patterns, especially those of keystone food resources (Rasamimanana and Rafidinarivo, 1993, Sauther et al., 1999; Rasamimanana, 2004). Because there are marked seasonal changes of food supplies and drought years occur quite frequently in gallery forests of south and southwestern Madagascar, ringtailed lemurs troops sometimes rely on low-quality food resources for long periods, and it is not uncommon in this context to observe high mortality rates in some years (Sauther, 1998; Gould et al., 1999;

Yamashita, 2002). It is thus necessary to investigate food constraints exerted on this species from the viewpoint of nutritional ecology in more detail. To date, little is known exactly of the levels of digestive inhibitors and toxins ingested by free-ranging lemurs, either *Lemur catta* or other Malagasy prosimians. There is evidence from semicaptive studies that ringtailed lemurs can ingest potentially toxic plants (Ganzhorn, 1986a; Mowry et al., 1997), but it remains unclear whether secondary metabolites consumed are innocuous—in the same way as caffeine ingested in moderate amounts each morning has no real toxic effect on a monogastric species—or whether ringtailed lemurs are able to counteract toxic effects. In this respect, analyzing taste sensitivity toward bitter or astringent taste stimuli allows one to define more objectively to what extent observed levels of allelochemicals in plants available in the environment could be deterrent to *Lemur catta*. Because taste perception is part of the digestive system, functioning to assess food edibility at a preliminary step of the ingestion/digestion process, measuring taste thresholds for various food-related compounds can provide evidence of psychophysiological adaptations to a specific diet or at least to a range of potential food resources.

In this paper, we aimed at defining whether ringtailed lemurs are widely exposed to plants potentially rich in toxins and antinutrients in the gallery forest of Berenty, southern Madagascar, and to what extent specific characteristics of taste perception of allelochemicals determine their food choices and feeding strategy. We analyzed concentrations of secondary metabolites in common plant species and in few less abundant plants at this study site as well as dietary levels of both these metabolites and gross nutrients in different seasons. Perception and avoidance of secondary metabolites considered unpalatable may vary according to their distinct biological activities (e.g., alkaloids among toxins with systemic effects on potential consumers at low concentrations versus tannins among digestibility reducers efficient at high doses). Accordingly, we determined taste thresholds for one tannin and one alkaloid, eliciting respectively tannic and bitter tastes in humans, in addition to responses to fructose. Experimental data on taste responses were obtained using a classical behavioral method on both captive animals from the prosimian colony of Thoiry in France and, for the first time, on free-ranging animals at the Berenty Reserve. As a complementary approach for assessing sensory adaptation to peculiar feeding niches, ringtailed lemur taste thresholds for food-related natural substances will be compared with taste sensitivity of other primate species.

11.2. Methods

11.2.1. Feeding Behavior and Food Chemistry

Diet typology and *ad libitum* observations of feeding behavior of lemurs in the gallery forest of Berenty Reserve (southern Madagascar) have been reported in a previous paper (Simmen et al., 2003b). We recorded food choices in three different seasons from May 22 to June 27, 1998 (middle of the dry season), from

October 10 to November 4, 1998 (late dry/early wet season), and from February 6 to March 17, 1999 (end of the rainy season). Results obtained by following focal individuals from four troops from the onset of morning activity to the nocturnal resting period, with observations being balanced between males and females, have been grouped to provide a broad picture of dietary trends, diet quality, and their seasonal variations. The diet was determined quantitatively, recording mouthfuls of each focal animal followed continuously and converting them into ingested matter after collecting and weighing food samples (Hladik, 1977). Details on animal observations and study site can be found in Simmen et al. (2003b). The seasonal variation of the diet can be summarized as follows: unripe fruits and mature leaves were the main food categories during the middle of the dry season (42% and 44%, respectively). In the late dry/early wet season, the proportion of fruits in the diet was 37% (with 19% ripe fruits), and leaves accounted for 63% of the ingested matter (young leaves: 52%). Ripe fruits were the main food category at the end of the rainy season (92%).

Chemical parameters investigated on plant samples (dried in oven at 60 °C) were soluble sugars (HPLC), crude lipids (HCl hydrolysis and extraction with petroleum ether), crude protein (N × 6.25; Kjeldahl method), crude fibers (Weende method), total phenolics (TP) using modified Prussian Blue assay (Price and Butler, 1977), condensed tannins (CT) using acid butanol assay (Porter et al., 1986), and total alkaloids (TA) using a titration method (Commission Nationale de Pharmacopée, 2002). Tannin analyses also included a biological activity test (Blue BSA; Asquith and Butler, 1985) to assess plants' ability to precipitate bovine serum albumin (BSAp), thus to get an idea of potential protein digestibility reducing effects of food items. Unpublished sources and an electronic database [Rasamimanana, 2004; FAO (www.fao.org/docrep/003/X6878E/X6878E00.htm#TOC)] provided some additional information on plant chemistry. Food chemical composition was then weighted by measured intake of the various plants to provide a quantitative estimate of gross nutrients, crude fibers and secondary compounds in the diet. Plant samples used for " diet reconstruction" accounted for a total of more than 97% of the dry matter ingested in the middle dry season (22 food items) and the early wet season (12 food items) and 79% in the late wet season (8 food items). A comparison was made of the chemistry of leaves eaten versus leaves not eaten. Analyses were performed on plant species that were ingested in significant amounts and on species that we did not observe to be eaten in any season, although these species, either trees or lianas, were common in the forest. We also contrasted the chemistry of leaves ingested against a random sample (which included a set of abundant and less abundant plant species).

11.2.2. *Determination of Taste Thresholds for Fructose, Quinine Hydrochloride, and Tannic Acid*

Taste perception was investigated on individual lemurs within one colony kept in the Parc Zoologique de Thoiry (France) as well as in the Berenty reserve (Madagascar) on wild *Lemur catta* from one group that was habituated to the presence of field observers and that could be quite regularly tested.

Taste thresholds for food-related compounds were determined using a classical two-bottle test derived from that of Glaser (1968). Compounds tasting sweet (fructose) or astringent (tannic acid; Fluka ref. 48811) to humans were tested in both the captive and the wild lemur population. In addition, responses to a bitter substance (quinine hydrochloride) were recorded in wild lemurs. The procedure described below for captive animals has been applied on Berenty lemurs, with some adjustments.

11.2.3. *Taste Thresholds of Captive Lemurs*

The two-bottle test is based on the spontaneous choice by animals between two liquids presented simultaneously, usually tap water (control solution) and a taste stimulus dissolved in the control solution, using a range of different randomized concentrations of the taste stimulus. Bottles are left for 1 minute as soon as a given individual starts drinking the solutions, yielding individual records. The respective location of the bottles (right or left side) is determined at random in each test to avoid a side-preference bias. The short duration is designed to reduce postabsorptive effects on taste perception and because tannins are not stable for long periods when dissolved in water. The threshold is defined as the lowest concentration of the taste stimulus that is preferred or rejected over the control solution. To standardize records among captive and wild animals, the time spent consuming the gustatory solution divided by total time spent consuming the two liquids is calculated in each test. The criterion for preference or rejection is reached if the consumption index differs from 66.7% for preference and 33.3% for rejection (Laska, 2000). This conservative criterion allows comparison with other primate species and is suited to small sample sizes. However, in most cases, we used a paired-sample t-test to search for significant differences, eventually grouping individual responses toward close concentrations of the taste stimulus (with $p < 0.05$ or lower).

When recording responses to distasteful stimuli versus the control solution during short tests, volumes of liquids ingested are not reliable measures, and a slightly attractive, fixed concentration of sugar must be added in each bottle. Because sweetness intensity of this solution depends on taste discriminative abilities, we determined the taste threshold for fructose in a first series of experiments, providing animals with fructose solutions versus water (11 different fructose concentrations in a two-bottle test). In a second series of experiments, animals were provided with one feeding bottle containing 20 mM fructose solution and the other containing the same sugar solution mixed with the deterrent substance (concentrations of tannic acid: either 0.1, 0.5, 1, or 2 g/L). The fixed concentration of the fructose solution at 20 mM was chosen according to the response–concentration curve determined in the first series of experiments: this concentration exerts a slight attraction to animals and corresponds with less than 5 times the threshold for this sugar (see "Results" below). Previous experiments showed that the rejection threshold for tannic acid determined using tannin/fructose mixtures did not vary significantly according to sugar concentration within this range of moderately sweet concentrations (Simmen et al., 1999).

Before starting the two series of experiments, animals were provided, first, with one single bottle containing a sweet solution (sucrose, 300 mM), and then one bottle with tap water and another with sucrose solution simultaneously. Habituation of a given individual to the two-bottle test was considered achieved when a marked preference for the sweet solution was recorded over four successive trials. Within the group, eight animals (seven females and one male) successfully reached this criterion and were subsequently used in the tests with fructose and tannins. However, because individuals could not be isolated from the group and were tested on the basis of their spontaneous motivation to approach the feeding bottles, the number of tests performed on each animal was not equal. Individual records for each concentration were averaged if several tests had been made on a given animal. When testing with distasteful compounds, animals were sometimes provided with attractive solutions to maintain high motivation for the test.

11.2.4. Taste Thresholds of Free-ranging Lemurs

The same protocol was used on wild animals (Figure 11.1) from a study group that ranged in the vicinity of the experimental area (at the edge of the forest).

FIGURE 11.1. A free-ranging *Lemur catta* female choosing among sapid solutions in a "two-bottle test" at the Berenty Reserve. Photo, B. Simmen.

Experiments were carried out during austral spring, between late September and early November (captive animals in the Northern Hemisphere were also tested in spring) as part of a program on ontogeny and diets of large sympatric lemurs within the Berenty Reserve.

Lemurs were initially attracted with pieces of bananas made accessible from the bottles, soon replaced by sweet solutions and water. As with captive animals, we first established a fructose concentration–response curve (testing with 10, 20, 40, 50, 100, 150, 200, 300, and 500 mM versus water). A slight but significant preference for fructose over water was observed only when the concentration was ≥40 mM (see "Results" below). Binary solutions were then prepared, mixing the 40 mM fructose solution with tannic acid (at 0.1, 0.5, 1, or 5 g/L) or with quinine hydrochloride (0.025, 0.05, 0.1, 0.2, 0.3, or 0.5 g/L). Mixtures and the control solution (fructose 40 mM) were prepared as soon as foraging animals were in sight of the observer. Tests lasted 5 minutes and were usually performed once a day in the morning or in the evening, mainly on females. Up to eight animals became habituated to the two-bottle test, but we had to discard some data, especially ingestion sequences of subordinate animals supplanted by dominant females.

11.3. Results

11.3.1. Diet and Food Chemistry

There were marked seasonal variations in the proportions of total phenolics (TP) and condensed tannins (CT) in the diet as well as in the amount of protein precipitated (BSAp; Figure 11.2a). Proportions of TP and CT did not change between the middle of the dry season and the late dry/early wet season but largely decreased toward the wet season. The precipitation ability (BSAp) decreased two fold between the middle of the dry season and the late wet season. Whether these dietary levels of antinutrients are high or low cannot be directly determined in the absolute because different standards and units are used for calibration curves associated with TP, CT, and BSAp. However, comparing the results with concentrations found in the pulp of one major fruit species consumed during the wet season (*Celtis bifida*, Ulmaceae; Figure 11.2a) may be helpful in this respect—*Celtis* fruits could be considered a baseline for low contents of antinutrients in attractive food items. The variations of TP, CT, and BSAp reported here reflect switches from a diet largely composed of leaves and unripe fruits during the middle of the dry season and the late dry/early wet season to a diet including predominantly ripe fruits during the wet season. Dietary concentrations of alkaloids decreased twofold during the dry season (Figure 11.2b). The concentration calculated for the late rainy season (0.058%; not figured) is presumably not representative of the diet because alkaloid contents of several major foods could not be determined. The fact that these items were mainly ripe fruit pulps nevertheless suggests that dietary proportions of alkaloids were probably low during this period.

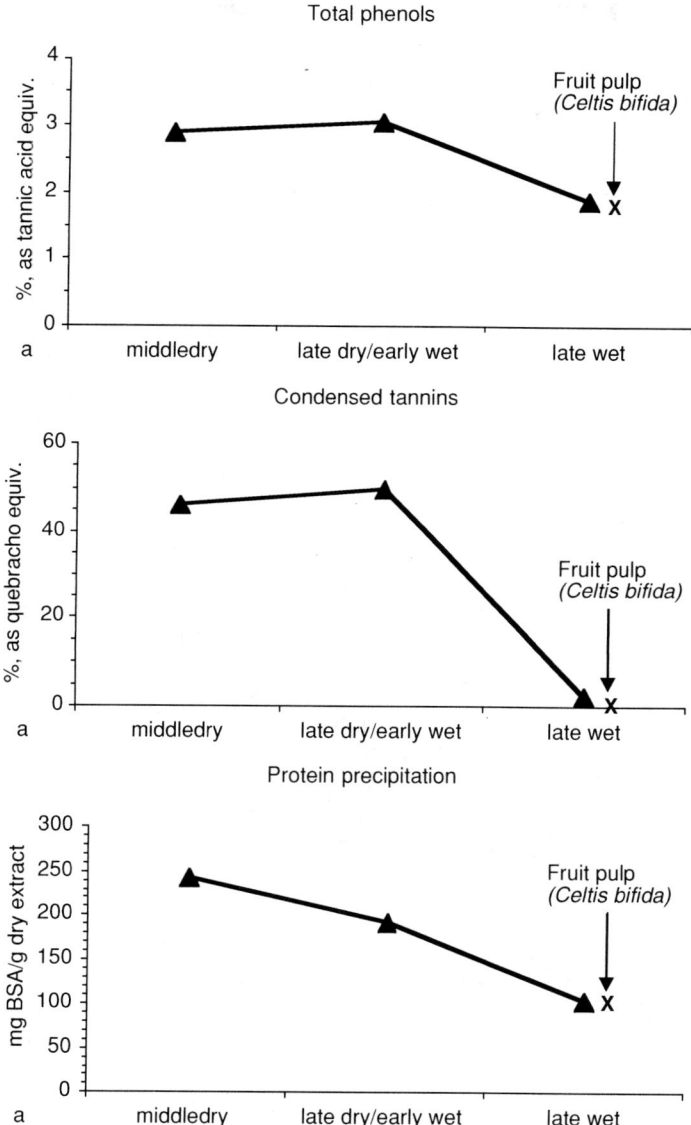

FIGURE 11.2. Seasonal variations of the quality of the diet of ringtailed lemurs at Berenty. Results are expressed in percent of the dry matter, except for bovine serum albumin (BSA) precipitated. Chemical analyses were made on antinutrients (a), alkaloids

In parallel, there was an increase in the protein/fiber ratio from the mid dry season to the wet season with crude fibers decreasing largely while crude protein was maximum during the early wet season (Figure 11.2c). Changes in protein proportion did not simply conform to expected variations based on diet typology: a similar protein content (ca. 12%) was observed in a diet mainly composed of ripe

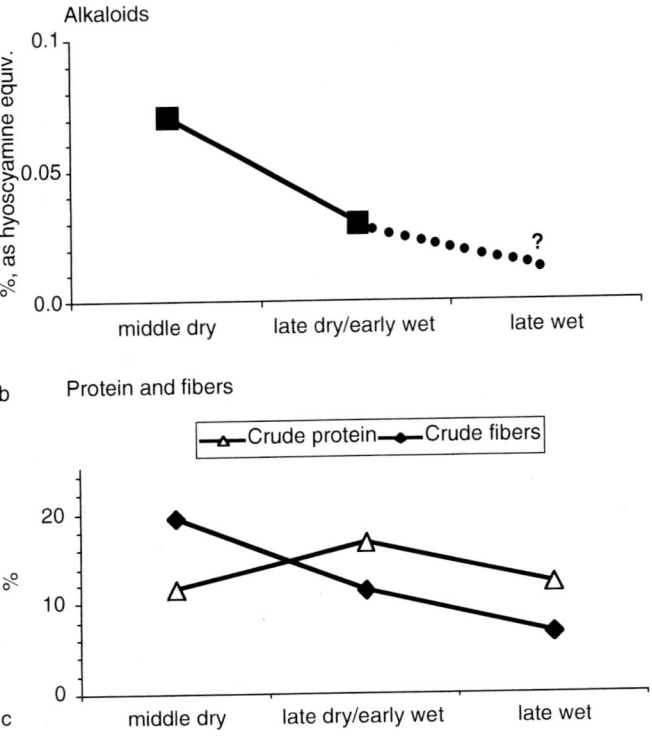

FIGURE 11.2. (*Continued*) (b), and protein and fibers (c). The antinutrient content of the ripe pulp of one main fruit species eaten during the late wet season (*Celtis bifida*) is indicated for comparison. Plants eaten during the late wet season, for which alkaloid concentration was measured, may not be fully representative of the diet, thus the dotted line (see "Results").

fruits (92% during the late wet season) and in a diet including equivalent proportions of mature leaves and unripe fruits (44% and 42% in the mid dry season; Simmen et al., 2003b). The relatively high nitrogen input observed during part of the wet season, when juveniles are just weaned, is largely due to the ingestion of *Celtis bifida* pulps, which contain unusually large protein concentration (25%; unpublished results) compared with other tropical fruits.

Leaves selected as food resources by ringtailed lemurs did not differ significantly from leaves not eaten for any of the antinutrients tested. Figure 11.3 shows results for mature leaves ($p < 0.4$ in all cases with a Mann–Whitney U-test). A lack of significant difference was also found for young leaves (with 10 species of ingested immature leaves analyzed) although there was a trend toward lower protein precipitation efficiency in plants eaten ($U = 12$; $p = 0.08$). A similar calculation, contrasting the chemistry of foliage ingested against a random sample of plants (27 species), confirmed the lack of a selective behavior with regard to the antinutrients tested. Lemurs for instance ingested leaves that were very

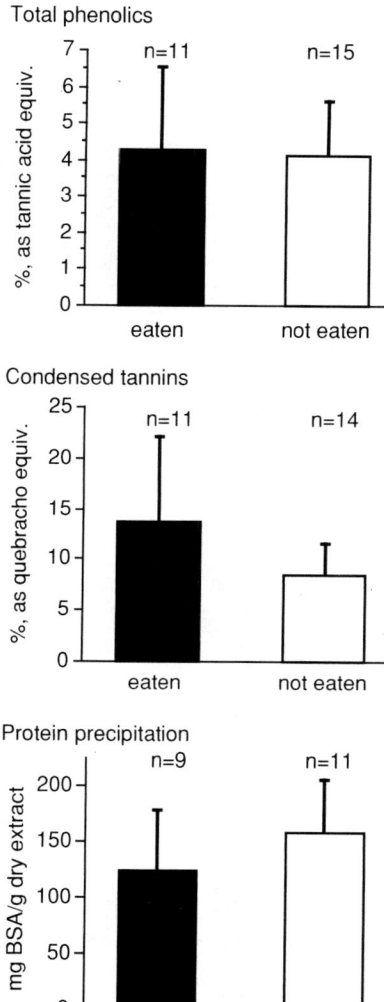

FIGURE 11.3. The average antinutrient content of mature leaves eaten by *Lemur catta* versus mature leaves not eaten.

efficient with respect to protein precipitation or that were very rich in phenolics or condensed tannins (as for instance, *Acacia rovumae* and *Tamarindus indica* among other legume species, and *Sclerocarya birrea* among Anacardiaceae). Similarly, alkaloid contents did not differ between leaves eaten and leaves not eaten (Figure 11.4). Likewise, lemurs consumed plants that, in some cases, were very rich in alkaloid (*Acacia rovumae*, Mimosoideae, and *Azadirachta indica*, Meliaceae). Because some of these leaves were major items in the middle of the dry season, they accounted for the high dietary concentration and significant amounts of alkaloids ingested by ringtailed lemurs during this period of low food availability (Figure 11.2b).

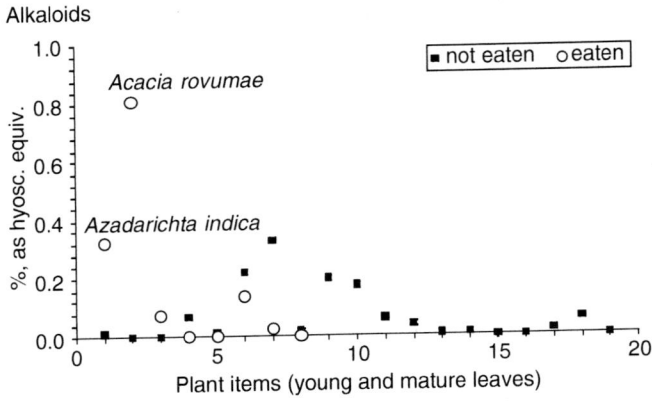

FIGURE 11.4. Plot of the alkaloid content of leaves eaten by *Lemur catta* and leaves not eaten.

11.3.2. Taste Perception of Captive Lemurs

Responses of captive ringtailed lemurs to fructose solutions versus water in a two-bottle test are presented in Figure 11.5A, showing the mean time that lemurs spent consuming the sweet liquid and water in each test as a function of increasing sugar concentrations. The lowest fructose concentration for which a significant difference is found, that is, the threshold for fructose, was 5 mM (with $p \leq 0.05$ in a paired sample t-test). Expressing these results in terms of the relative time spent drinking the sweet solutions (Figure 11.5B), indicates similarly a threshold value at 5 mM. A variation associated with consumption of slightly higher concentrations, at 10 and 20 mM (according to the 66.7% criterion), was observed, but grouping the consumption data recorded for these two concentrations yielded unequivocal preference for the sugar solutions over water, in agreement with significant differences obtained using a paired sample t test (Figure 11.5A).

Results of the second experiment, in which we recorded the effect on ingestion of adding various tannic acid concentrations to a moderately sweet fructose solution, namely at 20 mM (that is slightly above the threshold for fructose), are shown in Figure 11.6. The rejection threshold for tannic acid was 0.5 g/L.

11.3.3. Taste Perception of Wild Lemurs

A clear preference for fructose over water was observed for concentrations ≥ 40 mM (Figure 11.7a). Using a paired sample t-test, the mean difference of time spent consuming the sweet solution versus water was significant for this concentration ($p \leq 0.05$) and not significant for less sweet solutions. In the second series of experiments designed to determine rejection thresholds for tannic

11. Diet Quality and Taste Perception of *L. catta* 171

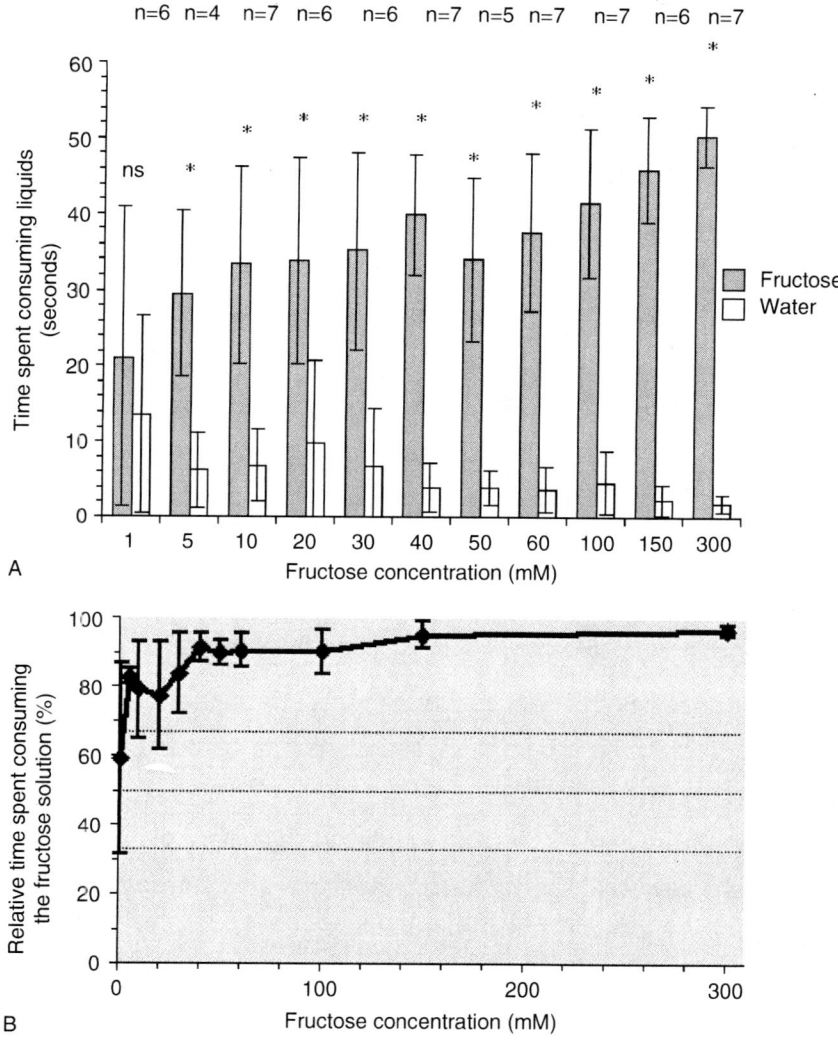

FIGURE 11.5. Results of the two-bottle test carried out on captive ringtailed lemurs using fructose solutions, as shown in absolute and relative terms. The first graph (A) shows the mean time (±SEM) spent drinking water versus sugar solution for each fructose concentrations. The number n of animals tested as well as the probability of a significant difference ($^*p < 0.05$) are indicated. The graph below (B) shows the mean relative consumption (±SD) of the sugar solution (i.e., time spent drinking the fructose solution divided by total time spent consuming water and the fructose solution in the two-bottle test). Lower and upper dotted lines correspond to 33.3% and 66.7%, respectively.

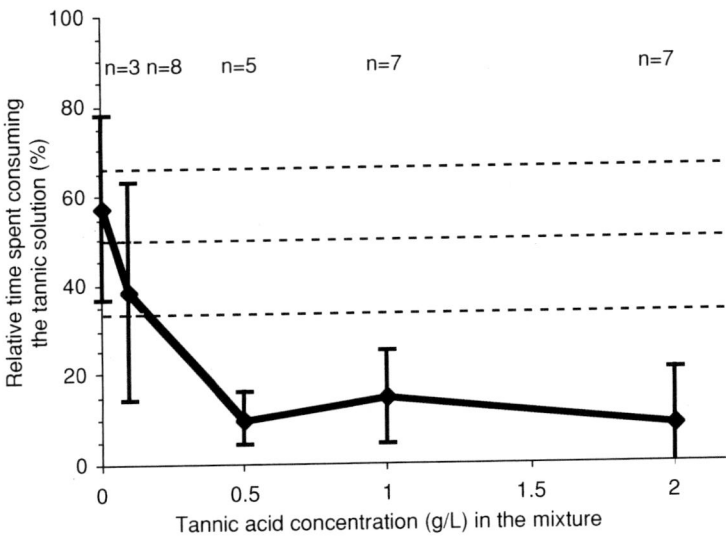

FIGURE 11.6. Results of the two-bottle test carried out on captive ringtailed lemurs using mixtures of increasing tannin concentrations. The bottles contain respectively a near-threshold concentration of fructose at 20 mM (as determined from tests presented in Figure 11.5) and the same fructose solution mixed with tannic acid the concentration of which is varied in each test. Lower and upper dotted lines correspond with 33.3% and 66.7%, respectively.

acid and quinine hydrochloride, the deterrent substances were mixed with a fructose solution at 40 mM. The rejection threshold for the tannin was 0.5 g/L (Figure 11.7b), exactly as was found in captive animals (with $p \leq 0.05$). When using the bitter compound (quinine), a consistent rejection of mixtures was observed for concentrations ≥ 0.2 g/L (Figure 11.7c; concentrations lower than 0.2 g/L were not significantly rejected over water with $0.8 < p > 0.3$ in a paired sample t-test). Clear behavioral expressions of distastefulness, such as head shaking, were observed for quinine concentrations reaching 0.3 g/L.

11.4. Discussion

Many behavioral and physiological characteristics of ringtailed lemurs can be understood as adaptive responses to the drastic variations of food availability and food quality that prevail in the dry climate of southern Madagascar. Although initially considered a predominantly frugivorous species (Sussman, 1974), the ringtailed lemur appears best described as a frugivorous–folivorous species that is able to feed predominantly on mature leaves and unripe fruits for long periods during the dry season (see Soma, this volume; Simmen et al., this volume). Ringtailed lemurs are not as specialized on leaves as are sifakas (*Propithecus*

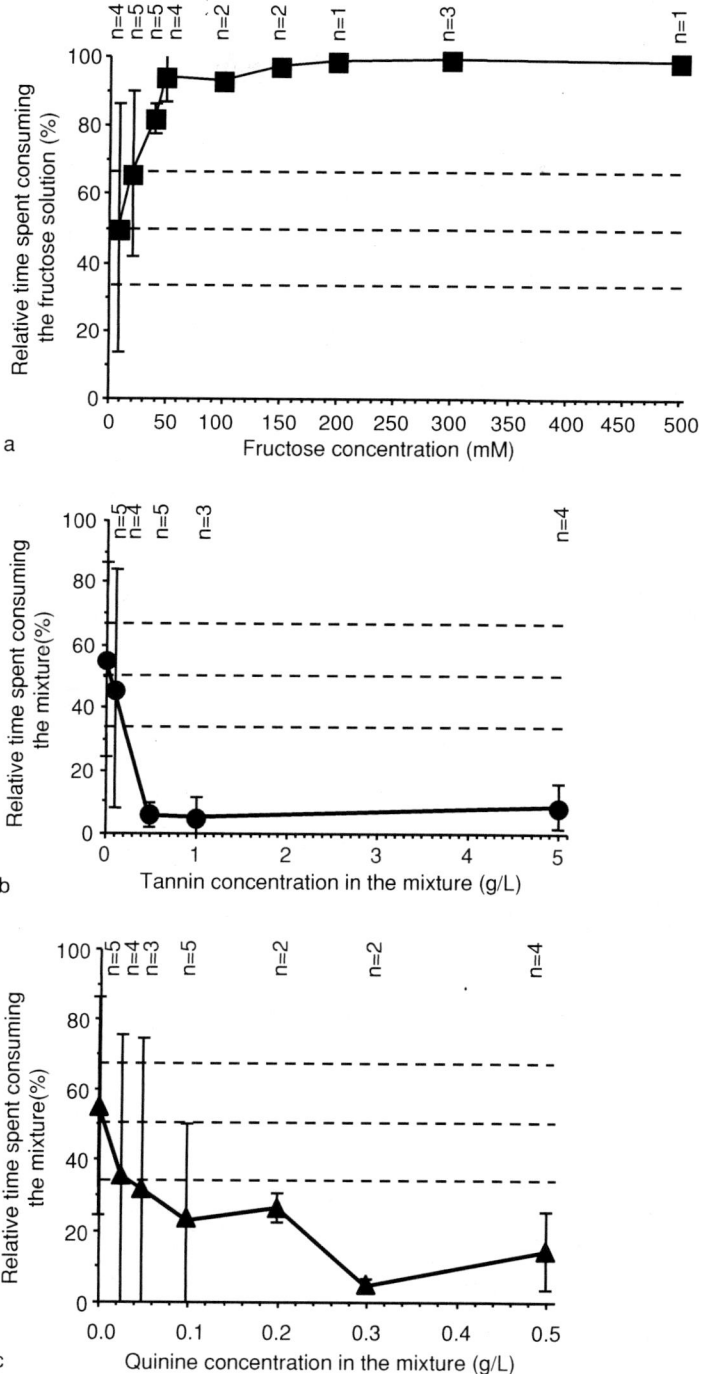

FIGURE 11.7. Results of the two-bottle test carried out on free-ranging ringtailed lemurs at the Berenty Reserve. (a) Responses to fructose. Same legend as Figure 11.5B. (b) Responses to increasing tannin concentrations mixed in a fixed near-threshold concentration of fructose at 40 mM (as determined from tests presented in Figure 11.7a). Same legend as Figure 11.6. (c) responses to quinine hydrochloride. Same protocol as for tannic acid.

verreauxi), with which they coexist in Berenty, but there is evidence that substantial amounts of cellulose can be digested by ringtailed lemurs in a somewhat enlarged cecum (Sheine, in Overdorf and Rassmussen, 1995; Campbell et al., 2000). Experimental data also suggest that gut transit time in *Lemur catta* is slow relative to body size or compared with *Eulemur fulvus*, a competing lemur species with similar body size (Ganzhorn, 1986b; Cabre-Vert and Feistner, 1995). Could we consider, in the same manner, that the gustatory characteristics measured in this study are psychosensory adaptations that facilitate the ingestion of poor-quality foods? Before interpreting sensory perception of *Lemur catta* in terms of adaptation to the biochemical environment, we have to examine the effects of the food context on similarities and differences observed in the taste responses between captive and free-ranging lemurs.

11.4.1. Taste Preference/Aversion Thresholds for Various Taste Stimuli According to Food Context

This study is the first one in which behavioral responses to taste stimuli have been recorded in free-ranging primates. We have been able to determine taste thresholds for compounds tasting sweet, tannic, or bitter to humans, or at least to identify the lowest concentrations of these compounds that were attractive or repellent to wild ringtailed lemurs. Taste thresholds determined from spontaneous choice between two solutions are assumed to provide a reliable reflection of discriminative abilities when using standard testing procedures. Indeed, it has been demonstrated that the "two-bottle test" and evoked potentials recorded on the chorda tympani of different primate species yielded similar thresholds (Glaser and Hellekant, 1977; Hellekant et al., 1993; Iaconelli and Simmen, 1999). However, because in our study the hunger status of animals was not controlled, it is likely that regulatory mechanisms like oral sensory-specific satiety—by which the hedonic perception of sweet stimuli decreases in relation to oral stimulation accompanying ingestion of energy-rich foods (Rolls et al., 1981)—explain the difference found between the threshold of free-ranging animals (40 mM) and that of captive lemurs (5 mM).

Whereas hunger-related variations of pleasantness have been described for sugars, there is evidence that satiety does not modify affective perception of distasteful stimuli (e.g., Berridge, 1991). Studies that investigated taste perception in distinct colonies of a same non-human primate species yielded comparable taste avoidance thresholds either for bitter substances or for tannins using the method of the two-bottle test (Gray mouse lemur *Microcebus murinus*; Pygmy marmoset, *Cebuella pygmaea*; lowland gorilla, *Gorilla gorilla*; Glaser, 1968, 1977; Simmen et al., 1999; Iaconelli and Simmen, 2002; Remis and Kerr, 2002; Simmen and Charlot, 2003). In some cases (*Microcebus murinus*), identical thresholds for quinine were obtained from 1-hour tests carried out on fasting animals versus 24-hour tests during which subjects had access to foods. Furthermore, the profiles of supra-threshold responses to distasteful stimuli, especially to bitterness, of different primate species generally indicate quasi total inhibition of the consumption for near threshold concentrations (i.e., for low stimulus intensities; Simmen, 1994; Iaconelli and Simmen, 2002). That the lowest tannin concentrations avoided by captive and wild animals were strikingly identical in our study suggests that the taste threshold for tannin (and presumably for other distasteful compounds) may be quite unaffected by different food contexts. Considering the survival value of immediately rejecting foods the unpalatable taste of which may signalize noxious substances, we suspect that the lowest concentrations of bitter and tannic substances that were avoided by free-ranging lemurs were close to taste discrimination thresholds.

Among stable, genetically determined behavioral traits associated with taste perception, the gusto-facial reflex of adults and newborn primates, including anencephalous human babies, is the most spectacular one (Steiner; 1977; Steiner and Glaser, 1984). Although studies of this reflex focused on facial stereotyped movements triggered by a limited range of stimuli eliciting the so-called basic taste stimuli (sweet, sour, salty, bitter), nothing was known so far of early reactions to potentially noxious compounds eliciting tannic tastes. During our experimental study in the field, an infant lemur 7 days old that had been separated from its troop for some unknown reason (and which was no longer accepted by its conspecifics) was saved and kept safe by a colleague (Alison Tew) for a few weeks. We benefited from this opportunity to record and videotape the facial and body movements of the infant (at the age of 15 days) in response to a pure solution of tannic acid being applied on its tongue. The concentration 1 g/L was chosen because it is in the lower range of concentrations eliciting avoidance in adult lemurs. Reponses of the infant to other substances tasting sweet or sour were also recorded for comparison (Simmen, 2004). Three observers determined whether videotaped facial and body reactions to the tannin evoked indifference, acceptation, or rejection. As expected, rejection of the tannin was unambiguous, in that case mirrored by head shaking in presence of the stimulus. Because the animal was not exposed to tannic substances after birth, it seems reasonable to assume that astringency, like bitterness (Scott and Mark, 1987), is encoded as a "toxic" stimulus in the taste system. Certainly, one cannot rule out potential influence of *in utero* exposure to unpalatable substances, and in no case does the gusto-facial reflex mean a fixed behavioral response throughout individual life. To some extent, free-ranging lemurs consuming tannin or alkaloid-rich foods might have acquired a tolerance for initially distasteful stimuli. Such preference reversal may be triggered by an alimentary deprivation or stress, by some curative properties for individuals with bad health or disease, or through social influences that play a great role in group-living primates. We reported in a previous paper that the infant *Lemur catta* rejected acid solutions (mixed with moderate amounts of sugar) that were preferred by adults (Simmen, 2004). In that case, avoidance reversal for initially distasteful stimuli not only results in a taste tolerance but even probably corresponds with an acquired preference for sour foods. These changes of perception occurred very early in the course of individual ringtailed lemur development because infants less than 2 months old were already licking and feeding on sour pods of *Tamarindus indica*. Positive conditioning toward sour stimuli was certainly favored by mother–infant interactions, a well-known psychophysiological process occurring during ontogeny. Such learned preference may also have resulted from individual learning and precocious exploration of the food environment irrespective of mother feeding activity, as described in brown lemurs (Tarnaud, 2004).

11.4.2. Adaptive Value of Taste Sensitivity and Feeding Strategy

We found that the minimum concentrations of tannic acid and quinine that were deterrent to wild *Lemur catta* were of the same order of magnitude (ca. 0.2 and 0.5 g/L). This similarity was unexpected given that toxins like alkaloids are considered efficient at very low doses, whereas much higher concentrations of digestibility reducers like tannins are required to prevent herbivores from feeding (Howe and Westley, 1988). At the ecosystem level, theories of plant chemical defenses predict distinct frequencies of plants containing either specific toxins or generalist defenses in a given environment (Feeny, 1976; Lebreton, 1982; Coley et al., 1985), which could theoretically lead to asymmetrical selective pressure for high sensitivity toward substances tasting bitter (as in many toxins) or astringent (as in polyphenols). In this context, could gustatory/avoidance characteristics measured here indicate adaptive sensory specialization to peculiar categories of allelochemicals?

The combined results of experimental studies of taste perception on the one hand and data on food chemistry and food choices of ringtailed lemurs on the other hand lead us to hypothesize that low taste sensitivity toward quinine (but not toward tannic acid) is an adaptive specialization allowing lemurs to ingest foods that may be extremely bitter and eventually toxic to unspecialized species. There is indeed both empirical and theoretical evidence that alkaloids, including toxic forms, are widespread in Berenty. Results of screening tests performed on mature leaves of a large sample of plants from different Malagasy forests are presented in Table 11.1. Focusing on plant species showing positive responses, there was a significantly larger proportion of species likely to contain alkaloids in gallery forest

TABLE 11.1. The occurrence of alkaloids in mature leaves in different forest sites of Madagascar.

Site	Percent (%) of plant species with positive response (screening)	Number of plant species with positive response (titration)
Gallery forest (Berenty)	57 (n = 104)[a]	10 (n = 13)[b]
Dry spiny forest (Berenty)	61 (n = 51)[a]	—
Evergreen rainforest (Andasibe)	38 (n = 128)[a]	—
Deciduous dry forest (Marosalaza)	15 (n = 127)[a]	—
Semi-deciduous forest (Ampasikely)	—	6 (n = 14)[c]

Data presented here for the screening tests only refer to plant samples with large precipitate (using a scale from 0 to +++) obtained when adding one of either two reagents to aqueous extracts.
n: number of plant species tested.
[a]Simmen et al. (2003).
[b]This study.
[c]Simmen et al. (2005).

and Didiereaceae forest in Berenty than in the eastern wet forest of Andasibé or the dry forest of Marosalaza. Results of alkaloid titration performed in our study confirmed that a majority of indigenous species contains alkaloids—although small, the sample included the most abundant plant species of the study site, and a higher figure was obtained when including introduced plants in the analysis. Several authors (Lebreton, 1982; Hladik et al., 2000) argued that abundant plant species in environments with low plant diversity (as in Berenty gallery and spiny forests) are predominantly damaged by specialized herbivores, which would favor the emergence of compounds of increasing toxicity (as many alkaloids) instead of more generalist antinutrients (tannins). That *Lemur catta* is relatively insensitive to quinine, a prototypical stimulus of bitterness (to human taste), is highlighted by a comparison with other primate species. Plotting the taste rejection threshold for quinine hydrochloride—as determined under testing procedures similar to those used in this study—in a range of species (Figure 11.8a), thresholds vary by a factor of more than 2000 between the least sensitive species (*Aotus trivirgatus*: 0.62 g/L) and the most sensitive one (*Callithrix argentata*: 0.0003 g/L). We note that ringtailed lemurs reject quinine at median/high concentrations (0.2 g/L), and indeed the bitterness of such solutions is really distasteful to humans. Taste thresholds for tannic acid are currently available for a small number of primate species (Figure 11.8b). This is largely because until recently, tannins were not considered by physiologists to elicit a taste but rather to elicit a tactile sensation involving dryness of the tongue. As a matter of fact, in 1993 it was demonstrated in a prosimian primate that the chorda tympani proper nerve—which mainly conveys gustatory signals—could be stimulated by tannic acid (Hellekant et al., 1993). We note from Figure 11.8b that the taste threshold for tannic acid of *Lemur catta* is similar to that of several other primate species including fruit specialists that are assumed to lack digestive specialization for food detoxification (e.g., spider monkeys). Folivorous species like gorillas appear much less sensitive to tannic acid than *Lemur catta*. We thus suspect that high dietary levels of phenolic compounds observed in some seasons in a context of food scarcity reflected a nutritional stress (mid dry season or gestation period) or resulted from a trade-off in the choice of foods with high nutrient/high antinutrient content (e.g., protein-rich new leaves of common plant species in the late dry season/early wet season corresponding with the early lactating period). Polyphenols overall are not as toxic as are some poisonous compounds, and primate species including ringtailed lemurs commonly exhibit behaviors such as geophagy, which is likely to reduce the antinutritional properties of tannins (Hladik and Gueguen, 1974; Johns and Duquette, 1991). One cannot rule out that tannin ingestion is associated with short-term beneficent effects on health (Athanasiadou et al., 2001), although a demonstration of antiabortive properties, as suggested for sifakas (Carrai et al., 2003), would require more experimental data (see Bouquet et al., 1967).

To conclude, *Lemur catta* appears to exhibit sensory characteristics of both folivorous species and frugivorous species. On the one hand, high taste sensitivity toward secondary metabolites helps minimize the ingestion of potentially noxious substances, which is expected in consumers with unspecialized digestive tract,

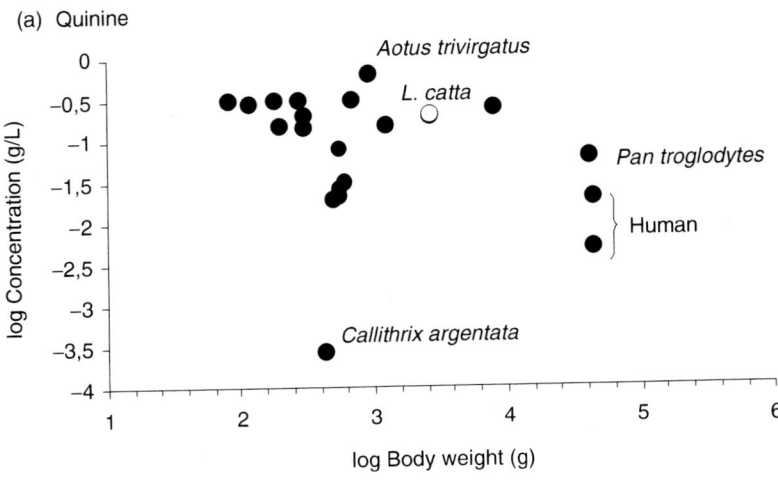
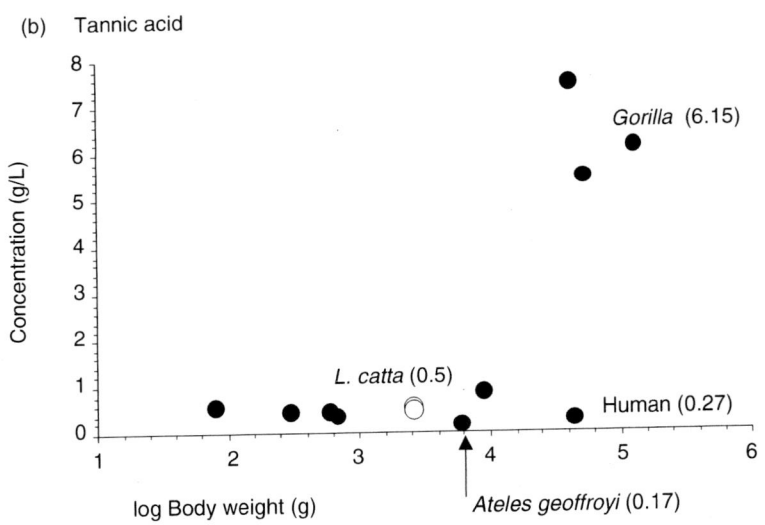

FIGURE 11.8. Plots of the rejection threshold for quinine hydrochloride (a) and tannic acid (b) in various primate species ranged according to their body mass. Taste thresholds values recorded in a two-bottle test are reviewed in Glaser (1986), Simmen and Charlot (2003), and Simmen (2004).

feeding predominantly on ripe fruits and other foods with readily digestible nutrients. But in *Lemur catta*, the taste response varies according to the type of secondary metabolites and could be considered, in the case of quinine, as an adaptation to ingest foods containing alkaloids and other toxins sharing bitter taste qualities (e.g., among cyanogenic glucosides or saponosides). At the opposite, a high sensitivity toward soluble sugars is not likely in folivorous species (e.g., the low sensitivity to sucrose in sifakas; Dennys, in Simmen and Hladik, 1998), which tend to exhibit feeding strategies of low energy input/low energy costs. In the case of *Lemur catta*, the threshold for fructose corresponds with one of the highest sensitivities found in primates tested to date. From these data, it is clear that categorizing species from their gross feeding tendencies (e.g., frugivore–folivore) actually refers to species showing distinct gustatory profiles and tells little about how these primates perceive their environments. Furthermore, there is suspicion that ringtailed lemurs within a group tolerate different concentrations of secondary metabolites in their foods. The extent to which avoidance levels vary according to individual gustatory sensitivity should be investigated further, even though, in this species, social factors such as female dominance are assumed to be one major determinant of the variation of diet quality among group members.

Acknowledgments. We are grateful to the director and staff of the Parc Zoologique de Thoiry who allowed us to carry out the experiments on taste in lemurs. We thank the Ministère des Eaux et Forêts, Madagascar, who provided research permits for fieldwork. We are also grateful to Jean and Philippe de Heaulme for giving permission to study ringtailed lemurs in the Berenty Reserve. Records of the infant lemur facial reactions in the field were made possible thanks to the help of Laurent Tarnaud and Alison Tew. We are indebted to Alison Jolly and Anne Mertl Millhollen who provided criticisms and improved the manuscript.

References

Asquith, T. N., and Butler, L. G. (1985). Use of a dye-labeled protein as spectrophotometric assay for protein precipitants such as tannin. *J. Chem. Ecol.* 11:1535–1544.

Athanasiadou, S., Kyriazakis, I., Jackson, F., and Coop, R. L. (2001). Direct anthelmintic effects of condensed tannins towards different gastrointestinal nematodes of sheep:in vitro and in vivo studies. *Vet. Parasitol.* 99:205–219.

Berridge, K. C. (1991). Modulation of taste affect by hunger, caloric satiety, and sensory-specific satiety in the rat. *Appetite* 16:103–120.

Bouquet, A., Debray, M.-M., Dauguet, J.-C., Girre, A., Leclair, J.-F., Le Naour, M., and Patay, R. (1967). A propos de l'action pharmacologique de l'écorce de *Combretodendron africanum* (Welw) Exell et particulièrement de son pouvoir abortif et perturbateur du cycle estral. *Thérapie*, XXII:325–336.

Cabre-Vert, N. and Feistner, A. T. C. (1995). Comparative gut passage time in captive lemurs. *Dodo, J. Wildl. Preserv. Trusts* 31:76–81.

Campbell, J. L., Eisemann, J. H., Williams, C. V., and Glenn, K. M. (2000). Description of the gastrointestinal tract of five lemur species:*Propithecus tattersalli, Propithecus verreauxi coquereli, Varecia variegata, Hapalemur griseus,* and *Lemur catta*. *Am. J. Primatol.* 52:133–142.

Carrai, V., Borgognini-Tarli, S. M., Huffman, M. A., and Bardi, M. (2003). Increase in tannin consumption by sifaka (*Propithecus verreauxi verreauxi*) females during the birth season:a case of self-medication in prosimians? *Primates* 44:61–66.

Coley, P. D., Bryant, J. P., and Chapin III, F. S. (1985). Resource availability and plant antiherbivore defense. *Science* 230:895–899.

Commission Nationale de Pharmacopée (2002). *Pharmacopée Française*, 10th ed. Agence Française de Sécurité Sanitaire de Produits de Santé, Saint Denis.

Critchley, H. D., and Rolls, E. T. (1996). Responses of primate taste cortex neurons to the astringent tastant tannic acid. *Chem. Senses* 21:135–145.

Danilova, V., Hellekant, G., Roberts, T., Tinti, J.-M., and Nofre, C. (1998). Behavioral and single chorda tympani taste fiber responses in the common marmoset, *Callithrix jacchus jacchus*. *Ann. N. Y. Acad. Sci.* 855:160–164.

Ehrlich, P., and Raven, P. H. (1964). Butterflies and plants:a study in coevolution. *Evolution* 18:586–608.

Faurion, A. (1988). Naissance et obsolescence du concept de quatre qualités en gustation. *J. Agr. Trad. Bot. Appl.* 35:21–40.

Feeny, P. (1976). Plant apparency and chemical defense. *Recent Adv. Phytochem.* 10:1–40.

Ganzhorn, J. U. (1986a). The influence of plant chemistry on food selection by *Lemur catta* and *Lemur fulvus*. In: Else, J. G., and Lee, P. C. (eds.), *Primate Ecology and Conservation*, 2. Cambridge University Press, Cambridge, pp. 21–29.

Ganzhorn, J. U. (1986). Feeding behavior of *Lemur catta* and *Lemur fulvus*. *Int. J. Primatol.* 7:17–30.

Ganzhorn, J. U. (1995). Low-level forest disturbance effects on primary production, leaf chemistry, and lemur populations. *Ecology* 76:2084–2096.

Glaser, D. (1968). Geschmacksschwellenwerte bei Callitrichidae (Platyrrhina). *Folia Primatol.* 9:246–257.

Glaser, D. (1977). Geschmacksleistungen bei nachtaktiven Primaten. *Z. Morph. Anthrop.* 68:241–246.

Glaser, D. (1986). Geschmacksforschung bei Primaten. *Vjschr. Naturf. Ges. Zürich* 131/2:92–110.

Glaser, D., and Hellekant, G. (1977). Verhaltens und electrophysiologische Experimente über den Geschmackssinn bei *Saguinus midas tamarin* (Callitrichidae). *Folia Primatol.* 28:43–51.

Gould, L., Sussman, R. W., and Sauther, M. L. (1999). Natural disasters and primate populations:the effect of a 2-year drought on a naturally occurring population of ringtailed lemurs (*Lemur catta*) in Southwestern Madagascar. *Int. J. Primatol.* 20:69–84.

Hellekant, G., Danilova, V., and Ninomiya, Y. (1997a). The primate sense of taste:behavioral and single chorda tympani and glossopharyngeal nerve fiber recordings in the rhesus monkey. *J. Neurophysiol.* 77:978–993.

Hellekant, G., Hladik, C. M., Dennys, V., Simmen, B., Roberts, T. W., Glaser, D., DuBois, G., and Walters, D. E. (1993). On the sense of taste in two Malagasy primates (*Microcebus murinus* and *Eulemur mongoz*). *Chem. Senses* 18:307–320.

Hellekant, G., Ninomiya, Y., and Danilova, V. (1997b). Taste in chimpanzees. II: Single chorda tympani fibers. *Physiol. Behav.* 61:829–841.

Hellekant, G., Ninomiya, Y., and Danilova, V. (1998). Taste in chimpanzees. III: Labeled-line coding in sweet taste. *Physiol. Behav.* 65:191–200.

Herrera, C. M. (1985). Determinants of plant-animal coevolution: The case of mutualistic dispersal of seeds by vertebrates. *Oikos* 44:132–141.

Hladik, A. (1980). The dry forest of the west coast of Madagascar:climate, phenology, and food available for Prosimians. In: Charles-Dominique, P., Cooper, H. M., Hladik, A., Hladik, C. M., Pagès, E., Pariente, G. F., Petter-Rousseaux, A., Petter, J. J., and Schilling, A. (eds.), *Nocturnal Malagasy Primates. Ecology, Physiology, and Behavior*. Academic Press, New York, pp. 3–40.

Hladik, A., and Hladik, C. M. (1977). Signification écologique des teneurs en alcaloïdes des végétaux de la forêt dense:résultats des tests préliminaires effectués au Gabon. *Rev. Ecol. (Terre Vie)* 31:515–555.

Hladik, C. M. (1977). A comparative study of the feeding strategies of two sympatric species of leaf monkeys:*Presbytis senex* and *Presbytis entellus*. In: Clutton-Brock, T. H. (ed.), *Primate Ecology: Studies of Feeding and Ranging Behaviour in Lemurs, Monkeys and Apes*. Academic Press, London, New York, pp. 323–353.

Hladik, C. M. (1993). Fruits of the rain forest and taste perception as a result of evolutionary interactions. In: Hladik, C. M., Hladik, A., Linares, O. F., Pagezy, H., Semple, A., and Hadley, M. (eds.), *Tropical Forests, People and Food: Biocultural Interactions and Applications to Development*. UNESCO-Parthenon Publishing Group, Paris, pp. 73–82.

Hladik, C. M., and Gueguen, L. (1974). Géophagie et nutrition minérale chez les primates sauvages. *C. R. Acad. Sc. Paris* 279:1393–1396.

Hladik, C. M., Pasquet, P., Danilova, V., and Hellekant, G. (2003). The evolution of taste perception:psychophysics and taste nerves tell the same story in human and non human primates. *C. R. Palevol.* 2:281–287.

Hladik, C. M., Pasquet, P., and Simmen, B. (2002). New perspectives on taste and primate evolution:the dichotomy in gustatory coding for perception of beneficent versus noxious substances as supported by correlations among human thresholds. *Am. J. Phys. Anthrop.* 117:342–348.

Hladik, C. M., Simmen, B., Ramasiarisoa, P., and Hladik, A. (2000). Rôle des produits secondaires (tannins et alcaloïdes) des espèces forestières de l'est de Madagascar face aux populations animales. In: Lourenço, W., and Goodman, S. (eds.), *Diversité et endémismes à Madagascar*, Mémoires de la Société de Biogéographie, Paris, pp. 105–114.

Howe, H. F., and Westley, L. C. (1988). *Ecological Relationships of Plants and Animals*. Oxford University Press, New York.

Iaconelli, S., and Simmen, B. (1999). Palatabilité de l'acide tannique dans une solution sucrée chez *Microcebus murinus*:variation saisonnière et implication dans le comportement alimentaire. *Primatologie* 2:421–434.

Iaconelli, S., and Simmen, B. (2002). Taste thresholds and suprathreshold responses to tannin-rich plant extracts and quinine in a primate species (*Microcebus murinus*). *J. Chem. Ecol.* 28:2315–2326.

Janzen, D. H. (1978). Complications in interpreting the chemical defenses of trees against tropical arboreal plant-eating vertebrates. In: Montgomery, G. G. (ed.), *The Ecology of Arboreal Folivores*. Smithsonian Institution Press, Washington, D.C., pp. 73–84.

Johns, T., and Duquette, M. (1991). Detoxification and mineral supplementation as functions of geophagy. *Am. J. Clin. Nutr.* 53:448–456.

Laska, M. (2000). Gustatory responsiveness to food–associated sugars and acids in pigtail macaques, *Macaca nemestrina*. *Physiol. Behav.* 70:495–504.

Lebreton, P. (1982). Tannins ou alcaloïdes: deux tactiques phytochimiques de dissuasion des herbivores. *Rev. Ecol. (Terre Vie)* 36:539–572.

Mowry, C. B., McCann, C., Lessnau, R., and Dierenfeld, E. (1997). Secondary compounds in foods selected by free-ranging primates on St. Catherines Island, GA. *Proceedings of the second conference of the nutrition advisory group/American zoo and aquarium association on zoo and wildlife nutrition*, Pap 5, 1–8 (Pp. Sess 2).

Oates, J. F., Whitesides, G. H., Davies, A. G., Waterman, P. G., Green, S. M., Dasilva, G. L., and Mole, S. (1990). Determinants of variation in tropical forest primate biomass: New evidence from West Africa. *Ecology* 71:328–343.

Overdorff, D. J., and Rasmussen, M. A. (1995). Determinants of nighttime activity in "diurnal" lemurid primates. In: Alterman, L., Doyle, G. A. and Izard, M. K. (eds.), *Creatures of the Dark: The Nocturnal Prosimians*, Plenum Press, New York, pp. 61–74.

Porter, L. J., Hrstich, L. N., and Chan, B. C. (1986). The conversion of procyanidins and prodelphinidins to cyanidin and delphinidin. *Phytochemistry* 25:223–230.

Price, M. P., and Butler, L. G. (1977). Rapid visual estimation and spectrophotometric determination of tannin content of *Sorghum* grain. *J. Agric. Food Chem.* 25:1268–1273.

Rasamimanana, H. (2004). *La dominance des femelles makis* (Lemur catta)*:quelles stratégies énergétiques et quelle qualité de ressources dans la réserve de Berenty, au sud de Madagascar?* PhD diss., Thesis of the National Museum of Natural History, Paris, France.

Rasamimanana, H. R., and Rafidinarivo, E. (1993). Feeding behavior of *Lemur catta* females in relation to their physiological state. In: Kappeler, P. M. and Ganzhorn, J. U. (eds.), *Lemur Social Systems and their Ecological Basis*, Plenum Press, New York, pp. 123–133.

Remis, M. J., and Kerr, M. E. (2002). Taste responses to fructose and tannic acid among gorillas (*Gorilla gorilla gorilla*). *Int. J. Primatol.* 23:251–261.

Rolls, B. J., Rolls, E. T., Rowe, E. A., and Sweeney, K. (1981). Sensory specific satiety in man. *Physiol. Behav.* 27:137–142.

Sauther, M. L. (1998). Interplay of phenology and reproduction in ring–tailed lemurs:implications for ring-tailed lemur conservation. *Folia Primatol.* 69(Suppl. 1): 309–320.

Sauther, M. L., Sussman, R. W., and Gould, L. (1999). The socio-ecology of the ringtailed lemur:thirty–five years of research. *Evol. Anthropol.* 8:120–132.

Schiffman, S. S., and Erickson, R. P. (1971). A psychophysical model for gustatory quality. *Physiol. Behav.* 7:617–633.

Schiffman, S. S., and Erickson, R. P. (1980). The issue of primary tastes versus a taste continuum. *Neurosci. Biobehav. R.* 4:109–117.

Scott, T. R., and Mark, G. P. (1987). The taste system encodes stimulus toxicity. *Brain Res.* 414:197–203.

Scott, T. R., Yaxley, S., Sienkiewicz, Z. J., and Rolls, E. T. (1986). Gustatory responses in the frontal opercular cortex of the alert cynomolgus monkey. *J. Neurophysiol.* 56:876–890.

Simmen, B. (1994). Taste discrimination and diet differentiation among New World primates. In: Chivers, D. J., and Langer, P. (eds.), *The Digestive System in Mammals: Food, Form and Function*, Cambridge University Press, Cambridge, pp. 150–165.

Simmen, B. (2004). Perception gustative des substances de défense chimique des végétaux: adaptation et plasticité des réponses des primates. *Primatologie* 6:149–170.

Simmen, B., and Charlot, S. (2003). Une étude comparative des seuils gustatifs de substances sucrée et astringente chez les singes anthropoïdes. *C. R. Biol.* 326:449–455.

Simmen, B., and Hladik, C. M. (1998). Sweet and bitter taste discrimination in primates:scaling effects across species. *Folia Primatol.* 69:129–138.

Simmen, B., Josseaume, B. and Atramentowicz, M. (1999). Frugivory and taste responses to fructose and tannic acid in a prosimian primate and a didelphid marsupial. *J. Chem. Ecol.* 25:331–346.

Simmen, B., Hladik, A., Hladik, C. M., and Ramasiarisoa, P. L. (2003a). Occurrence of alkaloids and phenolics in Malagasy forests and responses by primates. In: Goodman, S. M., and Bensted, J. P. (eds.), *The Natural History of Madagascar*. The University of Chicago Press, Chicago, pp. 268–271.

Simmen, B., Hladik, A., and Ramasiarisoa, P. L. (2003b). Food intake and dietary overlap in native *Lemur catta* and *Propithecus verreauxi* and introduced *Eulemur fulvus* at Berenty, Southern Madagascar. *Int. J. Primatol.* 5:949–968.

Simmen, B., Tarnaud, L., Bayart, F., Hladik, A., Thiberge, A.-L., Jaspart, S., Jeanson, M., and Marez, A. (2005). Richesse en métabolites secondaires des forêts de Mayotte et de Madagascar et incidence sur la consommation de feuillage chez deux espèces de lémurs (*Eulemur* spp.). *Rev. Ecol. (Terre Vie)* 60:297–324.

Steiner, J. E. (1977). Facial expressions of the neonate infant indicating the hedonics of food–related chemical stimuli. In: Weiffenbach, J. M. (ed.), *Taste and Development: The Genesis of Sweet Preference*. U.S. Department of Health, Education and Welfare, Bethesda, MD, pp. 173–189.

Steiner, J. E., and Glaser, D. (1984). Differential behavioral responses to taste stimuli in nonhuman Primates. *J. Hum. Evol.* 13:709–723.

Struhsaker, T. T., Cooney, D. O., and Siex, K. S. (1997). Charcoal consumption by Zanzibar red colobus monkeys: its function and its ecological and demographic consequences. *Int. J. Primatol.* 18:61–72.

Sussman, R. (1974). Ecological distinctions in sympatric species of *Lemur*. In: Martin, R. D., Doyle, G. A., and Walker, A. C. (eds.), *Prosimian Biology*. Duckworth, London, pp. 75–108.

Swain, T. (1979). Tannins and lignins. In: Rosenthal, G. A., and Janzen, D. H. (eds.), *Herbivores, Their Interaction with Secondary Plant Metabolites*. Academic Press, New York, pp. 657–682.

Tarnaud, L., Simmen, B., Marez, A., Thiberge, A.-L., Jeanson, M., and Hladik, A. (2004). Perception et choix des aliments par le jeune *Eulemur fulvus* en fonction des métabolites secondaires présents dans l'environnement et du comportement alimentaire de la mère. *Primatologie* 6:129–148.

Yamashita, N. (2002). Diets of two sympatric lemur species in different microhabitats in Beza Mahafaly Special Reserve, Madagascar. *Int. J. Primatol.* 23:1025–1051.

Part III
Social Behavior Within and Between Troops

12
Territory as Bet-hedging: *Lemur Catta* in a Rich Forest and an Erratic Climate

ALISON JOLLY, HANTANIRINA RASAMIMANANA, MARISA BRAUN,
TRACY DUBOVICK, CHRISTOPHER MILLS, AND GEORGE WILLIAMS

12.1. Introduction

Ringtailed lemurs (*Lemur catta*) impress scientists by differences in behavior between Berenty Reserve and the Beza Mahafaly Special Reserve. At Berenty, overt aggression between troops serves to defend the resources of circumscribed territories. At Beza there is indeed between-troop aggression as well as core areas for each troop, but intertroop encounters are less frequent and do not seem to mark out definable frontiers. A second difference is that Gould, Sussman, and Sauther (Gould et al., 1999) see the Beza population as r-selected, having watched the population crash by 40% during a major drought but then recover in 6 years to predrought levels. Observers at Berenty have been much more prone to fix on stability of range and even of population (Mertl-Millhollen, 1979; Jolly and Pride, 1999; Mertl-Millhollen, 2000).

The differences reflect the richer resources and higher lemur densities at Berenty, both within the tamarind-dominated gallery forest and recently in the super-provisioned Front and Ankoba zones. (See Jolly et al, this volume, for a description of Berenty as a study site.) However, the scientists' interpretations also carry heftier theoretical (or even philosophical) implications. Is the territorial aggression shown at Berenty evolved natural behavior or is it a pathological response to overcrowding? Has rich gallery forest in the past been such an important component of ringtail habitat that there would have been selective pressure to evolve behavior like that shown at Berenty? Does territoriality in rich habitat function to the lemurs' advantage, and if so, how?

The articles by Goodman and Wilmé and by Sussman and colleagues in this volume make it clear that most ringtail range seems to be in very dry regions where lemurs live at low population density. However, although gallery forest habitat is rare, gallery forests function as source populations of lemurs and many other species for the drier areas around. Rivers seem to have provided the major forest corridors for colonizing new range (Sussman et al., 2003; Goodman, this volume) Thus we can speculate that gallery forest behavior is important at the high end of the spectrum of natural lemur densities.

This paper falls into three parts. First, demography: the effect of the erratic climate of southern Madagascar on population, fertility, and 1-year survival in rich, poor, and super-rich habitats. In these habitats life-history response to environmental variability underlies year-to-year territoriality, ranging patterns and troop history.

Second, how is territoriality manifested in Berenty's range of habitats? Does the outcome of observed troop fissions suggest an advantage to long-term territoriality?

Finally, we suggest that territory which is conserved over generations is analogous to bet-hedging or K-selected life history: a means of bridging over bad times with competitive advantage to successful survivors, rather than success to fast reproducers and recolonizers. This has implications for the conservation of dense but bounded populations in forest fragments.

12.2. Demography

12.2.1. Methods

Methods are described in Jolly et al. (2002) and earlier publications. Censusing involves walking trails until a group is encountered, recognizing it by distinctive "marker" individuals, and following it until an apparently good count is obtained, usually when the troop walks down a path with their tails in the air. Most troops are counted on at least 4 different days; the clearest and most complete count is retained for the census. In early years we also located troops by transects between trails, or by switching troops when there was an encounter between troops off-trail, but on following these troops they always emerged to walk down a trail. Most Front and Gallery troops are known from year to year; Scrub troops are less well known.

From 1990 to 2000, there were counts of all Front and Gallery troops and from 1994 to 2000 the Scrub troops, with the aid of large numbers of students and Earthwatch volunteers. From 2001 to 2004, only a sample of troops were counted, (although in 2004 there was also a full count in Ankoba,) giving indicative data on juveniles per female. A full census was conducted again in 2005 (Table 12.1; see also Figure 3.2 in Jolly et al., chapter on Berenty Reserve, this volume).

A parallel study by Koyama and his students has followed the individuals of 7 troops at the Ankoba/Front interface since 1990 (Koyama et al., 2001, 2002). Their sample in part overlaps with the Malaza study, including C1, C2A and C2B, YF which is derived from A2 (all Front troops) and CX (a Gallery troop). They provide detailed reproduction and survival data from the super-rich forest area as compared with the broad-brush troop-based data of Front, Gallery, and Scrub.

12.2.2. Results

Figure 12.1 and Table 12.1 give population by habitat zone, 1974–2000. They should be compared with the yearly fluctuations in rainfall, Figure 3.4 in Jolly et al., this volume.

TABLE 12.1. Counts of 1-km Malaza study area.

	Front		Gallery		Scrub		Totals	
	Troops	N	Troops	N	Troops	N	G+S	F+G+S
1974	4	68	3	33	5	52	85	153
1975	4	57	3	37	5	61	98	155
1983	4	41	3	36	5	32	68	109
1984	4	46	3	40	5	39	79	125
1985	4	45	3	31	5	29	60	105
1989	4	53	3	44	1			
1990	5	58	5	58	1			
1991	5	86	5	73	5	49		
1992	6	92	5	64	6	80	144	236
1993	6	98	6	72	7	87	159	257
1994	7	109	7	67	7	53	120	229
1995	7	99	7	80	7	57	137	236
1996	7	101	7	97	7	65	162	263
1997	9	116	7	95	7	70	165	281
1998	9	123	8	80	8	59	139	262
1999	9	119	8	83	8	64	147	266
2000	11	138	9	78	6	46		
2001	8	84	6	46	3	19		
2002	7	109	8	62	7	66		
2003	7	106	7	65	6	53		
2004	9	127	5	40	6	46		
2005	10	152	7	50	8	90	140	292

Shaded cells: partial censuses. White cells: probable complete censuses. 1974: Budnitz and Dainis; 1983–1985, O'Connor and Pigeon; 1989–2004, Jolly, Rasamimanana et al.

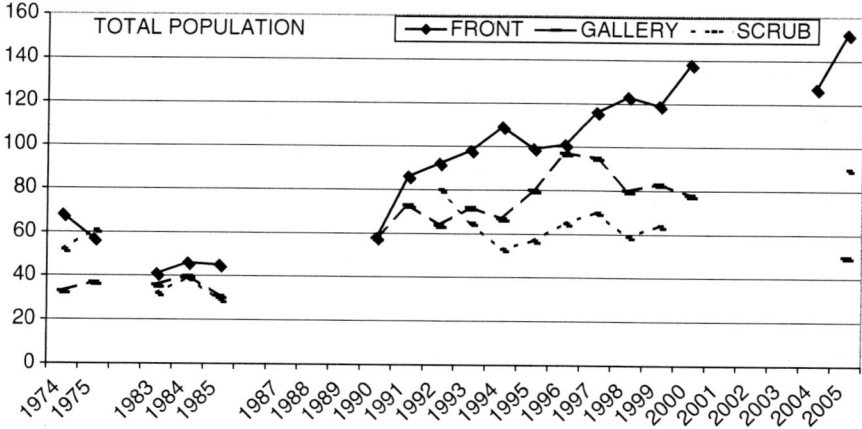

FIGURE 12.1. Total non-infant *Lemur catta* population for Malaza regions. Front population has increased 2.6 times since 1990. Scrub population outnumbered gallery population in years of high food stress: the drought and postdrought of 1992–1993, and during the forest-wide tamarind fruit failure of 2005. Total of scrub plus gallery has been relatively constant since 1990, averaging 146 non-infants.

The 1974–1975 total population of about 150 in the 100-ha Malaza study area seemed much the same density as in 1963–1964, though few troops were counted in the earlier decade (Budnitz and Dainis, 1975; Mertl-Millhollen et al., 1979). The lower density in the 1980s may have resulted both from the severe drought of 1983–1984 and just possibly from hunting during 1979 when the owner was absent from the reserve (O'Connor, 1987; Jolly, 2004) In the 1990s, the major population falls were in 1994 and 1998, respectively following the major droughts of 1991–1992 and 1997. The Scrub zone suffered worst, dropping 37% between 1992 and 1994, essentially the same degree as the 43% drop in total censused population at Beza during the same years. However, the Front troops at Berenty were buffered from the effects of bad years and grew fairly steadily. Gallery troops were intermediate between the buffered Front and vulnerable Scrub.

A major wind storm on October 2, 1999, damaged a quarter of the forest trees (Rasamimanana et al., 2000), in the same year that banana feeding was banned. Soon thereafter the increased consumption of planted leucaena trees began to turn from a protein benefit to a toxicosis (Crawford et al, this volume). Furthermore the brown lemur population presents increasing competition in all areas (Pinkus, this volume). Thus from 2000 on, the Front troops lost much of their buffering from environmental stress.

In 2005 there was a reserve-wide failure of tamarind fruit, normally the major component of the birth season diet. Many troops still slept in or near traditional territories, but ranged outward to the periphery, to feed on sisal flowers in fields far beyond the forest front, or in the spiny forest, or on a few remaining tamarinds outside the reserve. Lemurs observed in scrub outnumbered those in the gallery for the first time since the drought of 1992.

Looking at differences by age and sex (Figures 12.2 to 12.4) it is the adult and subadult females who remain within their troops and territories. Their population was fairly constant in the scrub, rose steadily in tourist, and fluctuated most widely in the gallery. The fall between 1991 and 1992 in the gallery forest reflects the loss of 10 females, five of them from one troop, which probably was a troop fission. Only two of the five were found again having fought their way in as dominants in a scrub troop. Some of the others may have survived, migrating out of the study area. If they died instead, this is a 21% mortality, comparable to the 20% mortality at Beza in the same period. The sharp rises in gallery female numbers, however, reflect cohorts of maturing female juveniles. not immigration.

Males do migrate, beginning during the October birth season when censuses are done. Males loosely associated with troops are included in Figures 12.2 to 12.4, but males seen only between troops are not. Sex ratios were equal in the front area, consistently female-biased in gallery forest, and fluctuated wildly in scrub.

Finally the number of 1-year-olds sums up the previous year's influence on both birth rate and survival. Figure 12.5, proportion of 1-year-olds to females, makes it clear that years of "lost cohorts" were 1985, 1993–1994, 1998–1999, and 2002. Each of these followed, rather than occurring within, years of drought (see Figure 3.4, yearly rainfall, in Jolly et al, this volume). This graph differs from

12. Territory as Bet-hedging   191

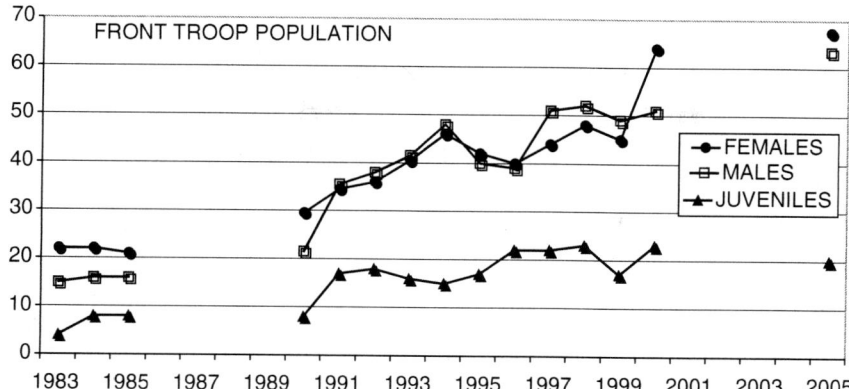

FIGURE 12.2. Non-infant population of front troops by age and sex. "Females" includes adult and subadult (2- and 3-year-old) females, with subadults of unknown sex assigned equally between males and females. Females rarely leave their troops, or if exiled their immediate region, so numbers reflect recruitment and mortality. Males includes adults and subadults as well as males loosely attached to troops while immigrating or emigrating, but not males that were seen wholly separate from troops. Juveniles refers to 1-year-olds only. Front troops had equal sex ratios taking the region as a whole and little fluctuation in juvenile numbers.

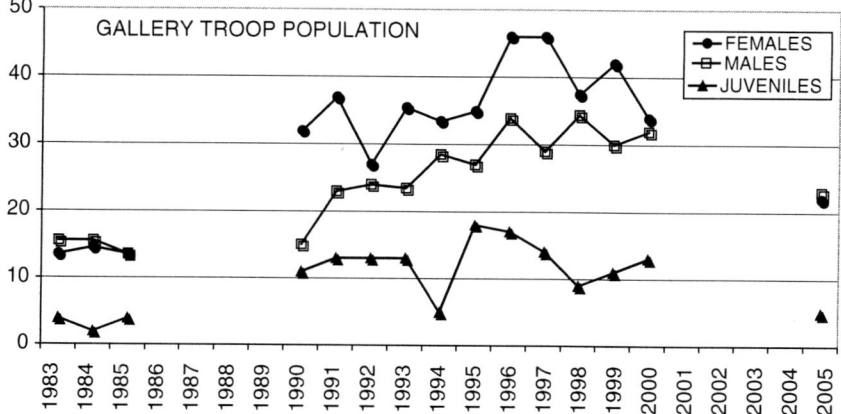

FIGURE 12.3. Gallery troop population by age and sex, as for Figure 12.2. The fall in females between 1991 and 1992, and the low numbers in 2005, may in part reflect daily movements toward scrub; the various rises reflect maturing female juveniles. Males consistently migrated either between troops or out of the region. Juvenile recruitment was erratic.

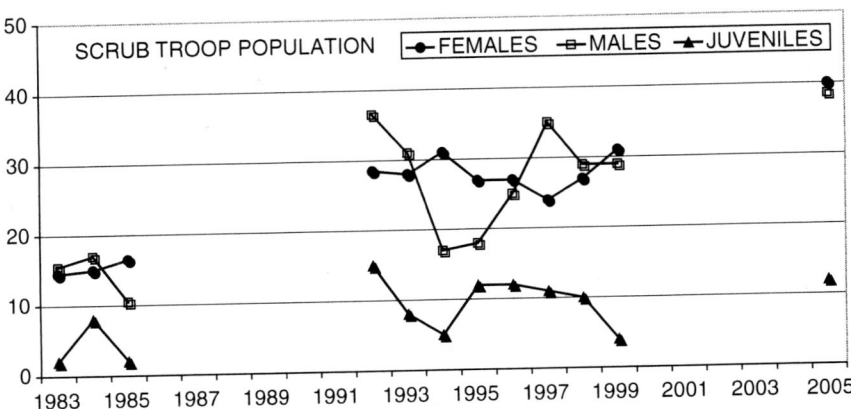

FIGURE 12.4. Scrub troop population by age and sex as for Figure 12.2. Juvenile and male numbers fluctuate even more widely than in gallery forest. Females remained more stable until the in-migration of 2005.

the one in (Jolly et al., 2002), which showed juveniles per adult female of the same troops of the year before, a more accurate picture of survival, and which makes it clear that the lows may be nearer 10% or 5% of females with 1-year-olds, not 20%. However, comparing juveniles per female of the same year allows inclusion of the 2001–2004 data where different years did not necessarily allow counts of the same troops. For juvenile survival, the Front troops do not seem to be buffered as they

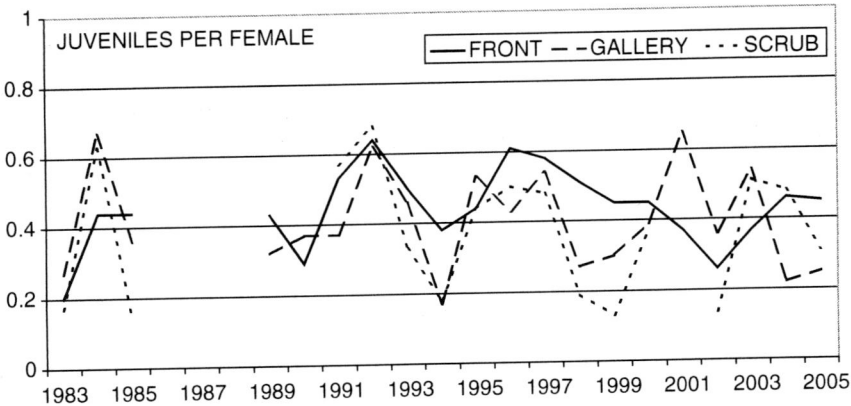

FIGURE 12.5. One-year-old juveniles per female of the same year, reflecting the sum of birth rate and survival. Low years were 1993–1994, 1998–1999, 2002, and 2005, each following (not during) drought years. Infant survival to 1 year at Beza is also around 50% in normal years but only 20% in the drought and postdrought years of 1992–1993.

are for births and for overall population (mainly adult) survival, especially in 2002. Finally, at Beza, which had suffered the same drought, infant mortality was 80% in 1992, which means a lost cohort of 1-year-olds in 1993: survivorship in 1994 is unknown. Survivorship of 1-year-olds in normal years at Beza is around 50% (Gould et al., 2003).

Figure 12.6 shows adult survivorship from Koyama et al. (2002) for the known females of Ankoba/Front junction, in comparison with sifaka from Beza and Ranomafana. Unfortunately there is no survivorship curve for Gallery or Scrub. Figure 12.7 compares ringtailed lemur survivorship in the wild with the greater longevity and different shaped curve of ringtails in European zoos.

12.2.3. Discussion

Southern Madagascar is among the top 25% of the world's most erratic rainfall climates (Dewar and Wallis, 1999). A swathe of variable weather reaches from Mauritius and Reunion islands through southern Madagascar including Berenty and Beza to the coast of Mozambique. These countries are alternately hit by cyclones that bring torrential downpour and flooding and by El Niño droughts that may effectively eliminate the wet season and the plant growth that depends on it. Some of the trees in western Malagasy forests may even depend on

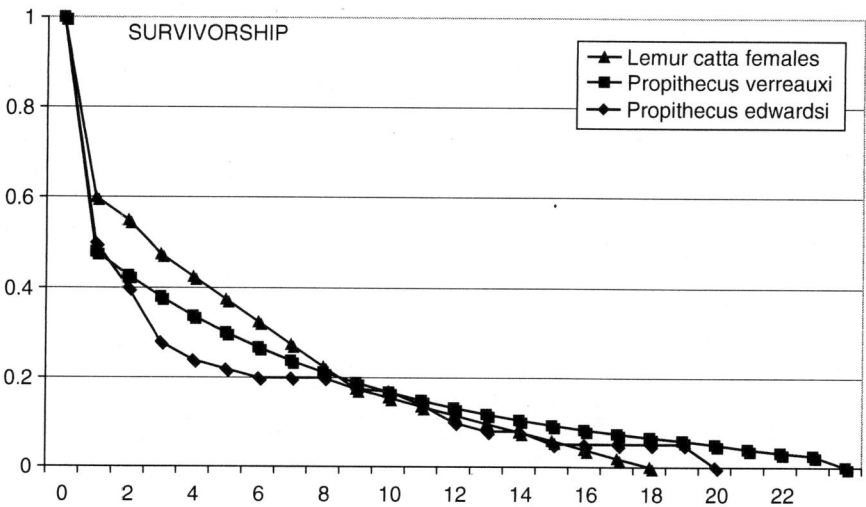

FIGURE 12.6. Survivorship: the curve for the first 9 years of Berenty females at the Front/Ankoba junction is higher and steeper, but not much steeper than the "slow lane" *Propithecus verreauxi* of Beza, or the similarly bet-hedging *Propithecus edwardsi* of Ranomafana. First 9 years of female *L. catta* data from Koyama et al. (2002), extrapolated to 18, which is maximum known age in the wild. Female *P. verreauxi*, Beza Mahafaly, from Richard et al. (2002). Female *P. edwardsi* at Ranomafana from Pochron et al. (2004).

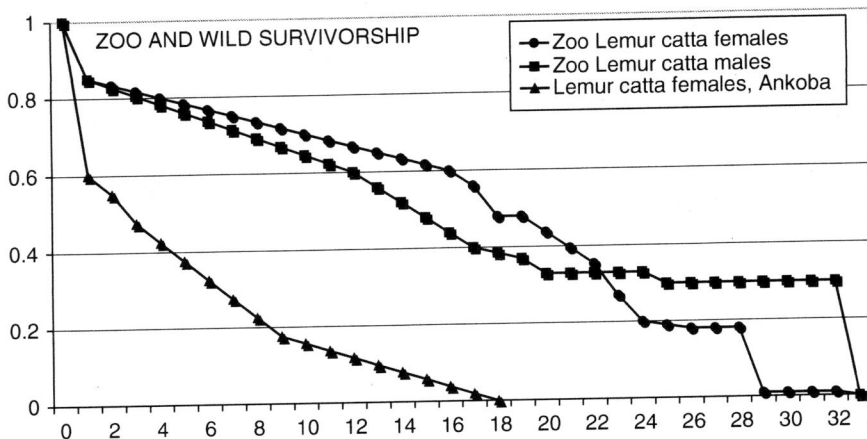

FIGURE 12.7. *L. catta* female survivorship from the European Zoo Studbook is an almost perfect straight line, typical of an r-selected species, except for first-year mortality and extreme old age. Females of Berenty's Front/Ankoba zones, which are among the most favored of wild populations, showed a more concave or "bet-hedging" survival curve. It is not known if more stressed populations have more or less concave curves. Zoo data courtesy of Elena Baistrocchi (pers. comm.); wild data from Koyama et al. (2002).

cyclones for enough water to germinate, and for light gaps that allow growth (Ganzhorn, 1995).

Berenty's region, Androy, has seen major droughts about every ten years, with the worst ones bringing significant loss of human life: 1932, 1943, 1984, and 1991–1992 (Jolly, 2004). Droughts in better economic times, such as 1970, and 1997 are buffered for the human population but not for plants and animals. The frequency is such that a ringtailed lemur with a life span of 10–15 years is almost certain to encounter one such drought, and a third of them likely to encounter two. Trees with lifespans in centuries like the tamarinds, the baobabs, and the *Alluaudia* of the spiny forest will encounter the most severe "drought of the century." It does not seem that drought itself is the crucial limit for the Berenty lemurs, but instead shortage of tamarind fruit (Koyama et al., this volume; Mertl-Millhollen et al., this volume.) During drought years, the tamarinds tend to bear fruit—a local Tandroy proverb says "The tamarinds take care of us during famine." However, one or two years later there seems to be a fruit failure with unknown triggers. The most serious such failure was in 2005, following cyclone-caused flooding and then total lack of rain in March–June, with the result that troops foraged outside the main forest.

Two major articles argue that a climate prone to catastrophe has shaped the evolution of the Malagasy fauna. Wright's (Wright, 1999) concept of "Energy Frugality Strategy" links erratic climate (and seasonality and poor soils) to a whole suite of lemur traits: female dominance, territoriality, and specifically, bet-hedging life style. The survivorship curves of Milne-Edward's sifaka, *Propithecus edwardsi*, shown in Figure 12.6 is strongly concave. Concave survivorship is usually an indication of "K selection" where juvenile mortality may be high but adults survive and compete in a fairly stable environment with low adult mortality. It also typifies "bet-hedging," where the environment is far from stable, but the population responds by a conservative life style, with slow growth, low reproduction and relatively high adult survival even through crises (Stearns, 1992; Pochron et al., 2004).

Richard et al's "Life in the Slow Lane" explores the bet-hedging strategy of white sifaka, *Propithecus verreauxi* at Beza in more detail (Richard et al., 2002). White sifaka females do not reach full fertility until 6 years of age, and may live more than 20 years, an extreme of slow life history for a primate of their 2.8-kg body weight.

Eisenberg (1981) had earlier pointed out an oddity about most mammalian life histories on Madagascar. The majority of mammals, including small rodents and carnivores, have only one young per year, extraordinary restraint for animals of their orders and their body size. If the litters are larger they are very much larger. Among those that opt for large litters are ruffed lemurs. *Varecia* are by far the largest primate that regularly twins, with quintuplets in captivity. The fossa *Cryptoprocta ferox* is the only Malagasy carnivore with several cubs, usually four at once. The tenrec *Tenrec ecaudatus* sets the mammalian record with a litter of 35. The summary table on p. 1174 in Goodman and Benstead's *Natural History of Madagascar* confirms Eisenberg's insight that for most other Malagasy mammals, litter size is low for body weight (Goodman et al., 2003).

The crucial question is the frequency of climatic catastrophes in relation to life span. If they occur on a multigenerational scale, the pressure is toward r-selection with build-up of total population, and then adult mortality when the population crashes. If they are as frequent as in Madagascar, most lemur adults must survive, while accepting the loss of cohorts of juveniles. The demographic pyramid of Richard's known-aged *Propithecus* shows missing cohorts from the bad years, like Figure 12.5 for the ringtails, and indeed like historic life tables of the Tandroy people of the region (Frère, 1958). *Lemur catta* are about on the edge of this divide. If extremely bad years come more often than once a decade, they should apparently veer toward bet-hedging; if less often, toward r-selection.

Gould observed 21% adult female mortality, 80% infant mortality, and the deaths of 4 of 7 juveniles, in and after the 1991–1992 drought at Beza Mahafaly, followed by a year of 100% birth rate. She concludes that as lemurs go ringtails are markedly r-selected, certainly so in comparison with Richard's white sifaka in the same reserve. At Berenty there was a similar population fall in the scrub, though much less in the gallery forest, and repeated loss of cohorts in both. In good years and areas, more than 70% of females gave birth; in one case 90%. In 1995 a Gallery female had twins; in 2001 a Front female, of which all survived (and see Koyama, 1992). In short, the parameters are the same in Beza and the Scrub area of Berenty, and there is no disagreement that ringtails are highly r-selected in comparison with sifaka. This is even more striking if one looks at the survivorship curves in captivity. The European Zoo Studbook shows a 10% infant mortality followed by an almost straight line fall in female survival from 90% at age 1 to negligible at age 29, with a few grandes dames persisting to 32, so in near-ideal conditions ringtails have a nearly ideal r-selected survival curve (Baistrocchi, pers. comm.)

However, if one changes one's frame of reference to include other primates or other small mammals outside of Madagascar, a single infant per year, full fertility at 4, and a life span of more than 15 years in the wild is more like K-selection/bet-hedging for a 2-kg animal. The survivorship curve of the Koyama groups in the most favored habitat during a period of rapid population growth is not so different in shape from the sifaka of Beza or Ranomafana. It is not clear whether in the poorer areas with no overall population growth the curve would become more concave like the bet-hedging sifaka curves, or steeper and more r-selected.

On the whole, the life-history strategy seems mixed, or indeed on the edge of the divide. Most ringtail adults survive major catastrophes that are indeed likely to occur within their life-times, even though at a high cost in the loss of a year or two of reproduction. They have an evolved "cautious" strategy of raising only one or very rarely two young even in good years. On the other hand, following catastrophes, lemurs like *L. catta*, the *E. fulvus* group and especially *Varecia* recover rapidly in comparison with indriids due to the high proportion of females who reproduce yearly and the potential to bear twins. It remains to be seen whether *L. catta* is even more r-selected than other lemurs, which one would expect from their dry, marginal habitat, but which pales beside the rapid population growth of *E. fulvus* at Berenty (Pinkus, this volume).

12.3. Territoriality

12.3.1. Methods

Day ranges for ringtailed lemur troops at Berenty are plotted on a grid of 25 m × 25 m quadrats originally devised by Koyama for Front/Ankoba troops. Williams' MAP computer program plots ranges, percent use polygons, point data and other information for 25 m × 25 m quadrats on his GPS-based map of Berenty (http://fontforge.sourceforge.net).

Observers followed each troop during the breeding seasons of 1995–1997, recording many parameters simultaneously, treating the troop as a whole, rather than focal individuals. The open forest of Berenty means that most individuals are in view if there are enough observers. Eight 12-hour day ranges were recorded per troop, spread over the changing phenology of September and October. In 1995, a team organized by Pinkus and Jolly recorded multiple data on matched *L. catta* and *E. fulvus* troops, with three watchers together per troop at all active times. In 1996–1997, pairs of students recorded troop data (Braun, 1996; Mills, 1997; Dubovick, 1998).

Troops are considered to meet when within 20-m range of each other. Beyond this range they show no behavioral reaction or altered movement. The first troop to leave beyond 20 m is defined as the loser, the other the winner, though meets may end as a draw. Observers recorded time each 10 minutes, grid square, individuals participating in aggression, maximum degree of aggression, and direction of retreat by winner and loser. Aggression both within and between troops is recorded as either 0, none; 1, staring and/or chewing at the opponents, which is easy to distinguish from the usual twitchy glances of ringtails; 2, moving feet to lunge, cuff at, or chase; and 3, physical contact including cuff, bite, grapple. Submission is similarly defined as 0, none; 1, look away, cringe; 2, retreat moving feet; 3, be cuffed, bitten, or grappled with physical contact. This scale is derived from the finer scale used by Pereira (see Pereira, this volume), but simplified to be used on whole troops during an encounter rather than just a focal individual.

Core area in this case is defined by a polygon including all quadrats with ≥2 female scent marks during the 8 days. Definitions of core area as 85% of time or of feeding time produces almost the same results, except in seasons when there is a long series of excursions outside the usual home range to feed on some scarce resource where there may be lowered scent marking (see Mertl-Millhollen, this volume).

12.3.2. Results

Table 12.2 gives results for 10 troops during the September–October breeding seasons of 1995–1997.

TABLE 12.2. Birth season intertroop encounters 1995–1997.

Troop	Area	Pop./ ha	Troop size	Meets/ day	Core meets	Core % wins	Peripheral meets	Peripheral % wins	Core vs. peripheral
SK 95	ANK	4	24	4.0	27	74%	3	33%	NS
A1 96	FRO	4	15	6.3	67	15%	9	0%	NS
A1 97	FRO	4	15	9.3	73	71%	30	70%	NS
A2 95	FRO	4	16	4.7	23	61%	17	65%	NS
A2 96	FRO	4	15	4.5	50	80%	6	67%	NS
A2 97	FRO	4	15	4.4	21	71%	8	50%	NS
D1 95	GAL	2.5	16	2.6	10	80%	3	0%	$p \leq 0.05$
VG 95	SCR	1.5	7	3.6	21	48%	12	0%	$p \leq 0.01$
YB 96	SCR	1.5	15	0.8	8	50%	5	20%	NS
SB 96	SCR	1	12	0.6	6	83%	0	—	—

1995: SK, A1, D1, VG: 8 full-day (8 × 12-hour) observations.
1996–1997: A2, A1, YB, and SB include additional meets seen on days beyond 8 full-day observations.
Meets per day is mean of 8 full-day (8 × 12-hour) observations.
Core defined by scent marking polygon enclosing quadrats with ≥2 scent marks.
Probability by Fisher's exact test.

12.3.3. Discussion

When does social repulsion between troops constitute territoriality? Kaufmann (1983) listed some 24 definitions of territoriality in use in the animal behavior literature. Even the *Oxford English Dictionary* is uncertain: "Ad. Latin territori-um: the land round a town . . . Etymology unsettled: a derivative of terra, earth . . . but the original form suggests terrere, to frighten, whence territorium, a place from which people are warned off." Thus the word itself has a double origin: either land reserved for exclusive use, or defense by the owner who "terrorizes" others. At Berenty and at Beza ringtail ranges have complete overlap, so the exclusive use definition does not hold. However they certainly attempt to defend their resources.

Ringtail intertroop behavior at Berenty might be summarized by three rules of thumb:

1. Dislike other troops. The proportion of troop encounters with level 2 (chasing) or level 3 (contact) aggression in the 1995–1997 data ranged from 60% to 100% per troop, but bore no relation to population density, though there were individual differences between pairs of troops. The spiny forest troop SB, with the lowest encounter rate, escalated all six encounters to contact aggression. Aggression occurs both within and far from the core: for instance in 2004 two troops had a encounter with high level chasing, each having traveled almost two km from their adjacent forest core areas to feed in the apparently endless expanse of a flowering sisal field.
2. Try to win encounters in your core area. Avoid or fight but be prepared to lose away from the core. This is crucial: it does assume that the troops know which is their core and expect to win there. Or looking at it another way, the core could be defined by the area where they generally win, rather than by high use or scent marking, and (see Takahata et al., 2005; Mertl-Millhollen, this volume). They may also show this knowledge by "going silent" (in Sauther's term) when in another troop's territory, and by precipitate retreat when they meet another troop on its home ground.
3. Let confrontation sites fall where they may. This is what reflects population density, as Dunbar (Dunbar, 1988) and Rubenstein (Rubenstein, 1981) explained in their theoretical models of costs and benefits of resource defense at different rates of incursion by others.

How to distinguish ownership of a contiguous area (territoriality) from ad hoc defense of particular food resources? Braun has suggested a series of criteria. First, day range length is larger than home range diameter, which means that a troop is able to patrol opposite edges of its range during a day (Mitani and Rodman, 1979). Not surprisingly this is true if the home range is determined by just 8 days of observation, but it is also highly characteristic in long-term Berenty studies, as pointed out for the gallery forest by Budnitz and Dainis (Budnitz and Dainis, 1975; Budnitz, 1978; see also Pride, this volume).

Second, encounters should be equally likely to occur at food sources or between them, proportional to the time spent at feeding or nonfeeding sites.

Third, there are equal rates of scent marking during encounters at food sources or encounters between food sources. Fourth, encounters are equally aggressive whether at food sources or between them. Fifth, scent marks should indicate core area borders (Mertl-Millhollen, 1988). Sixth, encounters should fall near core area borders rather than within the core. Finally, and crucially, encounters are won more in the high-use or high scent-marked core area, and lost more when peripheral to the core.

On most of these criteria in the 1995 study ringtails in all areas came out as territorial. Of course this could mean that any criterion that depends on finding no difference in measures might be falsified with larger sample size. The final two criteria varied between habitats. Only the Gallery and the Ankoba troop had encounters primarily on their borders. One Scrub troop and the Gallery troop won more in their cores than outside the core while the high-density troops showed no difference.

At low population densities, as in the Spiny forest or at Beza, few encounters happen and even fewer are seen. The defending troop may not patrol their whole range, and there are few incursions by others. There may be no frontier or one so weakly defended that an observer will not notice it.

At intermediate population density in the Gallery forest, there is more chance of troops meeting and thus more chance they will meet near the boundaries before the invading troop has penetrated the core. The gallery troop D1 and the troop based in scrub called VG that traveled into the gallery to drink had differences in their chance of winning confrontations in the core versus the periphery of their ranges. This produces classical territorial frontier zones. To observers at Berenty, the frontiers are locatable to within a few meters, often between specific trees, not just in a 25 m × 25 m quadrat, though observers often judge by subtle differences in demeanor rather than the rarer data from troop wins and losses. (Mertl-Millhollen, 1988, 2000).

At still higher populations, troops raid into each other's core areas. Total resource defense is difficult or impossible, particularly as two troops may invade from opposite borders at the same time. However, the "owners" continue to attempt defense, apparently because the expenditure of energy is still better than relinquishing the territory. With enough data observers can still determine frontiers by the distribution of wins and losses, though not always by which troop gains more of the resources (Takahata et al., 2005; Pride, this volume).

The Front troops also illustrate the complex interaction of troop history and territorial defense. In 1996, A1 lost consistently to A2. A2 had been A1's former subordinates, exiled in 1992, but which fought back and regained much of the parent troop range core (Hood and Jolly, 1995; Jolly, 2004). In 1997, A1 targeted and expelled a five-animal subgroup. Aggression toward that subgroup accounts for the high encounter rate and apparent success in winning, although A1 still lost to A2. A2 itself was a highly successful group in 1995 and 1996, but began to lose on its periphery in 1997 to a strong group, C1, which was expanding its range within the Tourist Front. Both aggression levels and overall success reflect particular troops' fortunes, not just stable territory.

In summary, territorial ownership appeared most clearly at the middle range of troop density at Berenty in the gallery forest, which is the richest known natural habitat. It is masked in the super-rich high density areas, where wins and losses may not correlate with ranging data and feeding success, even though wins and losses indicate quite sharp territorial frontiers. At the lowest density, the spiny forest troop rarely encountered any other. At such low density even year-long data (as in Pride, this volume) does not correlate wins and losses clearly with range occupancy. The full spectrum of resource defense outlined by Dunbar (Dunbar, 1988) is present at Berenty, with consistent differences between habitats and over years. This spectrum depends on a degree of aggression between troops that is consistent over all habitats and that at appropriate density serves as territorial defense.

It thus seems likely that territoriality is natural, evolved behavior, which would manifest itself in any ringtailed lemur troops living in gallery forest rich enough for moderately frequent encounter between troops.

12.4. Bequeathing Territory

The stability of troop ranges, and indeed of territorial frontiers at Berenty, has been a recurrent theme since Klopfer's observations of January 1969 (Klopfer and Jolly, 1970). He watched a troop that contained two marker animals, an old crop-eared male *L. catta* and the reserve's one *E. fulvus collaris*, both remaining from Jolly's 1963–1964 troop. The troop occupied only the eastern half of the 1963–1964 range. (Either this troop or its western sister group probably gave rise to the 1972–2004 D troops.) Klopfer pointed out that the northern, eastern, and southern range boundaries had been conserved during the 6 years: only on the west was the range divided. He saw hasty, silent flight from another troop, co-feeding and infant intertroop play with a third, but no intertroop aggression. He noted that the range boundaries seemed sharp, highly stable and that they fell in rich feeding areas, not at some externally given frontier. He suggested, then, that the boundaries were socially maintained. This conclusion has recurred in all subsequent studies (Budnitz, 1978; Mertl-Millhollen, 1979, 2000; Pride, this volume).

Jolly and Pride (Jolly and Pride, 1999) identified two Front lineages (A and G) and two Gallery troops (C and D) whose female descendents are known to have remained in essentially the same core areas since the Budnitz/Dainis census of Malaza in 1972. Six other core areas remain the same, and troops have been repeatedly re-identified in these areas, though there is always the possibility of complete replacement of one troop by another.

This does not mean that the ranges are fossilized. C troop, for instance, pushed an amoeboid lobe from the gallery forest to the Front during the 1990s as tourist feeding became a major factor. Both D1A, based in Gallery, and G3, based in the Front near the restaurant, travel deep into the scrub to feed on tamarinds when that is the only rich source of fruit (Mertl-Millhollen, 2000; Rasamimanana, 2004; Mertl-Millhollen, this volume). However, each of these troops kept their original home base.

Even the home base may be squeezed, as A1's range was compressed in 1996–1997 by the success of A2, although A1 later reclaimed ground. In short, territories and their frontiers are constantly maintained by confrontations and by scent marking, or else they can be lost to competitors.

Most troops have split over the years. Troop splits occur as the result of exiling one to four females. Harassment may last for three previous birth seasons, with the subordinates accepting targeted aggression from dominants rather than leave (Vick and Pereira, 1989). Eventually the dominant females escalate from aggression within the troop to actively chasing the subgroup beyond the troop boundaries. There follows a period of nomadic roaming by the exiled subgroup, attacked by both their former troop and all neighbors. The subordinate group may or may not achieve control of a territory. Following Koyama's definition, it should be called group, not a troop, until males join, usually at or after the point when the subordinate females begin to win confrontations and claim ground (Koyama, 1991; Hood and Jolly, 1995; Dubovick, 1998; Ichino et al, this volume). Female groups or single females may remain nomadic for up to 2 years, which means being chased by any troop they meet and having no secure food supply. These groups either gain range (and males) or disappear. Ichino, in this volume, gives twelve case studies of troop change. Table 12.3 adds nine cases of similar changes in other regions of the forest.

In all but one case, surviving daughter troops have remained in contact with each other as neighbors. The one case was when C2A, derived from Troop C2, inserted itself between the existing T1A and T1B, moving one range over from its sister troop—but still within 50 m in the compressed "bottleneck" of Ankoba/Tourist Front. Other leapfrogging has not been seen at Berenty though it might have passed unnoticed in the Scrub zone. Instead, observed daughter troops split the original range, claimed neighbors' range, or in four cases, completely took over the ranges of a neighbor troop, forcing them to flee in turn.

Figure 12.8 shows the outcome of troop fissions for the dominant (exiling) subgroup, for the subordinate (exiled) subgroup, and for 12 females in the three troops known to have lost their range completely, from the 14 troop fissions observed by the Ichino et al. (this volume), and for 9 observed by the Jolly et al.

TABLE 12.3. Troop fissions.

Troop	Year	Dom. female fate	Sub. female fate	Observer
A	1992	A1 div A range	A2 took A core	Hood, Jolly
A1	1997	A1 kept A1 core	"A3" died	Dubovick, Jolly
D1	2000	D1A kept range	D1B div E1 range	Pride, Mertl-Millhollen
E	1990	E1 div core	E2 div core	Jolly
E1	1994	E1 kept core	2 E1 F joined YB as doms	Koyama, Jolly
E2	1998	E2A kept core	E2B div E2 range	Jolly, Pinkus
F	1992	F1 kept core	"F3" div F range	Jolly
G	1989	G1 kept core	G2 div G and A ranges	Jolly
G1	1994	G1 div core	"G3" (G1B) div core	Rasamimanana, Jolly

F, females; range, total range; core, high use core of range.

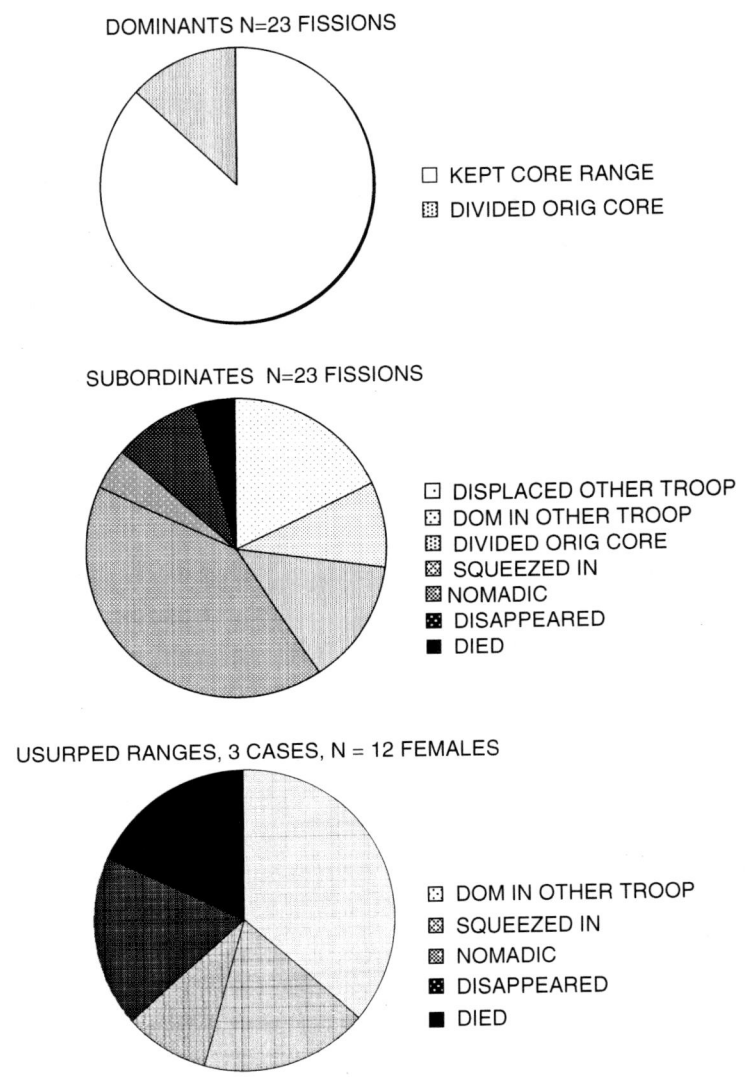

FIGURE 12.8. Fate of dominant female subgroups, subordinate female subgroups, and usurped females after troop fission or range takeover. The darker the shading, the worse the probable outcome. Moral: don't lose your range.

In 20 cases out of 23, the dominant group kept its core range. In just three cases the exiled group fought back to take the major part, though not all of the core.

The subordinate group, however, most often squeezed itself in on the periphery taking some of the original range and some from a neighbor (9 of 23 cases), Some displaced females joined another troop, fighting their way in as dominants over the resident females (two cases). Some disappeared, or are known to have died (three cases). Finally, in four cases the exiled females completely demolished another troop, driving the residents out of their range. The 12 females of the troops that were totally displaced either joined another troop, disappeared, or were killed in the takeover, one dying of wounds in a week, another of wounds and/or starvation later on after the takeover. (obs. of takeover and first death by Ichino, second death by Jolly). One displaced female remains nomadic and solitary within her former range.

Three troop fissions at Beza followed similar patterns of range division with daughter troops either remaining in the parent troop range or taking part of the parent range and part of a neighbor's. Two displaced females failed to join another troop, and were attacked, with an attempt by a resident female to kill the infant of a newcomer. However, interestingly, four of the nine original Beza Mahafaly study troops no longer exist, and the females from two separate troops joined to form a new one. In situations of population shrinkage, it seems even ringtail females can join together as is known in other primates (Gould et al., 2003).

The conclusion is that lemur females do not exactly bequeath territorial success but they can certainly bequeath territorial failure. That is, lineages endowed with territory must maintain their territorial defense, but this is apparently easier than breaking in to establish a new territory in a full environment. Exiled subgroups on the other hand face intense aggression from neighbors as well as from the original dominants, and may fail to gain territory at all. Finally, in the cases at Berenty where a group lost its entire territory they ceased to exist as a group, with pairs of females either joining another group or dying.

FIGURE 12.9. Between-troop aggression. Central females of A1 and A2 in 1992. Rival troops typically face outwards from their core: here A1 has their backs to us, A2 faces us. The face-off may escalate to lunges and feints, sometimes to contact aggression. In most troop encounters only a few females take an active role in escalated attack, often the alpha and/or her chief henchwoman. Photo, A. Jolly.

12.5. Conclusion and Conservation Implications

Bet-hedging life histories and multigenerational territories are both strategies that minimize risk. In the one case, animals reproduce at a lower rate than they might in the best years in order to safeguard adult survival. In the other case, adults invest energy, and sometimes put themselves and their infants in danger, to maintain access to resources even when this is not currently advantageous (Pride, this volume). This only makes sense to avoid still greater cost of establishing a new territory—or worse, the even greater risk of failing to establish a new territory (Pride, pers. comm.).

At Berenty, both slow life histories (in mammalian, though not lemur terms), and territoriality function to promote unequal survival for lineages of adult females that compete in a full environment. In spite of this paper's focus on exiled females and the consequences of failure, we should not forget that it is the close bonds within matrilines that lead to survival—and to what looks to human primates like social delight pervading ringtailed lemur life (Figures 12.9 and 12.10).

It seems entirely plausible that in dry, marginal habitat, *L. catta* populations fluctuate far more than they do at either Beza or the scrub area of Berenty. It may also be that whole troops are likely to die out from environmental stress rather than density-dependent competition. This would open new areas for occupation when good times return, which would eliminate the struggle to establish territory. If the

FIGURE 12.10. The close bonds within a matriline persist through much or all of a female ringtail's life, beginning in earliest infancy. Social contact, grooming, and support are the brighter—and much more frequent—side of the interplay between out-group aggression and in-group affiliation.

experience of the gallery forest is a guide, though, this will depend on the plants' response to drought and rain, rather than directly on rainfall. If the succulent plants of the spiny forest are themselves buffered against drought, they may even reduce the variation in lemur populations. This is a whole new question to explore.

Meanwhile there are conservation implications for lemurs as a whole and for lemur populations in small forest fragments—whether reserves, sacred forests, or the few remaining unprotected gallery forests still standing.

For lemurs as a whole, the extremely slow life histories of indriids, and the moderately slow ones of lemurids (and apparently most other Malagasy mammals) render them more vulnerable to changing environmental insults that highly r-selected animals on other continents. If local deforestation, lowered water tables, and global warming increase the frequency of catastrophic climate events, they may be under greater threat than more resilient mammals.

Within gallery forests, heritable territoriality is a form of resource competition which favors the haves against the have-nots and thus reduces the total gene pool. Small forest fragments that do not have male or troop immigration may preserve their populations at a cost in genetic narrowing mediated through lineage-group territoriality, rather than being more randomly spread among individuals.

A final speculation is the possible role of partially heritable reproductive success in favoring female dominance. If dominant female lineages are statistically more likely to succeed over successive generations, this might even increase the reproductive variance of females over that of the males who essentially start afresh in each generation. However, the ringtails would have to be compared to more clearly nepotistic male-dominant societies like that of baboons or many macaques, as well as across species, habitats, and generations of lemurs—data unavailable as yet.

Much of this train of argument is speculative and would benefit (or perhaps be contradicted) by more data, in particular comparative data on *Lemur catta* in sparse, spiny habitat. In spite of all we have learned, and the refinement of questions we ask, I cannot resist ending with Peter Klopfer's conclusion of 1970: "Lemurs remain lovely, and enigmatic!".

Acknowledgments. Thanks, as always, to the de Heaulme family and the people of Berenty for the creation and maintenance of the Berenty Reserves, as well as for their hospitality to visiting scientists. Thanks to the volunteers and students who contributed data: in 1995, Marisa Braun, Stuart Hall, Katherine Heavers, Hee-Joo Park, Susan Pinkus, Solofomaminiaina Rasamoelina, John Walker, George Williams, and Teresa Williamson; in 1996–1997, Tracy Dubovick, Jessica Flack, Oliver Maxwell, and Christopher Mills; and in 2005, Vonjy Andrianome, Paul Boucher, Deborah Custance, Francine Dolins, Veronica Kaiser, Lisa Lane, Josia Rasafinmandimby, Hajarimanitra Raveloarivony, and Jennifer Savage. Funding was provided by Earthwatch (Jolly, Rasamimanana, and 120 Earthwatchers), Wildlife Trust (Rasamimanana and her students from the École Normale Supérieur, University of Madagascar), and by students, volunteers, and their families.

References

Braun, M. A. (1996). *Variable Use of Territorial Behavior in Four Troops of Lemur catta at Berenty Reserve, Madagascar*. Unpublished BA, Princeton University, Princeton, N.J.

Budnitz, N., and Dainis, K. (1975). *Lemur catta:* Ecology and behavior. In: Tattersall, I. and Sussman, R.W. (eds.), *Lemur Biology*. Plenum, New York, pp. 219–236.

Budnitz, N. (1978). Feeding behavior of *Lemur catta* in different habitats. In: Bateson, P. and Klopfer, K. (eds.), *Perspectives in Ethology*, Vol. 3. Plenum Press, New York, pp. 85–108.

Dewar, R. E., and Wallis, J. R. (1999). Geographical patterning of interannual rainfall variability in the tropics and near tropics: an L-moments approach. *J. Climate* 12:3457–3466.

Dubovick, T. H. (1998). *A Historical, Social and Ecological Analysis of Three Tourist Ranging Troops of Lemur catta, Berenty Reserve, Madagascar*. Unpublished BA, Princeton University, Princeton, N.J.

Dunbar, R. I. M. (1988). *Primate Social Systems*. Croom Helm, London.

Eisenberg, J. F. (1981). *The Mammalian Radiations*. University of Chicago Press, Chicago.

Frère, S. (1958). *Panorama de l'Androy*. Ed. Aframpe, Paris.

Ganzhorn, J. U. (1995). Cyclones over Madagascar: Fate or fortune? *Ambio* 24:124–125.

Goodman, S. M., Ganzhorn, J. U., and Rakotondravony, D. (2003). Introduction to the mammals. In Goodman, S. M. and Benstead, J. M. (eds.), *The Natural History of Madagascar* University of Chicago Press, Chicago, pp. 1159–1186.

Gould, L., Sussman, R. W., and Sauther, M. L. (1999). Natural disasters and primate populations: the effects of a two-year-drought on a naturally occurring population of ring-tailed lemurs *(Lemur catta)* in southwestern Madagascar. *Int. J. Primatol.* 20:69–85.

Gould, L., Sussman, R. W., and Sauther, M. L. (2003). Demographic and life-history patterns in a population of ring-tailed lemurs (*Lemur catta*) at Beza Mahafaly Reserve, Madagascar: A 15-year perspective. *Am. J. Phys Anthrop.* 120:182–194.

Hood, L. C., and Jolly, A. (1995). Troop fission in female *Lemur catta* at Berenty Reserve, Madagascar. *Int. J. Primatol.* 16:997–1016.

Jolly, A., and Pride, R. E. (1999). Troop histories and range inertia of *Lemur catta* at Berenty: A 33 year perspective. *Int. J. Primatol.* 20:359–373.

Jolly, A., Dobson, A., Rasamimanana, H. R., Walker, J., O'Connor, S., Solberg, M., et al. (2002). Demography of *Lemur catta* at Berenty Reserve, Madagascar: Effects of troop size, habitat and rainfall. *Int. J. Primatol.* 23:327–354.

Jolly, A. (2004). *Lords and Lemurs: Mad Scientists, Kings with Spears, and the Survival of Diversity in Madagascar*. Houghton Mifflin, Boston.

Kaufmann, J. H. (1983). On the definition and functions of dominance and territoriality. *Biol. Rev.* 58:1–20.

Klopfer, P. H., and Jolly, A. (1970). The stability of territorial boundaries in a lemur troop. *Folia Primatol.* 12:199–208.

Koyama, N. (1991). Troop division and inter-troop relationships of ringtailed lemurs (*Lemur catta*) at Berenty, Madagascar. In: Ehara, A., Kimura, T., Takenaka, O., and Iwamoto, M. (eds.), *Primatology Today*. Elsevier, Amsterdam, pp. 173–176.

Koyama, N. (1992). Multiple births and care-taking behavior of ring-tailed lemurs *(lemur catta)* at Berenty, Madagascar. In: Yamagishi, S. (ed.), *Social Structure of Madagascar Higher Vertebrates in Relation to Their Adaptive Radiation*. Osaka City University, Osaka, pp. 5–9.

Koyama, N., Nakamichi, M., Oda, R., Miyamoto, N., and Takahata, Y. (2001). A ten-year summary of reproductive parameters for ring-tailed lemurs at Berenty, Madagascar. *Primates* 42:1–14.

Koyama, N., Nakamichi, M., Ichino, S., and Takahata, Y. (2002). The population and social dynamics changes in ring-tailed lemur troops at Berenty, Madagascar between 1989–1999. *Primates* (43):291–314.

Mertl-Millhollen, A. S. (1979). Olfactory demarcation of territorial boundaries by a primate–*Propithecus verreauxi*. *Folia Primatol.* 32:35–43.

Mertl-Millhollen, A. S. (1988). Olfactory demarcation of territorial but not home range boundaries by *Lemur catta*. *Folia Primatol.* 50:175–187.

Mertl-Millhollen, A. S. (2000). Tradition in *Lemur catta* behavior at Berenty Reserve, Madagascar. *Int. J. Primatol.* 21:287–298.

Mertl-Millhollen, A. S., Gustafson, H. L., Budnitz, N., Dainis, K., and Jolly, A. (1979). Population and territory stability of the *Lemur catta* at Berenty, Madagascar. *Folia Primatol.* 31:106–122.

Mills, C. N. (1997). *Lemur catta Behavior in Different Habitats*. Unpublished BA, Princeton University, Princeton, N.J.

Mitani, J. C., and Rodman, P. S. (1979). Territoriality: The relation of ranging pattern and home range size to defendability, with an analysis of territoriality among primate species. *Behav. Ecol. Sociobiol.* 5:241–251.

O'Connor, S. M. (1987). *The Effect of Human Impact on Vegetation and the Consequences to Primates in Two Riverine Forests, Southern Madagascar*. Unpublished PhD thesis, Cambridge University, Cambridge.

Pochron, S. T., Tucker, W. T., and Wright, P. C. (2004). Demography, life history and social structure of *Propithecus diadema edwardsi* from 1986–2000 in Ranomafana National Park, Madagascar. *Am. J. Phys. Anthropol.* 125:61–72.

Rasamimanana, H. R. (2004). *La dominance des femelles makis (Lemur catta): quelles stratégies énergétiques et quelle qualité de ressources alimentaires, dans la réserve privée de Berenty, au sud de Madagascar*. Unpublished PhD, Muséum Nationale d'Histoire Naturelle, Brunoy, France.

Rasamimanana, H. R., Ratovonirina, Jolly, A., and Pride, E. (2000). Storm damage at Berenty Reserve. *Lemur News* 5:7–8.

Richard, A. F., Dewar, R. E., Schwartz, M., and Ratsirarson, J. (2002). Life in the slow lane? Demography and life histories of male and female sifaka (*Propithecus verreauxi verreauxi*). *J. Zool., Lond.* 256:421–436.

Rubenstein, D. I. (1981). Population density, resource patterning, and territoriality in the Everglades pygmy sunfish. *Anim. Behav.* 29:155–172.

Stearns, S. C. (1992). *The Evolution of Life Histories*. Oxford University Press, New York.

Sussman, R. W., Green, G. M., Porton, I., Ony, L., Andrianasolondraibe, and Ratsirarson, J. (2003). A survey of the habitat of Lemur catta in Southwestern and Southern Madagascar. *Primate Conservation* 19:32–57.

Takahata, Y., Koyama, N., and Miyamoto, N. (2005). Inter- and within-troop competition of female ring-tailed lemurs: A preliminary report. *African Study Monographs* 26(1):1–14.

Vick, L. G., and Pereira, M. E. (1989). Episodic targeted aggression and the histories of Lemur social groups. *Behav. Ecol. Sociobiol.* 25:3–12.

Wright, P. C. (1999). Lemur traits and Madagascar Ecology: Coping with an island environment. *Yrbk. Phys. Anthropol.* 42:31–72.

13
Resource Defense in *Lemur catta*: The Importance of Group Size

R. ETHAN PRIDE, DINA FELANTSOA, TAHIRY RANDRIAMBOAVONJY, AND RANDRIAMBELONA

13.1. Introduction

Why do *Lemur catta* live in groups? And why does this species form groups of different sizes in different areas? Long-term demographic studies have demonstrated persistent differences in typical *Lemur catta* group size in different locations, both within and across field sites (Sussman, 1991; Jolly et al., 2002; Koyama et al., 2002; Gould et al., 2003). At Berenty Reserve, these differences coincide with differences in habitat and population density (Jolly et al., this volume), ranging from 9 animals per group in the southern open-canopy "scrub" to 14–16 animals per group in the northern "tourist front" and Ankoba sections (Jolly et al., 2002; Koyama et al., 2002). In this chapter, we explore the hypothesis that female *L. catta* adjust group size in response to the intensity of intergroup competition for food resources. We also examine the ecological conditions in which intergroup resource defense may provide foraging advantages and thereby promote increased group size.

A simple ecological model of group size can be derived from the assumptions that (1) larger groups gain foraging advantages by acquiring or defending high-quality food resources from smaller groups; and (2) animals in larger groups suffer foraging disadvantages due to intragroup feeding competition. Animals should seek to maintain membership in larger groups until the costs of intragroup feeding competition balance the benefits of intergroup resource defense. This has been considered as one of the possible ultimate causes of social grouping in primates (Wrangham, 1980).

The model assumes that large groups can supplant smaller groups from food resources due to their greater fighting ability, that there is variation in food resource quality such that animals in large groups can gain energetic benefits by doing so, and that these resources are defendable (Wrangham, 1980). The model predicts that group size will be proportional to intergroup conflict rate, and large groups will always occupy the highest quality habitats or food patches (i.e., those in which their daily energy intake is highest).

Here we examine how well this model describes the behavior *L. catta* of Berenty Reserve. In particular, we ask the following questions:

1. Are observed group sizes proportional to intergroup conflict rates?
2. Does membership in a large group increase the chance of winning intergroup conflicts and/or lower resource defense costs?
3. Does resource defense provide foraging benefits to individuals in large groups (occupying best habitat and gaining greater food intake rates)?

13.2. Methods

We report data collected on six *L. catta* groups, representating the full range of group sizes typically observed at Berenty (Jolly et al., 2002). Data were collected over a 1-year period (August 1999 to July 2000). Groups studied are presented in Table 13.1.

13.2.1. Seasonal Characterization

Because costs and benefits of resource defense may vary with resource availability, we compare behavior of these groups in two distinct ecological conditions ("typical" and "food-scarce"), as well as based on year-long averages. The conditions of food scarcity described here refer not to the annual dry season to which *L. catta* are adapted (Sauther and Sussman, 1993; Pereira et al., 1999), but to an atypical period of food scarcity, occurring at a time of year usually associated with high food availability (Sauther, 1998). During the "typical" birth/lactation season, all groups could exploit fruiting *Tamarindus indica* trees within their ranges. For 2 months during this period, all groups foraged heavily on *Rinorea greveana*, which was also found within the typical ranges of all groups. During the "food-scarce" weaning season (February–April 2000), *Azadirachta indica* was the only fruiting tree and was found only within the typical range of one group (A2). Figure 13.1 shows the location of the fruit trees exploited by these groups and the typical ranges of these groups.

The atypical food scarcity can be attributed to a tornado in late 1999 that had destroyed one-third of the forest canopy trees, severely damaged another third of the trees, and blown most of the fruit and leaves off the trees (Jolly et al., 2001) such that the *Tamarindus indica* fruit commonly exploited by *L. catta* was almost entirely absent by February 2000; a concurrent drought may also have lowered

TABLE 13.1. Size and composition of study groups.

Group	Total group size	Adult males	Adult females
A2	26 (26)	9	10
A1	19 (19)	6	6
D1	14.5 (13–16)	4–6	6–8
CX	9 (8–10)	1–3	4
SB	9 (9)	2	4
SE2	5.5 (4–8)	1–4	3

210 R.E. Pride et al.

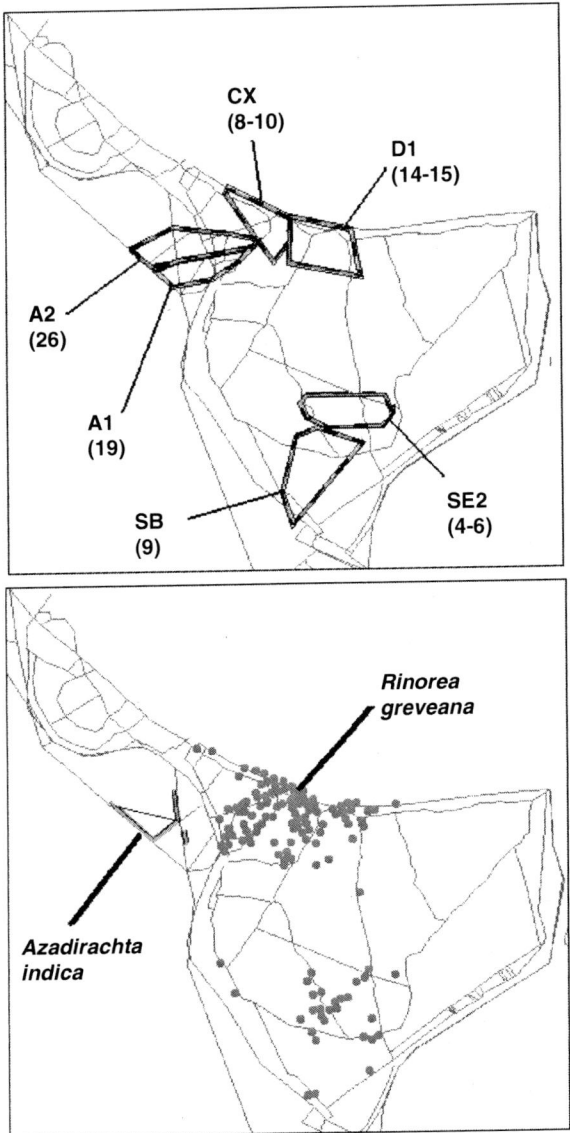

FIGURE 13.1. The top panel shows the sizes and typical ranges of the six study groups. Ranges shown are minimal convex polygons of the 85% most commonly used grid squares. The bottom panel shows the location of *R. greveana* trees exploited by these groups (the principal fruit source for 2 months in the "typical" season), and the location of the *A. indica* plantation (the only fruiting tree in the "food-scarce" season). Although all groups had access to *R. greveana* within their typical ranges, only the largest group maintained a range with *A. indica* trees.

productivity. Because high-quality food resources were scarce and female energetic demand was high, competition for food was likely to be intensified at this time.

13.2.1.1. Ranging

For each group, we recorded its location in the reserve (using a 25-m grid coordinate system) every 30 minutes during full-day follows (6:00 a.m. to 6:30 p.m.). We conducted 22 full-day follows for each of five groups (A2, A1, D1, CX, and SB); we conducted 13 full-day follows for the sixth group (SE2) under "typical" conditions but could not conduct follows in the "food-scarce" season because the group was ranging extensively outside of the reserve. We defined typical ranges for each group as the central 85% of locations recorded during full-day follows. We compared ranging in normal and food-scarce conditions by plotting day-ranges observed in normal conditions (N = 19 per group: August–November 1998, August–November 1999, May–July 2000) and food-scarce conditions (N = 3 per group: February–April 2000) using the range determination program Map (Williams, 1999). We then compared these to the distribution of available fruit trees in each season to determine the relative quality of ranges. Because food-scarce conditions were a brief aberration, only 3 day-ranges per group are available from this time period, and extent of ranging is likely to be underestimated at this time. However, *ad libitum* observation of troop locations during the food-scarce season (an additional 3–12 days per group) support the characterizations of each group's ranging determined from these day-range data.

13.2.1.2. Intergroup Conflict

We estimated intergroup conflict rates by recording all occurrences of intergroup encounters during full-day follows (typical season: August–November 1999; Food-scarce season: February–April 2000). We noted location, duration, which individuals participated at the maximum level of aggression for that encounter (displacement, lunges, physical contact), and which group withdrew/lost. We also determined the escalated defense rate, as this was found to correlate with high cortisol levels, and *L. catta* females may be particularly sensitive to this stressor (Pride, 2005a). Escalated defense encounters are those in which neither group immediately withdrew and which occurred inside the group's typical range. We then calculated these same rates at the individual level. We also calculated individual participation as the proportion of a group's observed conflicts in which a given individual participated at the maximum level for that encounter.

To determine if membership in large groups increases competitive ability, we compare proportion of conflicts won in large and small groups, indicating greater success, as well as individual participation per conflict in large and small groups, indicating lower *per capita* agonism costs. We considered number of adult females as well as group size as the independent variables in my analyses, because it is primarily the adult females that participate in intergroup agonism (Sauther, 1993). In each season, we determined each group's total intergroup conflict rate.

13.2.1.3. Food Intake

Feeding data are available on three adjacent groups (CX, A1, A2). We measured individual intake rates for two food items: *Rinorea greveana* fruit and *Azadirachta indica* fruit. *Rinorea greveana* was the primary fruit source exploited by *L. catta* for 2 months during the typical season (October–November: 165/580 feeding occurrences based on half-hourly scan samples); at this time, all groups had access to fruit-bearing trees within their typical ranges, as *R. greveana* was densely distributed throughout the gallery forest (Figure 13.1). *Azadirachta indica* was used heavily for 2 months in the food-scarce season (February–March: 228/574 feeding occurrences based on half-hourly scan samples); at this time, it was the sole fruit-bearing tree species. Most *L. catta* groups did not have fruiting trees within their typical ranges, as *A. indica* was planted only along roads as an ornamental tree near the hotel bungalows (Figure 13.1); among groups studied here, only the largest group (A2) had fruiting trees within its typical range at this time. *R. greveana* and *A. indica* provide good measures of individual differences in intake because the trees of both species are relatively small (~3–5m crown diameter), facilitating observation, and the fruit are small (1–2 cm) discrete items, consumed one at a time, so it was possible to obtain accurate fruit intake counts on the most heavily-used high-energy food items in both normal and food-scarce conditions.

For each individual, we calculated intake rates during foraging bouts on each species (the number of fruit consumed per unit time foraging on that species). During focal animal observations (Altmann, 1974), we recorded foraging duration by measuring the focal animal's start and stop times of active searching for fruit and counted the total number of fruit ingested during that time period. Because focal samples were taken randomly throughout the animals' entire daily active period, not exclusively while animals were foraging on these food sources, duration of foraging observations varied across individuals, and not all individuals or groups are evenly sampled (*A. indica* median: 16 minutes; range: 8.5–36 minutes; *R. greveana* median: 7 minutes; range: 4–16.2 minutes). However, there was no relationship between intake rate and duration sampled for *A. indica* (linear regression: $F_{1,20} = 0.073$, $r^2 = 0.00$, $p = 0.790$) or *R. greveana* (linear regression: $F_{1,9} = 0.292$, $r^2 = 0.04$, $p = 0.606$), and visual inspection of the data did not suggest that variance changed substantially with sample duration.

We estimated time spent foraging per day by recording behavior with instantaneous scan samples taken every 30 minutes from 6:00 a.m. to 6:30 p.m., one day per group per month, for the 2 months in which each fruit was eaten (*R. greveana*: October–November; *A. indica*: February–March). In a previous field season, it was determined that daily time spent foraging estimated from half-hourly scans correlate highly with that calculated from scans taken every 10 minutes (r = 0.89), and therefore give a fair estimate of daily time spent foraging.

For each individual in each season, we estimated food intake per day by multiplying fruit eaten per minute of foraging time times minutes spent foraging per day. We then compared food intake rates of animals in groups of different size in normal and food-scarce conditions to determine if animals gained foraging advantages through membership in larger groups in each condition.

13.3. Results

13.3.1. Are Group Sizes Proportional to Intergroup Conflict Rates?

Group size is directly proportional to intergroup encounter rate (linear regression: $F_{1,5} = 17.854$, $r^2 = 0.82$, $p = 0.013$). However, as number of adult males and offspring are likely to depend on the number of adult females in a matrifocal social system, and females are responsible for most intergroup conflicts, a comparison of number of females is more appropriate. Number of adult females per group increases with daily intergroup encounter rate (linear regression: $F_{1,5} = 17.631$, $r^2 = 0.82$, $p = 0.014$). Based on this regression, the number of adult females is predicted to be 1.5 + 2.5*(daily intergroup encounter rate). The relationship is similar if only escalated intergroup conflicts are considered (linear regression: $F_{1,5} = 19.679$, $r^2 = 0.83$, $p = 0.011$). Intergroup conflict rates from typical conditions may be more representative than the aberrant conditions of the atypical food-scarce season; if data from the atypically harsh weaning season are excluded, the relationship based on the 19 remaining appears non-linear (Figure 13.2).

Differences in intergroup conflict rates among Berenty's three habitat regions (Tourist = 2.7/day, Gallery = 1.7/day, Scrub = 0.4/day) parallel differences in group size among these regions observed in longterm demographic studies (Tourist = 14 to 16, Gallery = 13, Scrub = 9) (Jolly et al., 2002; Koyama et al., 2002). Similar variation in intergroup conflict rates among these regions have been noted in prior studies (Jolly et al., this volume).

13.3.2. Does Membership in Large Groups Increase Chances of Winning Intergroup Conflicts?

Proportion of conflicts won does not vary with group size (linear regression: $F_{1,5} = 0.058$, $r^2 = 0.01$, $p = 0.821$) or with number of adult females in group (linear regression: $F_{1,5} = 0.011$, $r^2 = 0.00$, $p = 0.920$), as shown by Table 13.2.

For the 188 observed conflicts in which the number of adult females in both rival groups is known, the larger group won in 103 instances, which was not

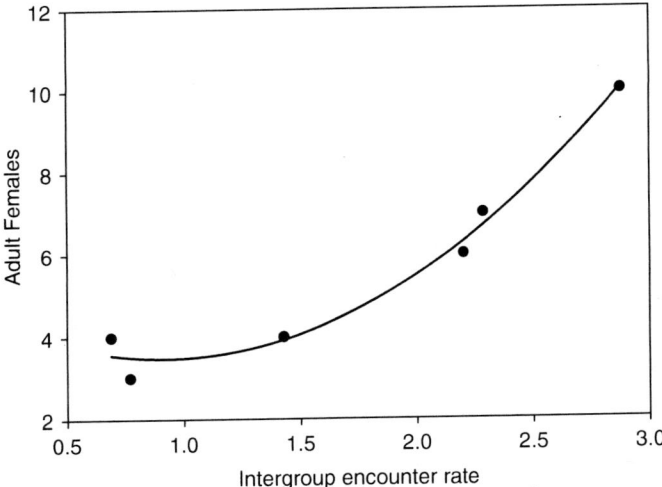

FIGURE 13.2. Number of adult females in a group is an increasing function of the intergroup conflict rate (encounters per day). Each point represents one of the study groups; intergroup conflict rate represents the average daily rate observed in typical seasons (August–November 1999; May–July 2000). The curve shows number of adult females predicted from a regression of number of adult females on daily intergroup encounter rate squared ($F_{1,4} = 73.605$, $r^2 = 0.95$, $p = 0.001$).

significantly greater than expected by chance (chi-square test: $\chi^2 = 1.785$, $p = 0.182$). These results agree with prior studies (Jolly et al., 1993). A group's probability of winning an intergroup conflict did not vary with the difference in number of adult females between it and its opposing group (logistic regression: $N = 188$, $\chi^2 = 1.068$, $r^2 = 0.00$, $p = 0.302$).

Outcome of intergroup conflicts depended not on group sizes but on location, with groups tending to win conflicts within their typical ranges (Figure 13.3). Groups won 92/141 conflicts observed within their typical range, but only 28/103 conflicts observed outside of their typical range (chi-square test: $\chi^2 = 35.520$, $p = 0.0001$). When location (inside or outside of typical range) and relative group

TABLE 13.2. Outcome of intergroup conflicts.

Group	Size	Females	Percent of conflicts won
A2	26	10	52% (43/83)
A1	19	6	42% (26/62)
D1	14.5	6–8	56% (18/32)
CX	9	4	54% (21/39)
SB	9	4	31% (5/16)
SE2	5.5	3	62% (8/13)

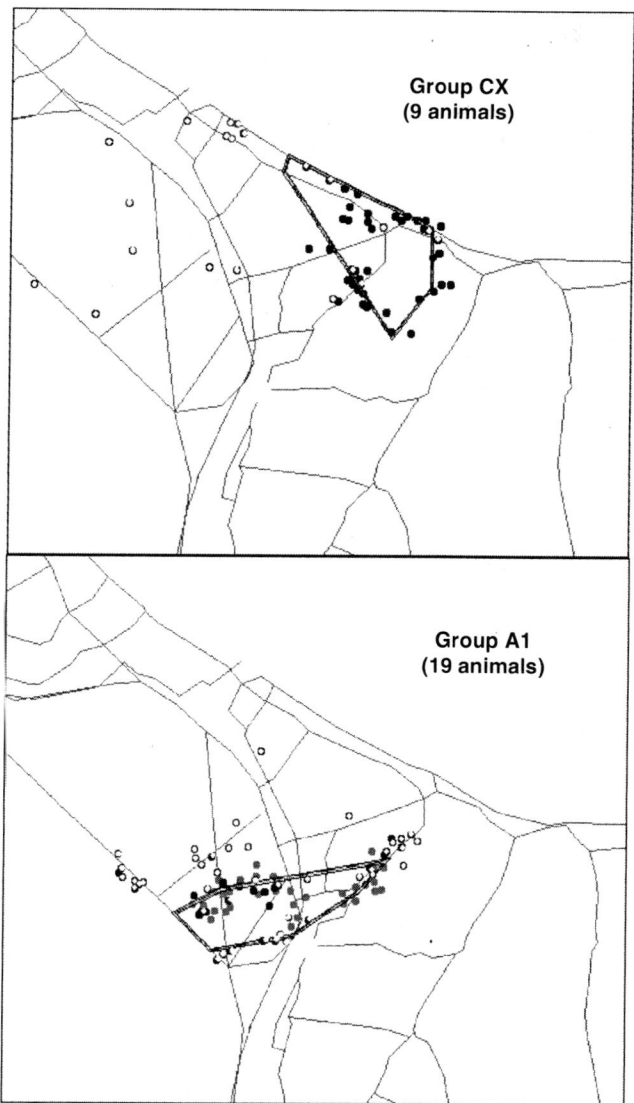

FIGURE 13.3. The outcome of intergroup agonism is spatially determined. Open circles indicate losses, closed circles indicate wins. Groups tended to lose conflicts outside their range, particularly in the high-quality habitat (A1, A2, CX) where conflicts were common and ranges were consistent. The two groups foraging on low-quality habitat (SB, SE2) exhibited less consistent ranging patterns, so typical ranges displayed here should be taken as approximations. They also had few intergroup encounters, possibly due to the poor quality of the habitat (supporting few groups at low density and few resources worth defending).

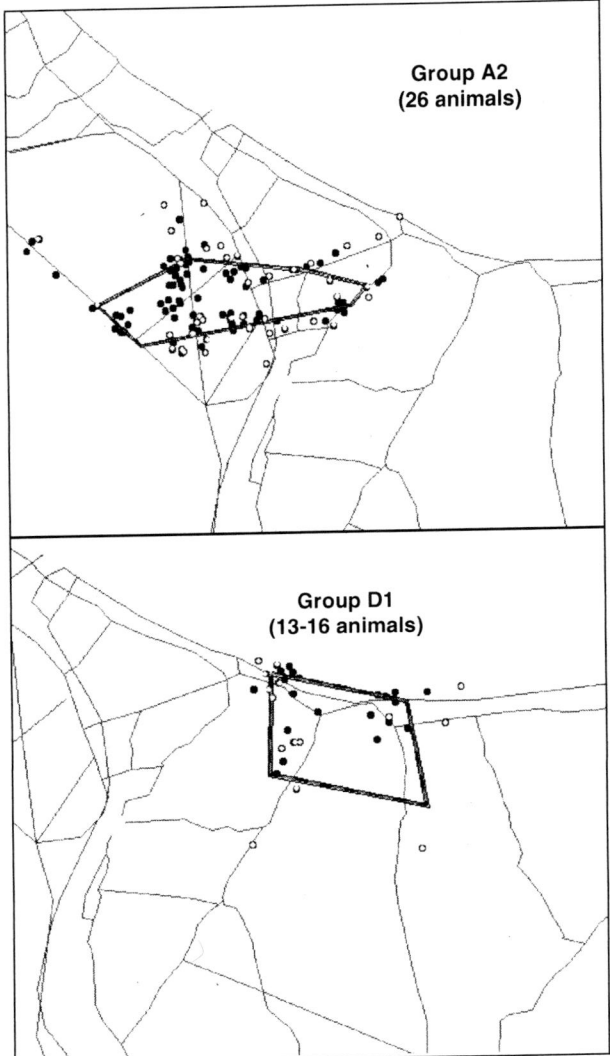

FIGURE 13.3. Cont'd.

size (larger or smaller than opponent) are both included as predictor variables, only location significantly predicts conflict outcome (nominal logistic regression: N = 188, χ^2 = 42.193, r^2 = 0.16, p = 0.0001; location χ^2 = 35.28, p = 0.0001; size χ^2 = 2.48, p = 0.115). Large groups will only tend to supplant smaller groups from food patches if the food patches are contained with large groups' typical ranges.

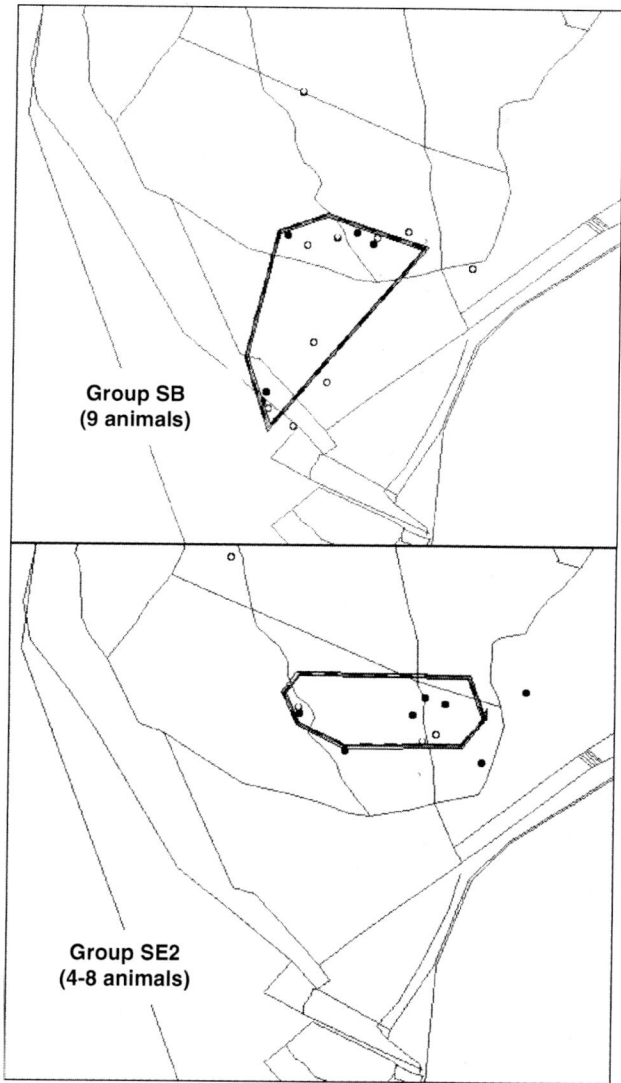

FIGURE 13.3. Cont'd.

13.3.3. Does Membership in Large Groups Reduce Costs of Intergroup Conflicts?

Although outcome of conflicts is independent of group size, the costs of attaining a given outcome are lower in large groups. Individual participation per intergroup conflict declined with group size in both typical and food-scarce seasons, indicating a lower cost per conflict for individuals in large groups (linear regressions; typical season: $F_{1,29} = 6.922$, $r^2 = 0.26$, $p = 0.014$; food-scarce season: $F_{1,19} = 9.286$, $r^2 = 0.33$, $p = 0.007$). Median individual participation (proportion of conflicts an individual participated in) is inversely related to the number of females in the group (linear regression of each group's median participation rate onto the reciprocal of number of females: $F_{1,5} = 88.922$, $r^2 = 0.95$, $p = 0.001$, Figure 13.4).

Participation in intergroup conflicts is not shared evenly among groupmates. Dominant females participated in a greater proportion of their group's encounters than subordinate females in both the typical season (mean$_{dom}$ = 63%, mean$_{sub}$ = 21%) and in the food-scarce season (mean$_{dom}$ = 61%, mean$_{sub}$ = 23%) (ANOVAS: typical season: overall $F_{6,29} = 7.194$, $r^2 = 0.51$, $p = 0.0001$; group $F_5 = 2.581$, $p = 0.048$; dominance $F_1 = 16.561$, $p = 0.0003$; food-scarce season: overall $F_{3,17} = 6.931$, $r^2 = 0.47$, $p = 0.003$; group $F_2 = 2.308$, $p = 0.130$; dominance $F_1 = 7.120$, $p = 0.015$). If participation in intergroup conflicts is costly, these costs are borne more heavily by dominant than subordinate females:

13.3.4. Do Large Groups Gain Advantages Through Resource Defense?

Large groups gained advantages through maintaining access to higher quality habitat. As noted previously, the largest group (A2) was the only group whose typical range contained fruiting trees even in the atypically harsh weaning season. Even in a typical weaning season, A2's habitat may be more productive than neighboring areas, due to the rows of contiguous *A. indica*, with their briefly

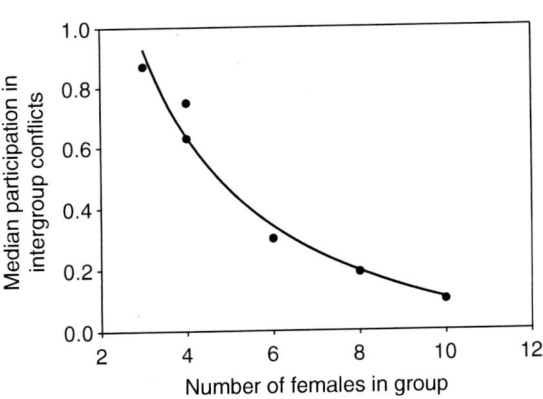

FIGURE 13.4. Individual participation per intergroup conflict is inversely proportional to the number of females in the group. The black line is a first-order inverse regression, and suggests that females can lower costs of intergroup agonism by grouping, as they may participate less often in intergroup conflicts.

TABLE 13.3. Participation in intergroup conflicts.

Group (no. of females)	Mean individual participation in intergroup conflicts	
	Dominant	Subordinate
A2 (10)	0.46	0.25
D1 (6–8)	0.34	0.15
A1 (6)	0.51	0.24
CX (4)	0.70	0.35
SB (4)	0.96	0.29
SE2 (3)	0.94	—

abundant clusters of fruit. More generally, though, the range size *per capita* decreases linearly with the size of the group (Figure 13.5), suggesting that the ranges of large groups are more productive, as less land is required to support each animal.

However, benefits of resource defense are expected to accrue only when resources are defensible as well as worth defending. Ranging data suggest that groups were able to exclude rivals from their habitats under typical conditions, leading to a territorial dispersion pattern; however, during the atypical food-scarce weaning season, groups ranged widely and ranges overlapped considerably (Figure 13.6). Resource defense could provide benefits to larger groups in the typical season if the groups they excluded were relegated to foraging on lower quality food patches, and in the food-scarce season if rival groups' foraging time within their range was reduced even when they did not have exclusive access to their food resources.

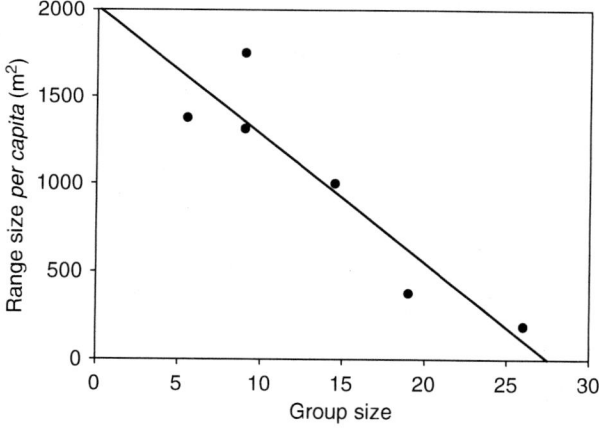

FIGURE 13.5. Range size per capita decreases with group size, suggesting that habitat of larger groups is more productive. Range size was calculated as the area of a minimal convex polygon of all grid squares in which a group was recorded during 22 observation days.

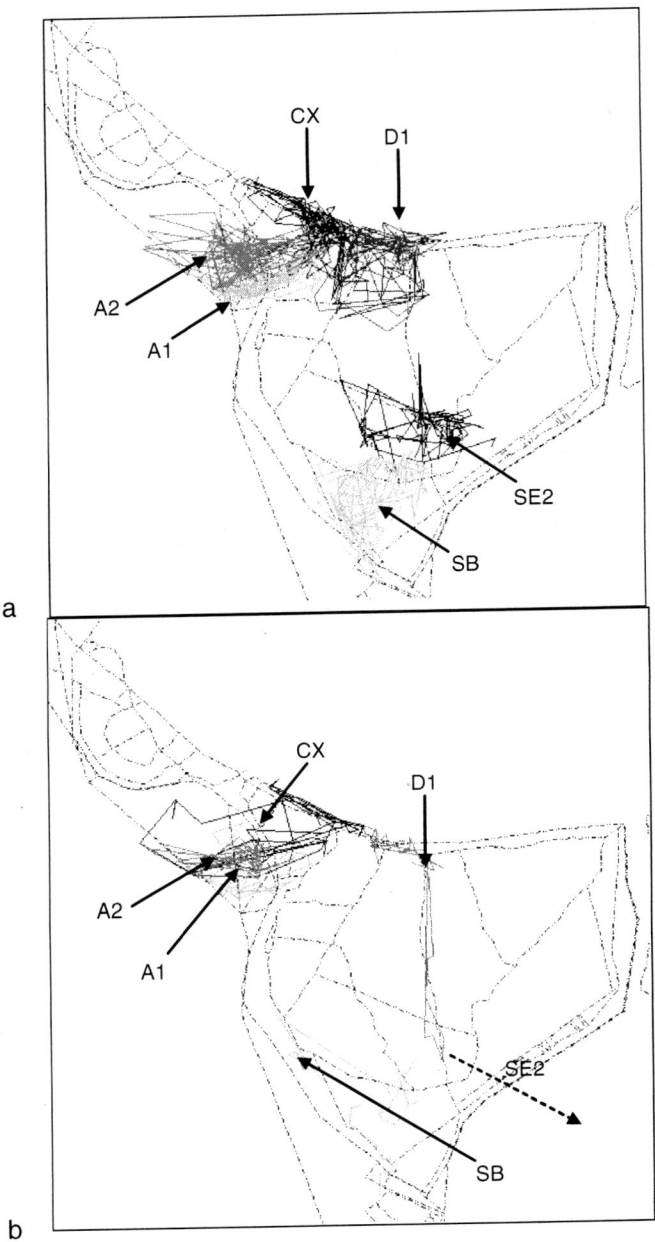

FIGURE 13.6. Ranging patterns in (a) typical conditions and (b) food-scarce conditions. When all groups have access to high-quality food sources, range overlap is minimal. When food is scarce, the large group (A2) with high-quality habitat defends the same range, while all other groups make excursions from their typical ranges (raid).

While foraging on *R. greveana* fruit (in the typical birth/lactation season), large groups attained greater intake rates. Females in the small group (CX; 9 animals) ingested 3.4 ± 1.1 *R. greveana* fruit per minute while foraging (mean ± SE) while those in larger groups ingested more (A1; 19 animals: 6.1 ± 0.8 and A2; 26 animals: 6.6 ± 1.0). If the larger groups are pooled, females in the smaller group obtained significantly fewer fruit per minute of foraging time (Student's t-test: N = 9, t = 2.42, p = 0.046). Proportion of time spent foraging did not differ among different groups (ANOVA: $F_{5,25}$ = 1.637, r^2 = 0.10, p = 0.187). Estimated daily fruit intake of females in the small group (254 ± 122 fruit/day) was considerably lower than the estimated intake of females in the larger groups (452 ± 100 and 420 ± 100 fruit/day), suggesting that large groups did obtain foraging advantages at this time (Figure 13.7a). However, this result should be taken with caution, as this possible advantage is too small to detect statistically given the low sampling intensity (ANOVA: $F_{2,5}$ = 1.142, r^2 = 0.31, p = 0.390).

In contrast, while foraging on *A. indica* (in the food-scarce weaning season), large groups had lower intake rates. Females in the small raider group (CX; 9 animals) ingested significantly more *A. indica* fruit per minute while foraging (mean ± SE: 3.9 ± 0.32) than those in the larger groups (A1 (19 animals): 2.8 ± 0.22; A2 (26 animals): 2.6 ± 0.20) (Tukey–Kramer: $F_{2,19}$ = 6.54, r^2 = 0.43, p = 0.008). This may have been due to crowding, as there were more animals per tree in larger groups (CX: 1.8 ± 0.2; A1: 1.6 ± 0.2; A2: 2.5 ± 0.2) (Tukey–Kramer: $F_{2,141}$ = 0.368, r^2 = 0.04, p = 0.697), and a lower proportion of animals eating at any given time during foraging bouts (CX = 78%, A1 = 70%, A2 = 56%) (Tukey–Kramer: $F_{2,141}$ = 3.376, r^2 = 0.05, p = 0.037), based on 144 *A. indica* trees in which foraging was observed during instantaneous scan samples. The large defender group (A2) maintained presence in or near fruit patches slightly longer than the raider groups (A2: 8 hours/day; A1 and CX: 6.5 hours/day, based on half-hourly scans of group location). However, this did not translate into differences in proportion of time spent foraging (means ± SE: A2: 18 ± 1%; A1: 19 ± 1%; CX: 18 ± 1%) (ANOVA: $F_{2,17}$ = 1.532, r^2 = 0.15, p = 0.244). Females in the small raider group had a significantly higher daily fruit intake than those in the large defender group (Tukey–Kramer: $F_{2,17}$ = 3.914, r^2 = 0.32, p = 0.040), with females in the large raider group having intermediate intake (large defender: 378 ± 31; large raider: 442 ± 36; small raider: 535 ± 47). The animals in the large groups that did not attain high intake rates may not have been meeting their daily energy requirements. Assuming a typical caloric value for *A. indica* fruit of ~67 calories per 100 g (Wu Leung et al., 1972), and assuming each fruit pulp ingested is approximately 1.5 g, then according to the general mammalian mass-metabolism equation in which FMR (kJ/day) = 4.82 mass$^{0.73}$ (Nagy et al., 1999), a 3- to 3.5-kg lemur would need to eat 400–450 *A. indica* fruit per day to meet its daily energy requirements. Females in the largest groups (subordinates in A1 and both dominant and subordinate females in A2) had intake rates lower than 400 fruit/day (Figure 13.7b). Although their diet also included foliage (*Cordia* and *A. indica* leaves), and their actual field metabolic rates are unknown, it is quite possible that these animals faced an energy deficit at this time.

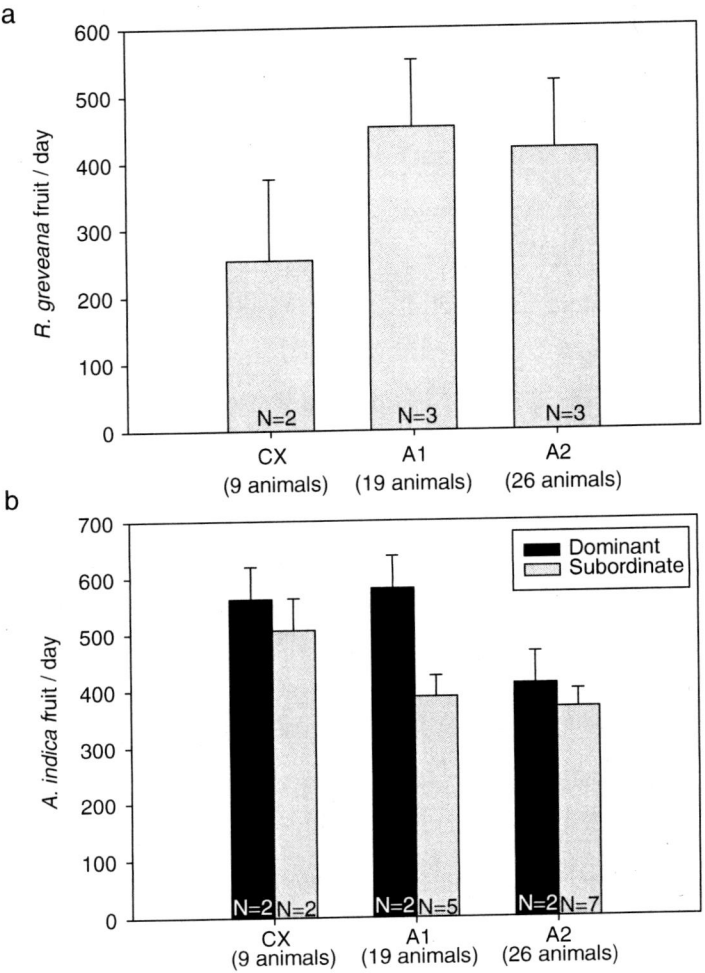

FIGURE 13.7. (a) Food intake rates for females in groups CX (9 animals), A1 (19 animals), and A2 (26 animals) foraging on *R. greveana* in the typical birth/lactation season. (b) Food intake rates for these same groups foraging on *A. indica* in the food-scarce weaning season. Although the *A. indica* exploited by all three groups was located within A2's typical range, and A2 evicted rival groups from this area, females in this group did not gain greater food intake by doing so.

13.4. Discussion

Intergroup competition for food resources is one factor that may promote larger group size in female-bonded primates such as *L. catta*. The importance of this factor in explaining variation in group size within the Berenty population is suggested by the remarkably close correlation between group size and intergroup conflict rate shown here. This relationship is predicted by the Wrangham (1980) model, which suggests that animals remain in larger groups because doing so enhances their ability to defend food resources from rival groups. The predicted values for group size based on a simple linear regression suggest that groups of four females should be found where intergroup conflict rate is low (e.g., <1/day) but can increase to eight females when conflict rates are high (3/day), which adequately reflects differences in Berenty's three habitat regions. Indeed, long-term differences in group size among the three regions (Jolly et al., 2002; Koyama et al., 2002) parallel differences in daily intergroup conflict rates observed in this and prior studies (Jolly et al., this volume). Smaller groups of ~4 adult females are typical at Beza Mahafaly Reserve (Gould et al., 2003), suggesting that intergroup competition for food resources may be less important there, as one may expect given the lower observed conflict rates (Sauther, 1992).

The benefit of membership in larger groups appears to be that a female's participation in intergroup conflicts decreases as she has more groupmates to share the burden of defense, lowering costs of attaining a given level of success in intergroup competition. Large groups' ability to maintain ranges in the most productive and stable habitats may be facilitated by the lower costs of defense *per capita*. Under typical conditions, the two largest groups maintained greater food intake rates than a smaller rival (since no animals in the larger groups had intake rates as low as those in the small group), suggesting that the spatial exclusion of rival groups observed in these conditions provided foraging benefits. Concurrent study shows that females in these large groups also exhibited lower cortisol levels at this time, suggesting lower stress and mortality risk (Pride, 2005b, 2005c).

However, results from the food-scarce weaning season highlight the importance of resource defensibility in determining the payoffs associated with group size, showing that benefits of being in a large group are eroded when resources cannot be monopolized. At this time, differences in food resource quality across ranges were extreme, and these conditions are expected to favor large groups (Wrangham, 1980), provided that they can monopolize the scarce resources. Females in the largest group (A2) were unable to prevent smaller adjacent groups from gaining access to the food resources, and they foraged less efficiently than their smaller rivals. Since they could not completely exclude rivals, benefits to defense depended on defenders limiting the raiders' foraging time enough to overcome the costs of their own relatively inefficient foraging, which they did not do. Females in large groups had highest cortisol levels at this time (Pride, 2005c).

Defensibility typically depends on the defender's ability to patrol resources in its day-range (Mitani and Rodman, 1979), the defender's sensory abilities to detect invaders (Lowen and Dunbar, 1994), and the aggregation of the resources (Rubenstein, 1981). In this case, the defending group's range was small relative to the distance it traveled in a typical day (range diameter = 375 m, daily path = 1250 m, based on 21 full-day follows), suggesting high defensibility on Mitani and Rodman's (1979) defensibility index ($D = 3.33$). Ability to detect intruders was likely to be greater in the *A. indica* plantation than in the forest, as tree density was much lower. However, the resources to be defended were not sufficiently aggregated to be continuously guarded against multiple groups that simultaneously made incursions to exploit them. Because defense behavior did not result in exclusive use of the resources in a range, and raiders were more efficient at harvesting resources because of their smaller group size, defense did not provide net foraging benefits.

The failure of resource defense to provide benefits may be understood in the framework provided by theoretical models showing that territoriality and resource defense are expected when there are intermediate levels of habitat variation (Carpenter and MacMillen, 1976; Pride, 2003). When habitat variation is low, resources are not worth defending because the gross benefit is too low, but when habitat variation becomes too pronounced, resources are not worth defending because gross costs of defense are too high. When the alternative habitat becomes poor enough, as in the food-scarce weaning season, many groups may raid the scarce high-quality food patches, and overwhelm the ability of the resident groups to mount effective defense. Thus even though females in large groups maintain access to high-quality habitats, have lower costs per conflict, and can evict rivals from resources within their typical range, they can be at a disadvantage because they are unable to repel multiple small raider groups, and have lower food intake than those small groups that gain access.

Given that, in the food-scarce season, large groups did not gain exclusive access to food resources from defense behavior, and may have suffered lower foraging efficiency as a result of their size, we must ask (1) why defense behavior was maintained at all; (2) why foraging time was not extended to compensate for the lower intake rate in large groups; and (3) why large groups did not fission, when doing so would give them greater foraging efficiency.

13.4.1. Why "Defend" When It Does Not Result in Foraging Advantages?

The defender group continued to allocate effort to resource defense even though doing so did not result in foraging advantages over raider groups, while defense costs could have been avoided by tolerating the presence of rivals. Three possible explanations for this can be made:

13.4.1.1. Resource Defense Is Maladaptive in Current Conditions, but Not Under Those in Which *L. Catta* Evolved

Although maladaptive behavior would be discouraged both by learning and natural selection, it is possible that the conditions in which resource defense was maladaptive are sufficiently uncommon that neither of these forces have shaped its expression. The conditions seen here are unlikely to have been prevalent throughout *L. catta*'s evolutionary history for several reasons. First, the habitat structure in the defender group's range was atypical. The defended resources were rows of exotic fruit trees planted along roads, and this contiguous fruit-tree monoculture would be uncharacteristic of most *L. catta* habitats prior to human arrival in Madagascar. Second, the invasion pressure the defending group faced was atypically high. Population density of *L. catta* at Berenty is much higher than anywhere else in Madagascar, possibly due to the high quality food and water usually available (Jolly et al., 2002). Where the habitat quality is lower, and population density is lower, fewer neighboring groups can attempt to exploit a defender group's resources. If groups are more widely dispersed across the landscape, a defender group will have to defend its resources far less often, and a fixed strategy of defense can be practical. While defense may not have been beneficial in this atypically harsh season in this atypical environment, a fixed strategy of resource defense—or antagonism toward all rival groups—could be the response best suited to the range of ecological conditions under which *L. catta* evolved. Characterizing behavior of *L. catta* outside of high-density gallery forest, that is, in most places this species is currently found (Sussman, this volume; Goodman, this volume), is an important next step towards understanding the costs and benefits of the behaviors previously found.

However, there are differences in *L. catta* intergroup agonistic behavior across seasons and among groups in different habitats (Jolly et al., 1993), which suggests that the effort devoted to resource defense is conditional rather than fixed. For example, at Beza-Mahafaly, a site with lower *L. catta* population density and little human modification of food supply, groups exhibit greater tolerance of rival groups (L. Gould, pers. comm.), at least in some seasons, and maintain overlapping typical ranges (Sauther and Sussman, 1993). If resource defense is not a fixed strategy, then its occurrence suggests that it was either less costly than alternative behavior in the short term, or actually provided net benefits over a longer term.

13.4.1.2. Resource Defense Is Less Costly Than Permitting Rivals to Deplete Resources

In the immediate or short-term timescale, the large defending group A2 had lower daily food intake than the smaller raider groups during the food-scarce months. However, it is not known what their food intake would have been if they had not exerted effort to evict raiders. If incursions by neighboring rival groups were not repelled, the resident group would face lower food availability as rival groups depleted the food supply. With no defense effort, costs to raiders will decrease, and more groups are expected to invade the highest quality food patches. Furthermore, since *L. catta* maximize food intake during the wet season (Pereira, 1993b), the expected depletion of food resources by unchallenged raiders can be substantial, particularly given the relatively efficient foraging of small raider groups. If this exploitation competition lowers the foraging efficiency of the resident group more than the cost of defense, then defense behavior is beneficial (Gill and Wolf, 1975). The large defender group faced at least 8 rival groups whose typical ranges were close to the *A. indica* plantation (comprising >100 animals), and it was observed evicting at least 5 of these groups in the food-scarce season. Assuming that these animals ingested food at a comparable rate to the observed groups (3–4 fruit/individual per minute), this represents a loss of 18,000–24,000[CE2] fruit per hour, which is approximately twice the observed daily intake the large group. Given the high potential depletion of food resources, the large defender's response can be seen as "the best of a bad situation," choosing to pay high defense costs instead of even higher costs of food scarcity, a strategy of converting exploitation competition to interference competition. If the payoff of raiding is greater than the payoff of defending, and this in turn is greater than the payoff of tolerating rivals, this will result in a "quasi-territoriality" (Sauther et al., 1999) in which territorial behavior is observed but exclusive range use is not.

13.4.1.3. Maintaining Spatial Dominance Is Less Costly Than Establishing It

Even if paying greater defense costs is not beneficial in the short term, there may be long-term consequences to permitting invasion that make it more costly than maintaining defense efforts. When the cost of maintaining spatial dominance is much lower than the cost of establishing it, for example, then greater investment in evicting rival groups even at times when it is not immediately advantageous will be favored. As shown here and in prior studies (Sauther and Sussman, 1993), *L. catta* maintain spatial dominance relationships in which the resident group wins intergroup conflicts; when far from their core areas, raiding groups often retreat immediately from residents without contest, suggesting a "resident wins" convention. This strategy reduces costs of intergroup agonism by eliminating escalation when challenges are unlikely to change the *status quo* (Davies, 1978). By maintaining a "resident wins" convention, costs of maintaining spatial dominance are kept low, and *L. catta* intergroup encounters rarely escalate to physical contact (Jolly, 1966). However, if two (or more) rival groups acquire familiarity with habitat such that residency is contested, both groups may value the resources more highly, conflicts would escalate, and the original resident would have to pay high costs to reestablish its spatial dominance, or lose its dominance status (Tobias, 1997). It may be cheaper to exert small effort to continuously evict rivals through periods when defense does not provide immediate net benefits than to exert great effort to displace fellow residents only when it does. In this study population, high *escalated* intergroup defense rates were associated with high cortisol levels (Pride, 2005a), while overall conflict rates were not, suggesting that the costs of establishing spatial dominance when contested may be higher than routine displacement of subordinate raider groups that immediately retreat. If reestablishing spatial dominance is stressful and maintaining it is not, then the duration for which nonbeneficial defense should be maintained may be quite high, presuming that defense benefits accrued at other times will ultimately make defense worthwhile. Given that periods food scarcity of the magnitude seen here occur as seldom as once in an animal's lifetime, even a slight benefit conferred by spatial dominance in typical conditions may outweigh the substantial costs accrued to maintain it through these atypical periods.

13.4.2. Why Did Large Groups Not Compensate for Lower Intake Rates by Foraging Longer?

Large groups could have compensated for their lower foraging efficiency if they had foraged longer, which is one way in which resource defense could provide benefits to large groups. Given that the small raider group obtained 50% more fruit per minute foraging than the larger groups, however, the two larger groups would have needed to forage 50% longer to compensate. Time individuals spent actively foraging on *A. indica* was small (e.g., 3 hours/day), but this figure does not include waiting for access to food patches, or movement among or within patches other than when directly searching for food, so actual time required to meet food requirements was considerably greater.

One possibility is that all groups were already foraging as much as was energetically feasible. *L. catta* maximize food intake in the rainy season, when resources are usually most abundant, and ingest up to 50% more food than they do in other seasons (Pereira, 1993b). At this time, the large defender group A2 devoted less time to resting than it did in the rest of the year (food-scarce season: 4.75 hours/day; other seasons: 6–7 hours/day; based on half-hourly scan samples of modal group activity during monthly full-day follows), suggesting that they were already maximizing active time. Active time may have been constrained due to thermal limitations, an important factor shaping time budgets in many taxa (Porter et al., 1973; Dunbar, 1996). The period of food scarcity occurred in the hottest season (daytime shade temperatures commonly >40°C), and foraging activity in the middle of the day (particularly in the exposed roadside *A. indica* plantation) could have been more costly than the expected gain from foraging, rendering extension of foraging time unfavorable.

Comparison with other groups supports this interpretation, as the groups that foraged on lower quality food sources at this time (i.e., foliage, dry fallen tamarind fruit) took longer midday siestas (A2, A1, CX: 2–2.5 hrs/day; SB, D1: 4–4.5 hrs/day). This would be expected because having a lower foraging payoff lowers the threshold temperature at which foraging payoff exceeds metabolic costs of being active, lengthening siestas. If large groups could not reduce their midday "siestas" beyond a certain threshold due to thermal constraints, then their foraging time would be limited. This could impose a cost on groups that are least efficient foragers, as they must either suffer low food intake or diminish other activities. Alternatively, the longer siestas taken by groups foraging on lower quality food sources may be considered a mechanism for coping with lower energy or moisture intake by reducing metabolic demand. Even if these groups obtained less food, they would be able to maintain positive energy balance by sleeping more (they also travelled less: A2, A1, CX: 1.1–1.7 km/day; SB, D1: 0.7–1.0 km/day). This option may not have been available to the groups foraging on *A. indica*; given the potential for depletion of resources by multiple groups, there may have been strong benefits to harvesting resources as fast as possible. Regardless of whether siestas limited foraging time of large groups or provided a way for groups foraging on uncontested resources to lower their demand, it is clear that large groups did not forage long enough to compensate for their lower intake rate.

13.4.3. Why Did Animals Stay in Large Groups?

Although large group size may provide benefits in typical conditions, this does not explain why large groups do not fission when large group size becomes unfavorable. Although *L. catta* do actively constrain the size of their groups (Vick and Pereira, 1989; Koyama, 1991; Hood and Jolly, 1995), they do not attempt to match short-term environmental fluctuations that alter optimal group size (Pride, 2005c). A likely explanation is that costs associated with the process of evicting rivals from a group (or establishing one's own group after being evicted), as well as costs of reassessing and reestablishing dominance relationships within and among unstable groups, exceed costs of being suboptimally large for a short time period. In other primate species, periods of dominance assessment and changes in group composition are accompanied by increases in glucocorticoid levels and agonism rates (Sapolsky, 1983; Alberts et al., 1992). In *L. catta*, establishment of dominance relations within and among groups can involve intense or sustained fighting (Vick and Pereira, 1989; Hood and Jolly, 1995), and therefore can be energetically costly and involve risk of injury. While group fissions sometimes are gradual processes of increasingly segregated ranging by subgroups—as is usually observed at Beza Mahafaly (R. Sussman, pers. comm.)—they are not always so, with very high intergroup conflict rates following a fission (Hood and Jolly, 1995; pers. obs.). Given that the food-scarce season when large groups were disadvantageous was atypical and brief, the benefit of reducing group size to match it could be less than the "energy of activation" associated with the reduction.

Due to these potentially high costs, group size modulation could be constrained to occur only in certain times of the *L. catta* annual cycle, when they are least likely to interfere with competing demands such as lactation or seasonal replenishment of fat reserves. Fissions that have occurred during early lactation have resulted in infant deaths (Jolly et al., 2000), suggesting immediate fitness costs in addition to energetic burdens that may constrain fissions. No large groups split during the food-scarce season (even though this was when the penalty for being in a large group was greatest due to low food intake); however, two groups (A2 and D1) did eventually fission at the end of this study. This occurred during the typical dry season, when variation in habitat quality was minimal, intergroup competition was likely to be low, and group size would not be expected to greatly affect food intake. It has been argued that seasonal patterns in targeted aggression (intragroup dominance reversals that can result in evictions) result from seasonal differences in intensity of competition (Pereira, 1993a); the timing of fission here provides anecdotal support for this idea, but suggest that the events are either (1) constrained by fixed annual cycles (like much of *L. catta* physiology (Pereira et al., 1999) such that fissions are unlikely to occur at certain times of year even if atypical conditions make them favorable, or (2) constrained to times when costs of eviction are borne most easily, not simply at times when eviction would be most favorable. If these or other constraints limit the ability of groups to modulate group size, then observed behavioral strategies may be considered "the best of a bad situation."

13.4.4. Extensions and Other Considerations

Other factors not examined here, such as predation and infanticide avoidance, can promote larger primate group sizes. The importance of one factor—intergroup competition for food resources—does not diminish the potential importance of other selection pressures. When the resource distribution causes lower variation in intergroup conflict rates, the capacity of intergroup conflict to explain group size variation may be negligible. This is expected to occur when there is little variation in resource availability among groups (Pride, 2003). Furthermore, intergroup conflict rates depend not only on the distribution and abundance of resources in groups' habitats, but also on group history (Jolly et al., this volume). Recently fissioned groups may have higher conflict rates as they modulate their ranging patterns or establish spatial dominance relations with their former groupmates and neighboring groups [e.g., A1/A2 (Hood and Jolly, 1995), D1A/D1B (Mertl-Millhollen et al., 2003), CX/SH and T2/U2 (Koyama et al., 2002)]. Group history may interact with resource patchiness to produce "high-confrontation pairs" (Jolly et al., 1993) as former subgroups contest resources worth defending. Thus we would expect to find variation in the conflict-groupsize relationship, but mainly when dominance relationships among groups are perturbed by changes in group composition or the resource base. Finally, it should be noted that modulations of conflict rate and group composition offer two solutions, but by no means the only solutions, to ecological problems associated with sharing habitat with conspecifics.

13.5. Conclusion

Large group size conferred resource defense advantages by allowing members to participate in a smaller proportion of intergroup conflicts, and large groups maintained access to the highest quality habitat in all seasons. This may lead to foraging advantages and promote membership in larger groups in conditions where intergroup conflict is common. The correlation between group size and intergroup conflict rate is consistent with the idea that intergroup competition for food resources drives increases in group size, and therefore may explain the longterm differences in group size observed among Berenty's different habitat regions. However, defense behavior does not always allow groups to maintain exclusive access to food resources. As a result, large groups do not always gain foraging benefits in spite of their occupying the highest quality habitat and exerting spatial dominance over rivals. Alternative strategies for large groups (tolerance of rivals, increasing time spent foraging, or fissioning) were not adopted. Defense behavior may be maintained because it diminishes the impact of the raiders and provides long-term advantages, across seasons and possibly generations.

References

Alberts, S. C., Sapolsky, R. M., and Altmann, J. (1992). Behavioral, endocrine, and immunological correlates of immigration by an aggressive male into a natural primate group. *Hormones Behav.* 26:167–178.

Altmann, J. (1974). Observational study of behavior: Sampling methods. *Behaviour* 49:227–265.

Carpenter, F., and MacMillen, R. (1976). Threshold model of feeding territoriality and test with a Hawaiian honeycreeper. *Science* 194:639–642.

Davies, N. B. (1978). Territorial defense in the speckled wood butterfly (*Pararge aegeria*): The resident always wins. *Anim. Behav.* 26:138–147.

Dunbar, R. I. M. (1996). Determinates of group size in primates: A general model. In: *Evolution of Social Behaviour Patterns in Primates and Man.* Oxford University Press, Oxford, pp. 33–57.

Gill, F. B., and Wolf, L. L. (1975). Economics of feeding territoriality in the golden-winged sunbird. *Ecology* 56:333–345.

Gould, L., Sussman, R., and Sauther, M. L. (2003). Demographic and life-history patterns in a population of ring-tailed lemurs (*Lemur catta*) at Beza Mahafaly Reserve, Madagascar: A 15-year perspective. *Am. J. Phys. Anthropol.* 120:182–194.

Hood, L. C., and Jolly, A. (1995). Troop fission in female *Lemur catta* at Berenty Reserve, Madagascar. *Int. J. Primatol.* 16:997–1015.

Jolly, A. (1966). *Lemur Behavior: A Madagascar Field Study.* University of Chicago Press, Chicago.

Jolly, A., Rasamimanana, H. R., Kinnaird, M. F., O'Brien, T. G., Crowley, H. M., and Harcourt, C. S. (1993). Territoriality in *Lemur catta* groups during the birth season at Berenty, Madagascar. In: Kappeler, P. M., and Ganzhorn, J. U. (eds.), *Lemur Social Systems and Their Ecological Basis.* Plenum, New York, pp. 85–109.

Jolly, A., Caless, S., Cavigelli, S., Gould, L., Pereira, M. E., Pitts, A., Pride, R. E., Rabenandrasana, H. D., Walker, J. D., and Zafison, T. (2000). Infant killing, wounding, and predation in *Eulemur* and *Lemur. Int. J. Primatol.* 21:21–40.

Jolly, A., Rasamimanana, H., and Pride, R. E. (2001). Storm damage at Berenty Reserve. *Lemur News* 5:13.

Jolly, A., Dobson, A., Rasamimanana, H. M., Walker, J., O'Connor, S., Solberg, M., and Perel, V. (2002). Demography of *Lemur catta* at Berenty Reserve, Madagascar: Effects of troop size, habitat, and rainfall. *Int. J. Primatol.* 23:327–355.

Koyama, N. (1991). Troop division and inter-troop relationships of ring-tailed lemurs (*Lemur catta*) at Berenty, Madagascar. In: Ehara, A. (ed.), *Primatology Today: Proceeding of the XIIIth Congress of the International Primatological Society.* Elsevier Science Publishers, New York, pp. 173–176.

Koyama, N., Nakamichi, M., Ichino, S., and Takahata, Y. (2002). The population and social dynamics changes in ring-tailed lemur troops at Berenty, Madagascar between 1989–1999. *Primates* 43:291–314.

Lowen, C., and Dunbar, R. I. M. (1994). Territory size and defensibility in primates. *Behav. Ecol. Sociobiol.* 35:347–54.

Mertl-Millhollen, A. S., Moret, E. S., Felantsoa, D., Rasamimanana, H., Blumenfeld-Jones, K. C., and Jolly, A. (2003). Ring-tailed lemur home ranges correlate with food abundance and nutritional content at a time of environmental stress. *Int. J. Primatol.* 24:969–985.

Mitani, J. C., and Rodman, P. S. (1979). Territoriality: The relation of ranging pattern and home range size to defendability, with an analysis of territoriality among primates. *Behav. Ecol. Sociobiol.* 5:241–251.

Nagy, K. A., Girard, I. A., and Brown, T. K. (1999). Energetics of free-ranging mammals, reptiles, and birds. *Ann. Rev. Nutr.* 19:247–277.

Pereira, M. E. (1993a). Agonistic interaction, dominance relation, and ontogenic trajectories in ringtailed lemurs. In: Pereira, M. E. and Fairbanks, L. A. (eds.), *Juvenile Primates: Life History, Development, and Behavior*. Oxford University Press, New York, pp. 285–305.

Pereira, M. E. (1993b). Seasonal adjustment of growth rate and adult body weight in ringtailed lemurs. In: Kappeler, P. M., Ganzhorn, J. U. (eds.), *Lemur Social Systems and Their Ecological Basis*. Plenum Press, New York, pp. 205–221.

Pereira, M. E., Strohecker, R. A., Cavigelli, S. A., Hughes, C. L., and Pearson, D. D. (1999). Metabolic strategy and social behavior in Lemuridae. In: Rakotosamimanana, B., Rasamimanana, H., Ganzhorn, J. U., and Goodman, S. M. (eds.), *New Directions in Lemur Studies*. Kluwer Academic/Plenum, New York, pp. 93–118.

Porter, W. P., Mitchell, J. W., Beckman, W. A., and DeWitt, C. B. (1973). Behavioral implications of mechanistic ecology. *Oecologia* 13:1–54.

Pride, R. E. (2003). The socio-endocrinology of group size in *Lemur catta* (PhD). Princeton University, Princeton, N. J.

Pride, R. E. (2005a). Foraging success, agonism, and predator alarms: Behavioral predictors of cortisol in *Lemur catta*. *Int. J. Primatol.* 26:295–319.

Pride, R. E. (2005b). High faecal glucocorticoid levels predict mortality in ring–tailed lemurs (*Lemur catta*). *Biol. Lett.* 1:60–63.

Pride, R. E. (2005c). Optimal group size and seasonal stress in ring-tailed lemurs (*Lemur catta*). *Behav. Ecol.* 16:550–560.

Rubenstein, D. I. (1981). Population density, resource patterning, and territoriality in the Everglades pygmy sunfish. *Anim. Behav.* 29:155–172.

Sapolsky, R. M. (1983). Endocrine aspects of social instability in the olive baboon (*Papio anubis*). *Am J Primatol.* 5:365–379.

Sauther, M. L. (1992). Effect of reproductive state, social rank, and group size on resource use among free-ranging ringtailed lemurs (*Lemur catta*) of Madagascar (PhD). Washington University, St. Louis, Mo.

Sauther, M. L. (1993). Resource competition in wild populations of ringtailed lemurs (*Lemur catta*): Implications for female dominance. In: Kappeler, P. M., and Ganzhorn, J. U. (eds.), *Lemur social systems and their ecological basis*. New York: Plenum Press; 135–152.

Sauther, M. L. (1998). The interplay of phenology and reproduction in ringtailed lemurs: implications for ringtailed lemur conservation. In: Harcourt, C. S., Crompton, R. H., and Feistner, A. T. C. (eds.), *Biology and Conservation of Prosimians. Folia Primatol.*, 69(Supp 1):309–320.

Sauther, M. L., and Sussman, R. W. (1993). A new interpretation of the organization and mating systems of the ring-tailed lemur (*Lemur catta*). In: Kappeler, P. M., and Ganzhorn, J. U. (eds.), *Lemur Social Systems and Their Ecological Basis*. Plenum Press, New York, pp. 111–121.

Sauther, M. L., Sussman, R. W., and Gould, L. (1999). The socioecology of the ringtailed lemur: Thirty-five years of research. *Evol. Anthr.* 8:120–132.

Sussman, R. W. (1991). Demography and social organization of free-ranging *Lemur catta* in the Beza Mahafaly Reserve, Madagascar. *Am. J. Phys. Anthropol.* 84:43–58.

Tobias, J. (1997). Asymmetric territorial contests in the European robin: The role of settlement costs. *Anim. Behav.* 54:9–21.

Vick, L. G., and Pereira, M. E. (1989). Episodic targetting aggression and the histories of Lemur social groups. *Behav. Ecol. Sociobiol.* 25:3–12.

Wrangham, R. W. (1980). An ecological model of female-bonded primate groups. *Behaviour* 75:262–300.

Wu Leung, W. T., Butram, R. R., and Chang, F. (1972). Food composition table for use in East Asia. Rome: Food policy and Nutrition Division, Food and Agriculture Organization of the United Nations.

14
Social Changes in a Wild Population of Ringtailed Lemurs (*Lemur catta*) at Berenty, Madagascar

SHINICHIRO ICHINO AND NAOKI KOYAMA

14.1. Introduction

Social changes are important social phenomena required to understand primate social systems. Troop fission has been reported in many primate species, particularly multimale, multifemale cercopithecine species (e.g., Japanese macaque: Koyama, 1970; Rhesus macaque: Malik et al., 1985; Toque macaque: Dittus, 1988; Chacma baboon: Ron, 1996; Anubis baboon: Nash, 1976). Troop fission generally occurs in a large-sized troop, and it is argued that it functions to decrease the high costs of female competition for local resources within the troop (Wrangham, 1980; van Schaik, 1983; Dittus, 1988). On the other hand, several studies have reported that adult males initiated the fission process, and male sexual competition might be an important factor affecting the fission pattern (Yamagiwa, 1985; Ron, 1996; Kuester and Paul, 1997).

Ringtailed lemurs (*Lemur catta*) form a multimale, multifemale troop, which is relatively similar to the social group of the cercopithecine species (Jolly, 1966; Sauther et al., 1999). The ringtailed lemurs have several unique social characteristics. For example, females are socially dominant over males (Jolly, 1966; Kappeler, 1990; Pereira et al., 1990) and bear the responsibility for range defense (Budnitz and Dainis, 1975). Agonistic interactions among individuals are almost dyadic and agonistic support for other individuals is rare (Nakamichi and Koyama, 1997; Pereira and Kappeler, 1997).

Two forms of major social changes have been reported in ringtailed lemurs. Troop fission has been reported in two major study sites in Madagascar—in Berenty and Beza Mahafaly (Berenty: Koyama, 1991; Hood and Jolly, 1995; Jolly and Pride, 1999; Koyama et al., 2002; Ichino, 2006; and Beza Mahafaly: Sussman, 1991; Gould et al., 2003). Female eviction is another common social phenomenon in this species (Vick and Pereira, 1989; Koyama et al., 2002) and may result in troop fission. In the semi-free-ranging population at the Duke University Primate Research Center, persistent aggression occasionally occurred among females, and this resulted in female eviction (Vick and Pereira, 1989; Pereira, 1993).

Range takeover is a unique social phenomenon observed in this species in Berenty. Koyama (1991) reported that an evicted female group took over the entire range of a neighboring troop. The troop lost their range and eventually disappeared from the study area. This social phenomenon (range takeover) is rare (Jolly and Pride, 1999; Koyama et al., 2002), but it may be important in Berenty, because it affects a particular population structure within a larger population.

Thus, troop fission, female eviction, and range takeover are major social changes in ringtailed lemurs in Berenty. In this paper, we summarize these three social changes that occurred in the 14.2-ha study area in Berenty for 12.5 years. The aim of this paper is to reveal the characteristics of social changes in ringtailed lemurs.

14.2. Methods

14.2.1. Study Population

The population of ringtailed lemurs reported in this paper inhabits the 14.2-ha study area at Berenty, Madagascar, where, since 1989, Koyama and his colleagues conducted a continuous, long-term research on the basis of individual identification. The age and kin relationships of all the individuals who were born after 1989 have been recorded (Koyama et al., 2001, 2002).

The study was conducted for a period of 12.5 years—from September 1, 1989, to January 19, 2002. A summary of the social changes during 10 years—from 1989 to 1999—and several other reports on the detailed process of social changes have already been published (Koyama, 1991; Koyama et al., 2002; Ichino, 2006). In this paper, we have used these published data to prepare a 10-year summary (Koyama et al., 2002) with more detailed information recorded by N. Koyama and newly collected data by S. Ichino for the following study periods: (1) from September 1 to 30, 1999; (2) from August 18 to November 19, 2000; and (3) from March 28, 2001 to January 19, 2002.

We have described each case of social changes for a period of 2.5 years—from September 1, 1999, to January 19, 2002; this is followed by a summary of the data for the 12.5-year period.

14.2.2. Definition of Terms

We have defined three social phenomena, that is, troop fission, female eviction, and range takeover, as social changes. "Troop fission" is a case of creation of permanent new social groups with secure ranges by the division of troop members. "Female eviction" is a case of persistent attack of one or several females by other females, eventually leading to their eviction from the other troop members (Vick and Pereira, 1989; Koyama et al., 2002). "Range takeover" is a case wherein one troop completely takes over another troop's secure range (Koyama, 1991).

We have also defined the terms "nomadic troop" and "nomadic group." A nomadic troop is a troop that does not maintain an exclusive core area (Jolly and Pride, 1999). Jolly and Pride (1999) considered a group that only comprised females as a troop. However, in this paper, we have referred to a nomadic female group without a single adult male as a nomadic group and the group with at least one adult male as nomadic troop.

Individuals who were 2.5 years of age or more were considered as adults.

14.3. Results

14.3.1. Brief Description of New Cases

In this paper, we present new cases of social changes for 2.5 years—from September 1, 1999, to January 19, 2002. During this period, three cases of female eviction (including one case of troop fission) and two cases of range takeover were observed.

14.3.1.1. Female Eviction from Troop T1 (Troop Fission)

On September 19, 2000, there was intense aggression among females occurred and this consequently resulted in troop fission. First, two adult females, the second ranked female (HIT-90195♀) and her mother (HIT-901♀), were evicted from troop T1. At that time, troop T1 consisted of 20 individuals, including nine adult females. Subsequently, two adult females and their immature offspring and one adult male immigrated into the group (Ichino, 2006). Finally, they established a new range by March 2001. As a result of this fission, troop T1 got divided into troop T1A (12 individuals including five adult females) and troop T1B (eight individuals including four adult females) (Figure 14.1). Consequently, all females of one kin group (HIT-kin group) left their troop and formed troop T1B.

14.3.1.2. Female Eviction from Troop C1

On September 22, 2000, four females, including three adults, were evicted from troop C1 as a result of intense aggression among females. At that time, troop C1 consisted of 20 individuals, including eight adult females. The victims of this aggression were the subordinate kin group (MK- and MW-kin groups) females and the aggressors were the dominant kin group (ME-kin group) females. As a result of the eviction, females of only one kin group (ME-kin group) remained in the troop C1. The evicted females (MK92-group) were nomadic until November 19, 2000. However, they had disappeared from the study area by the beginning of April 2001 (Figure 14.1).

14.3.1.3. Female Eviction from Troop C2A

On September 1, 2000, troop C2A consisted of 16 individuals, including six adult females. The females consisted of two kin groups (OD- and SI-kin groups) and two immigrant females (KN and KM; see Koyama et al., 2002) and their offspring.

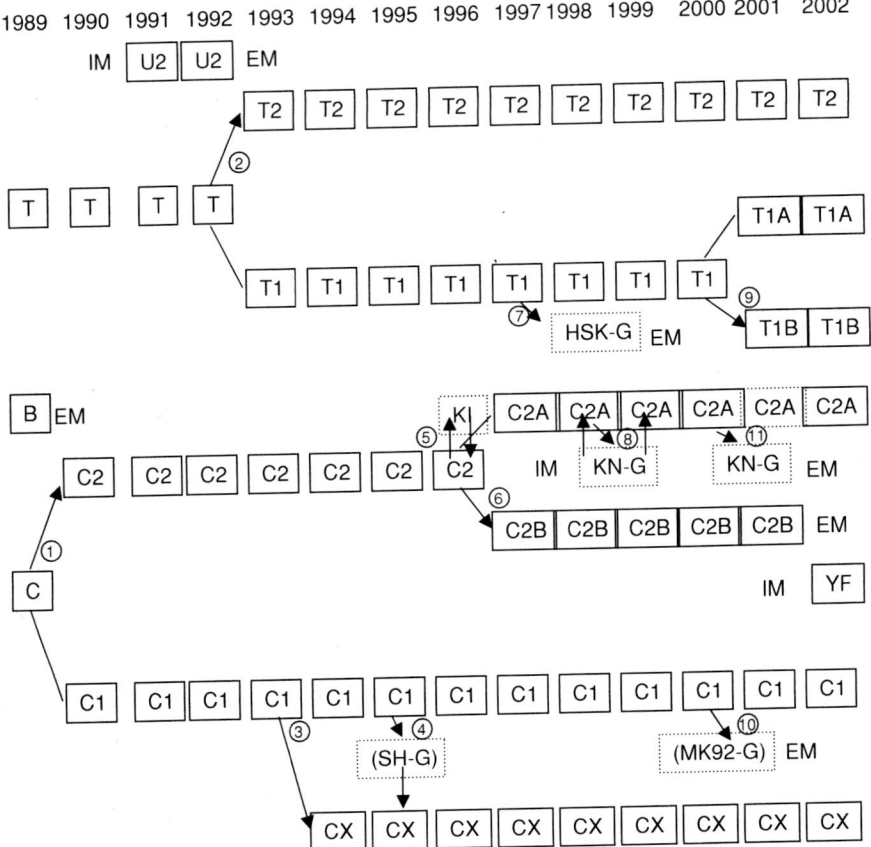

Figure 14.1. Troop histories of a ringtailed lemur population within the 14.2-ha study area for 12.5-years—from September 1989 to January 2002. Each square shows troops living within the study area on September 1, during each year (but January 19 for 2002). The dotted square shows nomadic groups/troops. The circled numbers correspond with the case number in Table 14.1. IM shows a shift of troop to the study area. EM shows a shift of troop to the outside of the study area.

On November 4, the top ranked female (OD) persistently chased an immigrant female (KN) for seven minutes. After the chase, the immigrant females began to travel, maintaining a safe distance from the OD-kin group females, although they continued to travel together as one troop till at least November 19. However, they (KN-group) traveled independently from troop C2A by the end of March, 2001 (Figure 14.1). However, by this time, the number of adult females in the troop C2A, including the evicted females (KN-group), had increased to seven as one female had reached the age of 2.5 years. The KN-group left the study area in mid-May.

14.3.1.4. Takeover of Troop C2A's Range by Troop T1B

Troop T1B, which was nomadic until November 19, 2000, took over the entire range of troop C2A by the end of March 2001. After the range takeover, troop C2A was nomadic for at least six months. During this period, three individuals, including two adult females, died, and three males immigrated to troop T1B. Although the troop size decreased rapidly during these six months, troop C2A successfully established a new range in the boundary area between troops T1A and T1B by October 2001.

14.3.1.5. Takeover of Troop C2B's Range by Troop YF

Troop YF was a branch troop of troop A2 (Pride, pers. comm.), which shared a boundary with several of our study troops. Troop YF shifted to the study area by September 2001. It consisted of six individuals, including four adult females who conflicted frequently and intensively with our study troops (troops C1, C2A, and C2B) in September and October 2001.

On November 22, 2001, we first observed that troop YF drove out troop C2B from a feeding tree. Following this, by the beginning of December 2001, troop YF completely took over the entire range of troop C2B. Subsequently, troop C2B left the study area by mid-January 2002. Troop C2B males stopped following the females within the group. The top ranked old female of troop C2B died shortly after they left the study area (Jolly, pers. comm.).

14.3.2. Summary for a 12.5-Year Period

14.3.2.1. Troop Fission and Female Eviction in the Study Area

During the 12.5-year period, a total of 11 cases of female eviction were recorded in the study area (cases 1–11 in Table 14.1; Figure 14.1). Seven cases (63.6%) of female eviction were directly observed. All the evictions occurred in September, corresponding with the birth season at Berenty (Koyama et al., 2001).

Female eviction occurred when the troop comprised 16 or more individuals (mean 22.5, range 16–28) and the number of adult females was seven or more (mean 8.4, range 7–10). The mean number of evicted adult females was 2.7 (range 1–5). Females were generally evicted due to intense and persistent aggression (targeted aggression: Vick and Pereira, 1989).

All the females belonging to the subordinate kin group were evicted (see "Brief Description of New Cases," above). When the females were evicted from their troop, they did not secure a core range at once. Except for the case of troop T2 (case 2 in Table 14.1; Koyama et al., 2002), the troop males did not follow the evicted females immediately. Thus, in most cases, evicted females began to travel around their former range as a nomadic group (Figure 14.2).

Only in five out of the 11 cases of female eviction (45.5%), the evicted females established a new range within the study area (i.e., troop fission). Three nomadic troops established new ranges by taking over the entire range of the neighboring

TABLE 14.1. Nomadic groups/troops observed in the 14.2-ha strudy area during the 12.5-year period.

Case	Year	Group/troop	Nomadic period	Cause	Consequence	Source
1	1989	C2	1 month	Eviction	Established new range (RT)	1
2	1993	T2	?	Eviction	Established new range (RT)	1
3	1993	CX	?	Eviction?	Established new range	1
4	1995	SH-G	1 month	Eviction	Transferred to another troop	1
5	1996	KI[a]	3 to 11 months	Eviction	Rejoined troop	1
6	1997	C2B	?	Eviction	Established new range	1
7	1997	HSK-G	14+months	Eviction	Left the study area	1, 2
8	1998	KN-G	5 months	Eviction	Rejoined troop	1, 2
9	2000	T1B	4 to 7 months	Eviction	Established new range (RT)	This study
10	2000	MK92-G	3+months	Eviction	Left the study area	This study
11	2001	KN-G	2 to 6+months	Eviction	Left the study area	This study
12	1991	U2	?	Shift to the study area	Established new range	1
13	1997	KN-G	7 months	Shift to the study area	Transferred to another troop	1, 2, 3
14	2001	YF	4+months	Shift to the study area	Established new range (RT)	This study
15	1990	B	?	Range takeover	Left the study area	1
16	1993	U2	?	Range takeover	Left the study area	1
17	2001	C2A	7 to 11 months	Range takeover	Established new range	This study
18	2001	C2B	1+months	Range takeover	Left the study area	This study

RT, by range takeover.
[a] KI is a female ID of a nomadic individual.
Source: 1, Koyama et al. (2002); 2, personal observation; 3, Miyamoto, personal communication.

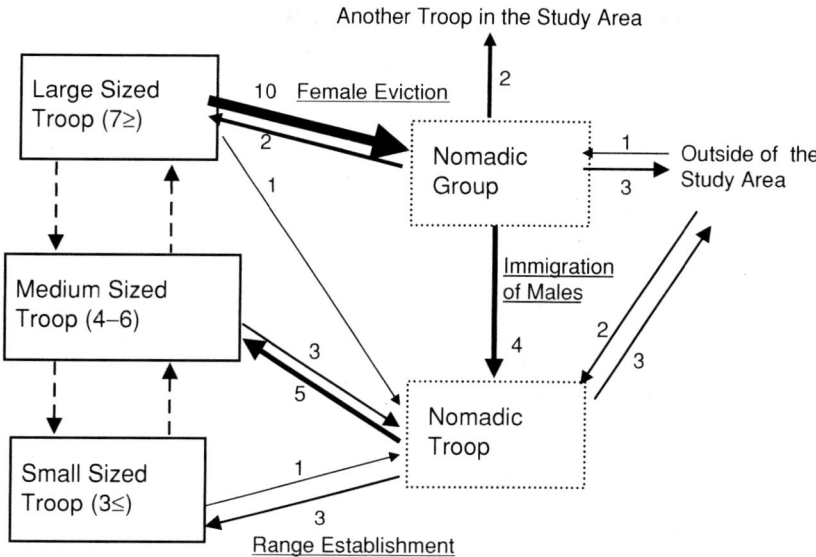

Figure 14.2. The process of social changes and phases of social groups in a ringtailed lemur population in Berenty, Madagascar. The thickness of each line and the numbers on the lines show the relative value and the number of observed cases. The number in the squares shows the number of adult females in the troop.

troops. Another two nomadic troops established small ranges within the study area by acquiring a part of their former troop or the neighboring troop's ranges. The number of adult females in the nomadic troops that successfully established new ranges was 3–6.

In all the five cases, the evicted female groups established their ranges after they accepted at least one male. Except for the case of troop CX (case 3 in Table 14.1), the males got separated from the former troop of evicted females. These immigrant males were not dominant in their former troops; all of them ranked third in their troops (troops C2, T2, C2, and T1).

Two evicted female groups rejoined their troops, while one joined another troop. Three other evicted female groups could not establish their new ranges, and hence, they left the study area. They were relatively small-sized groups that only 2–3 adult females and no males.

14.3.2.2. Range Takeover in the Study Area

During the 12.5-year period, a total of four cases of range takeover were observed in the study area. This means that 50% of the new troop's ranges were established by range takeover because eight nomadic troops established new ranges in the study area (Table 14.1). The number of adult females in the troops whose ranges were taken over was more than three (Figure 14.2). Two of them comprised more adult females than the troops that took over them. Two troops established a new range within or around the study area immediately after their entire ranges were taken over. However, two other troops lost troop members after the range takeover and consequently disappeared from the study area.

14.3.2.3. Consequences of Social Changes

Nomadic groups and nomadic troops were some phases of a social group and were associated to social changes in our study population in Berenty (Figure 14.2). During the 12.5-year period, a total of 18 nomadic troops/groups (including one nomadic individual) were recorded in the study area (Table 14.1). At least 50.0% of these had no secure core areas for more than two months (Table 14.1). This was due to female eviction from their troop in 11 cases (61.1%), range takeover in four cases (22.2%), and a shift to the study area in three cases (16.7%).

Eight cases (44.4%) established new ranges within the study area, four cases (22.2%) rejoined their troops or transferred to another troop, and six cases (33.3%) left the study area, 33.3% (6 cases). As a result, the number of troops increased and troop composition in the study area changed each year (Figure 14.1).

14.4. Discussion

The first characteristic of a social change in ringtailed lemurs is the first step of the fission process. Female eviction may be a typical process of troop fission in Berenty (Figure 14.2). Three previous detailed studies on the process of troop fission in Berenty have shown that a dominant subgroup persistently attacked a subordinate subgroup at the beginning of the fission process (Koyama, 1991; Hood and Jolly, 1995; Ichino, 2006). In another wild population in Beza Mahafaly, Gould et al. (2003) reported that a pair or a small subgroup of females had been forced out of the larger parent group. Furthermore, many cases of female eviction due to intense and persistent aggression (i.e., targeted aggression) have also been reported in the semi-free-ranging population at the Duke University Primate Center (DUPC; Vick and Pereira, 1989). On the other hand, in cercopithecine species, subordinate females generally leave their troop spontaneously and this results in troop fission (e.g., Dittus, 1988). Thus, the first step of the fission process in ringtailed lemurs was different from that in the cercopithecine species.

Many previous studies have suggested that ringtailed lemurs have a matrilineal society in which a troop consists of close kin females and non-kin mature males (Taylor and Sussman, 1985; Vick and Pereira, 1989; Sussman, 1992; Jolly, 1996; Nakamichi and Koyama, 1997; Sauther et al., 1999; Koyama et al., 2002). Our study supported these suggestions. Troop fission and female eviction occured between matrilineal kin groups in our study population (see "Brief Description of New Cases," above). Furthermore, female immigration is rarer than male immigration in ringtailed lemurs (Jones, 1983; Sussman, 1992; Koyama et al., 2002), although it was observed in both Berenty and Beza Mahafaly (Jolly and Pride, 1999; Koyama et al., 2002; Gould et al., 2003). In our study population, one female group (KN group) that is unlikely to have kin relations to our study troop females, immigrated into a troop (Koyama et al., 2002). However, they were evicted within a few years (Figure 14.1). Thus, ringtailed lemur troops may generally consist of close kin females. On the other hand, Gould et al. (2003) reported that a new troop had formed from two formerly different troops and established a range in Beza Mahafaly. When troop size reduces, non-kin females might form a troop.

Female evictions may function to decrease the intense competition within the group among females. Evictions occurred in troops that contained relatively more individuals (16 or more), particularly more adult females (7 or more) (Figure 14.2). As the number of adult females increases, the competition within the group among females may become critically high. Jolly et al. (2002) have shown that the birth rate is low in large-sized troops in Berenty. Takahata et al. (2005) also noted that in large-sized troops with six or more adult females, the birth rate, infant survival rate, and number of surviving infants within one year after birth in the low ranking adult females were lower than those in the other females.

Occurrence of female eviction may correspond to the reproductive season in ringtailed lemurs. Female eviction occurred primarily in September, the birth season in the study population in Berenty (Koyama et al., 2001). On the other hand, a large majority of targeted aggression among females at the DUPC occurred not only in the birth season (47%) but also immediately before and during the mating season (37%; Vick and Pereira, 1989). Because most of our observation period is biased, that is, they have been carried out during the birth season and lactation season, we cannot conclude whether this is a difference between populations; however, we have never observed female eviction during the mating season in our study population.

The second characteristic of the fission process is the recessive role of males. In the cercopithecine species, males sometimes initiated the fission process (e.g., Yamagiwa, 1985; Ron, 1996); however, we could not find clear evidence for the male playing a dominant role in the lemur fission process during our 12.5-year period. Except for one case, no males followed the evicted females immediately after an eviction. However, at least one adult male immigrated into the group before they established new ranges. This suggests that males decide the timing of the immigration into nomadic groups by monitoring them. The dominance rank of males who joined the nomadic groups was neither first nor second in their former troops (Koyama et al., 2005). Dominant males may avoid the risk of immigrating into the nomadic groups which may sometimes fail to establish new range. In contrast with dominant males, middle ranking males may take a risk of immigrating into nomadic groups because by becoming first-ranking males in these nomadic groups, they may have a better opportunity to mate with the females.

The third characteristic of the fission process is the nomadic phase of social groups (Figure 14.2). Nomadic troops/groups were caused not only by female eviction but also by a shift to the study area and by range takeover (Table 14.1). Although Jolly and Pride (1999) have reported nomadic troops to be a "rare event" in Berenty, it was not rare in our study area. During our 12.5-year study,

18 nomadic troops/groups were observed and at least 50% of these troops/groups stayed within the study area for more than two months (Table 14.1). In contrast, in Beza Mahafaly, new fission groups could establish a new range immediately after the fission. The difference between the two populations suggests that the nomadic phase of social groups is a specific phenomenon, which reflects the ranging pattern in high-density population areas. In Beza Mahafaly, the population density was lower than that in Berenty and the range of troops largely overlapped each other (Sauther and Sussman, 1993; Jolly and Pride, 1999; Jolly et al., this volume). On the other hand, in a high-density area in Berenty, the competition between troops may be intense (Pride et al., this volume) and ringtailed lemurs behave as territorial (Jolly et al., this volume). Therefore, at Berenty, immediately after the eviction, newly evicted females may have difficulty in establishing new ranges, particularly with fewer adult females. The nomadic groups that had left the study area comprised three or less adult females.

Nomadic troops established their new ranges by range takeover in four cases (50.0%) of the observed cases. Nomadic troops/groups in our study area have probably increased because of the intense competition between the groups. In our study area, as the population increased steadily (at 2.7% per year for 1989–1999), the troop size and the range of each troop decreased than before (Koyama et al., 2002; Koyama et al., this volume). In such a situation, the nomadic troops may have an opportunity to completely take over another troop's range. Several troops whose ranges were taken over were not always smaller than the troops that took over. This suggests that the defense of a range may depend on the individual ability or bonding between females rather than the number of adult females present. In two out of the four cases of range takeover, the troop females divided into two groups immediately after the range takeover (troops B and C2A). This suggests that the instability of female-female relationships within the troop might increase the risk of loss of range.

14. 5. Summary

We summarized social changes (troop fission, female eviction, and range takeover) in a wild population of ring-tailed lemurs within a 14.2-ha study area at Berenty, Madagascar. During a 12.5-year period, total 11 cases of female eviction (including four cases of troop fission) were recorded. Female evictions occurred primarily around birth seasons, particularly in September, in large-sized troops with 16 or more individuals; among these, it particularly occurred in troops in which seven or more individuals were adult females. All the subordinate kin group females were evicted. When the females were evicted from their troop, they did not secure an exclusive core area at once. Such groups (i.e., nomadic groups) emerged as a result of female eviction (n = 11) as well as their entire ranges being taken over by other nomadic troops (n = 4) or a shift to the study area (n = 3). The evicted female groups (n = 11) resulted in individual females (1) establishing a new secure range by taking over a part of or an entire range that originally belonged to their former troop or the neighboring troop (i.e., troop fission; n = 5), (2) rejoining their troop (n = 2), (3) immigrating to another troop (n = 1), and (4) disappearing from the study area (n = 3). The nomadic groups established new ranges only after accepting at least one adult male.

Acknowledgments. We thank Jean de Heaulme and his family for their hospitality at Berenty Reserve and their kind permission to carry out this study; A. Randrianjafy, a former director of the Botanical and Zoological Park of Tsimbazaza, and the government of the Malagasy Republic for their kind permission to perform our research in Madagascar. We are grateful to A. Jolly, M.A. Huffman, G. Yamakoshi, Y. Takahata, and M. Nakamichi for their comments on the manuscript. This work was supported by Grants in Aid for Scientific Research from the Ministry of Education, Culture, Sports, Science and Technology, Japan, to S. Yamagishi (no. 01041079) and N. Koyama (no. 06610072 and 05041088) This work was also supported by the Kyoto University Foundation to S. Ichino.

References

Budnitz, N., and Dainis, K. (1975). *Lemur catta*: Ecology and behavior. In: Tattersall, I., and Sussman, R. W. (eds.), *Lemur Biology*. Plenum Press, New York, pp. 219–235.

Dittus, W. P. J. (1988). Group fission among wild toque macaques as a consequence of female resource competition and environmental stress. *Anim. Behav.* 36:1626–1645.

Gould, L., Sussman, R. W., and Sauther, M. L. (2003). Demographic and life-history patterns in a population of ring-tailed lemurs (*Lemur catta*) at Beza Mahafaly Reserve, Madagascar: A 15-year perspective. *Am. J. Phys. Anthropol.* 120:182–194.

Hood, L. C., and Jolly, A. (1995). Troop fission in female *Lemur catta* at Berenty Reserve, Madagascar. *Int. J. Primatol.* 16:997–1015.

Ichino, S. (2006). Troop fission in wild ring-tailed lemurs (*Lemur catta*) at Berenty, Madagascar. *Am. J. Primatol.* 68:97–102.

Jolly, A. (1966). *Lemur Behavior*. University of Chicago Press, Chicago.

Jolly, A., and Pride, E. (1999). Troop histories and range inertia of *Lemur catta* at Berenty, Madagascar: A 33-year perspective. *Int. J. Primatol.* 20:359–373.

Jolly, A., Dobson, A., Rasamimanana, H. M., Walker, J., O'Connor, S., Solberg, M., and Perel, V. (2002). Demography of *Lemur catta* at Berenty Reserve, Madagascar: Effects of troop size, habitat and rainfall. *Int. J. Primatol.* 23(2):327–353.

Jones, K. C. (1983). Inter-troop transfer of *Lemur catta* males at Berenty, Madagascar. *Folia Primatol.* 40:145–160.

Kappeler, P. M. (1990). Female dominance in *Lemur catta*: More than just female feeding priority? *Folia Primatol.* 55:92–95.

Koyama, N. (1970). Changes in dominance rank and division of a wild Japanese monkey troop in Arashiyama. *Primates* 11:335–390.

Koyama, N. (1991). Troop division and inter-troop relations of ring-tailed lemurs (*Lemur catta*) at Berenty, Madagascar. In: Ehara, A., Kimura, T., Takenaka, O., and Iwamoto, M. (eds.), *Primatology Today*. Elsevier Science Publishers, Amsterdam, pp. 173–176.

Koyama, N., Nakamichi, M., Oda, R., Miyamoto, N., Ichino, S., and Takahata, Y. (2001). A ten-year summary of reproductive parameters for ring-tailed lemurs at Berenty, Madagascar. *Primates* 42:1–14.

Koyama, N., Nakamichi, M., Ichino, S., and Takahata, Y. (2002). Population and social dynamics changes in ring-tailed lemur troops at Berenty, Madagascar between 1989–1999. *Primates* 43:291–314.

Koyama, N., Ichino, S., Nakamichi, M., and Takahata, Y. (2005). Long–term changes in dominance ranks among ring-tailed lemurs at Berenty Reserve, Madagascar. *Primates.* 46:225–234.

Kuester, J., and Paul, A. (1997). Group fission in Barbary macaques (*Macaca sylvanus*) at Affenberg Salem. *Int. J. Primatol.* 18:941–966.

Malik, I., Seth, P. K., and Southwick, C. H. (1985). Group fission in free–ranging rhesus monkeys of Tughlaqabad, Nothern India. *Int. J. Primatol.* 6:411–422.

Nakamichi, M., and Koyama, N. (1997). Social relationships among ring-tailed lemurs (*Lemur catta*) in two free-ranging troops at Berenty Reserve, Madagascar. *Int. J. Primatol.* 18:73–93.

Nash, L. T. (1976). Troop fission in free-ranging baboons in the Gombe stream National Park, Tanzania. *Am. J. Phys. Anthropol.* 44:63–78.

Pereira, M. E. (1993). Agonistic interactions, dominance relation, and ontogenetic trajectories in ringtailed lemurs. In: Pereira, M. E., and Fairbanks, L. A. (eds.), *Juvenile Primates.* Oxford University Press, New York, pp. 285–305.

Pereira, M. E., and Kappeler, P. M. (1997). Divergent systems of agonistic behaviour in lemurid primates. *Behaviour* 134:225–274.

Pereira, M. E., Kaufman, R., Kappeler, P. M., and Overdorff, D. J. (1990). Female dominance does not characterize all of the lemuridae. *Folia Primatol.* 55:96–103.

Ron, T. (1996). Who is responsible for fission in a free-ranging troop of baboon? *Ethology* 102:128–133.

Sauther, M. L., and Sussman, R. W. (1993). A new interpretation of the organization and mating system of the ringtailed lemur (*Lemur catta*). In: Kappeler, P. M., and Ganzhorn, J. U. (eds.), *Lemur Social Systems and their Ecological Basis.* Plenum Press, New York, pp. 111–121.

Sauther, M. L., Sussman, R. W., and Gould, L. (1999). The socioecology of the ringtailed lemur: Thirty-five years of research. *Evol. Anthropol.* 8:120–132.

Sussman, R. W. (1991). Demography and social organization of free–ranging *Lemur catta* in the Beza Mahafaly Reserve, Madagascar. *Am. J. Phys. Anthropol.* 84:43–58.

Sussman, R. W. (1992). Male life history and intergroup mobility among ringtailed lemurs (*Lemur catta*). *Int. J. Primatol.* 13:395–413.

Takahata, Y., Koyama, N., Ichino, S., and Miyamoto, N. (2005). Inter- and within-troop competition of female ring–tailed lemurs: A preliminary report. *African Study Monographs* 26(1):1–14.

Taylor, L., and Sussman, R. W. (1985). A preliminary study of kinship and social organization in a semi-free-ranging group of *Lemur catta. Int. J. Primatol.* 6:601–614.

van Schaik, C. P. (1983). Why are diurnal primates living in groups? *Behaviour* 87:120–144.

Vick, L. G., and Pereira, M. E. (1989). Episodic targeting aggression and the histories of *Lemur* social groups. *Behav. Ecol. Sociobiol.* 25:3–12.

Wrangham, R. W. (1980). An ecological model of female-bonded primate groups. *Behaviour* 75:262–300.

Yamagiwa, J. (1985). Socio-sexual factors of troop fission in wild Japanese monkeys (*Macaca fuscata yakui*) on Yakushima Island, Japan. *Primates* 26:105–120.

15
Obsession with Agonistic Power

MICHAEL E. PEREIRA

"Status striving"... [and] "dominance contest"... imply a goal-directed process... [but] there is no compelling evidence that the achievement of high status exists as an end in itself.... animal conflict is not about status.

—W. A. Mason 1993, pp. 20–21, 25, 41

How to explain the incredible energy put into rank reversals... and the abrupt change in attitude once the other submits, other than as [a process] aimed at *forcing the other into recognizing a new order?*

—F. B. M. de Waal 1996, p. 100 [emphasis his]

15.1. Introduction

With a fine volume, co-edited with Sally Mendoza (Mason and Mendoza, 1993), a seminal voice in American primatology began capping his career with a review of conflict among nonhuman primates. William A. Mason (1993) presented a full concept of psychosocial conflict as the context for his prescription for future work—identification and analysis of proximate causes of social conflict, with an eye toward ameliorating oversimplifications pervading the literature. Mason saw biological competition assumed without a moment's reflection to cause virtually all primate conflict and aggression uncritically considered primates' general solution. These two premises combined to help us all miss badly the pervasive, subtle, and complex influences of conflict in primates' lives and societies. We should instead treasure, craft, and pursue every opportunity to analyze competition, aggression, and types of conflict as the separate, if related, phenomena that they are. Much primate aggression occurs in no direct relation to competition, and competitors respond in at least as many nonaggressive as aggressive ways.

Animals are *simply hedonic*, Mason reminded us, their aggression manifesting whenever divergent priorities cannot be satisfied simultaneously (Table 15.1). Unfamiliarity, misunderstanding, and indifference all help to generate aggression

TABLE 15.1. Animal conflict.[a]

Entails opposition, resistance, or incompatibility
Occurs intrapersonally and interpersonally
Results from simultaneous arousal of incompatible behavioral tendencies
Inevitably accompanies social living
Occurs in every characteristic primate social context
Occurs interpersonally whenever one's efforts to achieve desired outcome is thwarted by another's interference, resistance, withdrawal, indifference, or other failure to complement

[a] Mason, W.A. 1993. The nature of social conflict: A psycho-ethological perspective. In: W.A. Mason and S.P. Mendoza (eds.), *Primate Social Conflict*. SUNY Press, Albany.

and subjects are often responding to fear, confusion, or pain unrelated to actual opponents. Aggression "redirected" among monkeys like the "pinball" in a mechanical game machine is one of many well-known agonistic phenomena meriting reconsideration here. With competition thought to cause all conflict, many primatologists came to understand most aggression as competition for dominance itself. Terms like "status striving" and "dominance contest" envision preprogrammed drives of the sort Bill had helped sweep from animal behavior's house. Putative lust for power lures especially students of relatively aggressive primates like baboons, some macaques, and male chimpanzees. Bring in the reporters, Mason seemed to huff; given expanding scientific literacy among them, one might even notice the chance for *actual* research that exists here on scientists' behavior.

15.1.2. Striving, Achievement, and Reward

Mason's (1993) proximate focus regarding social conflict contributes the important alternative to any hypothesis of adaptive status striving: a clear null hypothesis. His "minimax" model casts forth self-focused, strictly hedonic primates, each seeking to move about freely, interact with whomever whenever, and access all gratifications with minimal work, discomfort, or impediment. Wind them up, set them loose, and witness inevitable conflict as each strives independently to obtain the same satisfactions best contributing also to others' well-being. No task is any more or less urgent today than it was yesterday or will be tomorrow, all else equal, while everyone routinely seeks to minimize costs.

Thinkers backing this view range from Buddha, 2.5 millennia ago, to the most prolific utilitarian ethicists of today (e.g., Singer, 1999). For Epicurus, pure pleasure was the removal of all pain (and, thus, any means of protection against others was necessarily toward the good!), and the modern West acknowledged rational self-interest as basic even before Darwin's birth, let alone his explanation for selfishness across entire kingdoms lacking even neurons to convey the pains and pleasures with which we feel familiar. Famously, Bentham (1781) used only hedonics to define goodness and explain individual action, offering that "Pain and pleasure . . . alone . . . point out what we ought to do [and] . . . determine what we shall do." Society's morality hammer (justified or not) combined with every other cause-and-effect in our universe account for all we think, say, and do.

Knowing his friend's unpublished papers, J.S. Mill (1833) accepted as given both self-interest and reason toward its ends, citing Bentham's (1824) claim that "in every human breast . . . self-regarding interest is predominant over social interest, each person's own individual interest over [those] of all other persons taken together."

From adaptive emotion in daily pursuit of simple gratifications, contentedness, and minimized discomfort, Mason inferred that competition among nonhuman primates can never be for dominance itself, *species differences notwithstanding*. And, with that last bit, his argument parted ways from my own.

All behavior inspires two simple if enormous questions: *What is salient to the animal?* And, *Why?* Accordingly, we *expect* answers to vary in accord with phylogeny. My own work on agonistic interaction has been conducted much but not entirely with gregarious primates in which significant aggression occurs daily in virtually all subsets of group members. Among baboons (*Papio hamadryas cynocephalus*), macaques (*Macaca fuscata*), vervet monkeys (*Cercopithecus aethiops*) and ringtailed lemurs (*Lemur catta*) patterned systems of social dominance contribute to groupmates' essential experiences with one another and some aggression (and other behavior) seems advanced specifically to establish, reverse, and stablilize dominance directionality within relationships (Pereira 1995). These primates seem to recognize the status issues that research identifies at the hearts of their societies, seeming actually to obsess over them, showing every evidence of adaptive motivation or *appetite* (appetitive behavior) to address them *with particular social partners* (figure 15.1), via specific, functional, stepwise social initiative (Hinde 1981, 1987).

Sex and species differences in juvenile rank acquisition across Old World papionin monkeys support a robust status-striving hypothesis: *All else equal, maximization of social status, effected and maintained differently by males and females across developmental phases, maximizes for members of each sex prospects for high reproductive success across long lives.* In the relevant socioecological contexts, dominance works well generally to keep high-status individuals best-supplied with resources, including reproductive opportunity, often working very well under recurring sets of conditions.

Taking development as selection's general target while accounting for recognizable short- and long-term costs and benefits, support was obtained from (a) age-related and (b) sex-specific patterns of (c) juvenile aggression toward and (received) from (d) unrelated versus related (e) peers and adults and (f) juvenile delivery and receipt of (g) supportive versus punishing aggression (h) as/from third-parties (Pereira, 1989). This trying level of detail, in my view, is unavoidable in any genuine effort to understand social interactions among complex gregarious mammals. Though this will surprise many primatologists, for example, the hierarchical detail mentioned above has only ever been provided, even among best-studied anthropoids, by my report on savanna baboons, to my knowledge (while data can be extracted from literature similar to one partial degree and another for Japanese, rhesus, and longtailed macaques). The ineluctable complexity of the work, the analysis, and the reading (and the waxing and waning

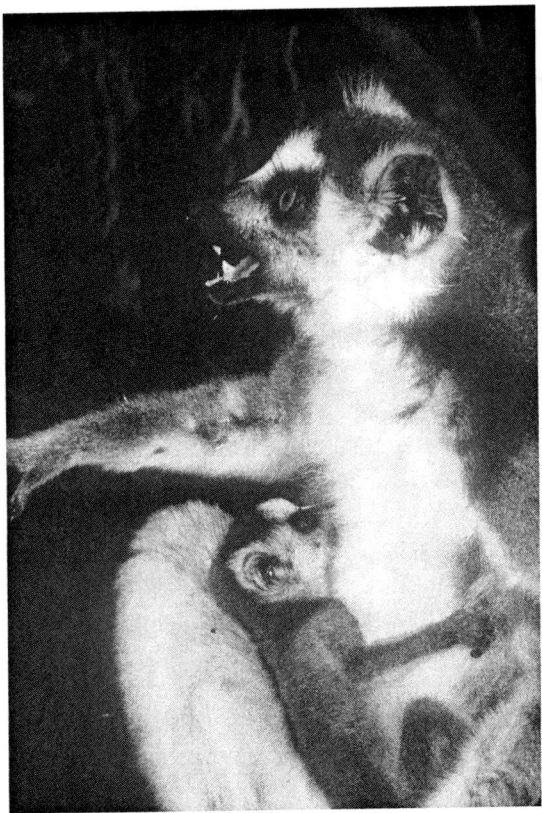

FIGURE 15.1. After an austral spring equinox, one mother of a new infant at Berenty Reserve extended an arm wounded by her aggressor while issuing ringtailed lemurs' formal signal of subordination, the spat-grimace display (Pereira & Kappeler 1997), thus losing dominance to a mature female groupmate that had targeted her for intense aggression specifically to over-turn her in dominance (photo: A. Jolly). As adult ringtails' dominance relations are not invariably transitive (Pereira & Kappeler 1997), the aggressor did not necessarily next occupy "alpha" or despotic dominance status in this group. Adolescent females successfully overturning dominant adults were almost were almost never evicted that same season, however, over my decade of near-daily study of Lc1 and Lc2 Groups (Nunn & Pereira 2000; Pereira & Leigh 2003).

of topics' popularity in any marginalized research industry with far too many questions per practitioner) keeps most primatologists ignorant of what has been demonstrated regarding dominance acquisition among primates or remains, rather, actually unknown. In large part, Mason (1993) sought to share this same, simple message: We know far less about patterns of primate behavior than today's publications too often suggest.

Mason's (1993) views of primates and our literature diverge from my own within two perceptual domains: What was salient to the different primate species we each studied, and what was salient to each of us as we have done so (different-aged adult males of different U.S. coastlines and personal histories, briefly if systematically observing a few primates of different species and particular ages, sexes, social groups, and individual histories). Here, I convey data that seem to me to document competition for dominance itself quite clearly. In a behavioral syndrome occurring consistently twice each year, mutually familiar, distantly-related, mature female groupmates neither cercopithecine nor even anthropoid generated these data, representing, instead, the most social, anatomically "primitive" extant nonhuman primate, comprising perhaps 750,000 individuals, in captivity across the world and remaining endemic forest and spiny desert in southern Madagascar (Sussman et al., this volume). I expect to convince most readers that the pattern documented in North American forest enclosures certainly functions adaptively among females of fast-growing groups living at high density in the ecological source of ringtailed lemurs: Madagascar's southern gallery forests.

15.2. Methods

15.2.1. Subjects, Housing, and Management

Throughout each of 40 months between 1988 and 1994, all juvenile and adolescent members of the two groups of ringtailed lemurs occupying adjacent forest ranges at the Duke University Primate Center (DUPC) were subjects for daily focal-animal sampling (Altmann, 1974; Table 15.2). I managed these and co-ranging groups of lemurs (*Eulemur fulvus rufus*, *Varecia variegata variegata*) from 1985 to 1995, expanding enclosure area in 1986 to establish second study groups of each species in 1987. After mature females' consistently selective targeting of particular peers for intense seasonal aggression was described for the DUPC colony, including patterns over years within Lc1 Group before and after its permanent placement into outdoor enclosures (Taylor and Sussman, 1985; Taylor, 1986; Vick and Pereira, 1989), Lc2 Group was founded using three

TABLE 15.2. Hours of focal-animal sampling of 14 adolescent subjects.[a]

Year	Lc1 group		Lc2 group
1989	KT 17.7		ML 17.1
1990	NI 19.3		CO 17.4
	ER 21.6		CH 29.0
	AA 25.5		
	PH 19.4		
1991	BU 23.3		DO 23.8
	SE 16.7		
	BI 16.6		AL 23.7
1992	NS 22.5		
TOTAL		292.6	

[a] Nine underscored subjects provided data not only before but also after reversals of dominance with adult female groupmates.

unrelated and mutually unfamiliar females released simultaneously, each supporting weanlings sired by different males (one set of twins). Four months later (April 1988), unrelated males also unfamiliar to the females were admitted to new study area (Pereira and Izard, 1989). Only one female Lc2-Group founder, CL, was related, distantly, to females in Lc1 Group (Pereira and Izard, 1989), making initial relatedness between Lc1 and Lc2 Groups similar to that between nonadjacent groups at Berenty and Beza-Mahafaly Reserves.

Note that primarily female CH represented Lc2 Group in present analyses, first as adolescent status-striver and then as adult target (see Table 15.7). To postpone merging of groups' gene pools and enable Lc2 Group to establish territorial ownership, fencing was retained between the two ranges for Lc2 Group's first three years. But, on the (1987) estrous day of one founding female, five-year-old natal Lc1 male PL discovered a hole beneath intermediary fencing created by another passer-through (e.g., fox, raccoon, skunk, opossum). Despite genetically homogenizing present data a bit, PL's disregard for management concerns (and not CH's estrous mother) ended up being good luck overall. Precluding all other male transgression that autumn, PL's few hours in the new area generated CH, our second group's main contributor here. AL (born the next year) ultimately reversed CH in dominance, while nearly all succeeding infants succumbed to predation or infanticide over the next three years. Lc1 and Lc2 Groups resumed exchanging members and genes naturalistically when barriers were removed between territories in 1990 (Pereira and Leigh, 2003) and many accounts of DUPC husbandry and research methods are available (e.g., Bogart et al., 1977; Glander and Rabin, 1983; Taylor and Sussman, 1985; Ganzhorn, 1986; Pereira, 1993a, 1993b; Pereira et al., 1999).

15.2.2. *Observational Procedures and Definitions*

All behavior recorded for present analyses had previously been operationally defined (Pereira, 1993a; Pereira and Kappeler, 1997). Focal samples were distributed comparably, via random permutation, among subjects and hours of morning and afternoon sessions, ruling out diurnal and short-term seasonal effects. *Ad libitum* samples of agonistic interaction and social grooming (Altmann, 1974) and all occurrences of high-intensity fighting were recorded throughout all time spent with groups over the five-year period. All social approaches and departures across a personal 1.5 m radius and initiations of huddling, social play, and other non-agonistic touching were recorded that involved focal subjects as actors or recipients, alongside detailed records for all fighting, grooming, and scent-marking. Point samples at five-minute intervals recorded subjects' nearest neighbors' identities and the identities of all "neighbors" within 1.5 m, as well as subjects' ongoing activities and groups' *levels* of activity (5-point scale), the latter two data enabling restriction of analyses to samples of comparable wakefulness or motivation (e.g., foraging, territorial, social predator avoidance) per subject or social group.

Agonistic interaction was detailed using seminal ethological methods first to identify species-typical aggressive and submissive acts and signals, including formal signals of subordination (Pereira and Kappeler, 1997; cf. Sade, 1967; van Hooff, 1970; de Waal and Luttrell, 1985). Conflicts began when either of two lemurs within 5 m of one another directed aggressive or submissive behavior toward the other and ended when both terminated agonistic behavior for at least 3 seconds. Initiators' identities and whether each participant issued aggressive or submissive behavioral elements, or both or neither, was recorded, summarizing contributions with an "A," an "S," an "AS," or an "O." Eight meaningful *types* of agonistic interaction are rendered, excluding S-S (rarely witnessed among dominance-oriented primates) and O-O, a by-product describing no agonistic interaction. (Asymmetric contributions to six of the types suggest six additional "types" of agonistic interaction only from either opponent's specific vantage.) Initiators first emitted agonistic behavior. Decided interactions were cleanly "won" and "lost," only one participant expressing only submissive behavior while the other emitted only aggressive behavior or no agonistic behavior at all. Barring rare reversal attempts by subordinates, entirely unidirectional "AS" and "OS" events constitute more than 90% of all agonistic interaction for any pair of ringtailed groupmates throughout almost any study period (Pereira and Kappeler, 1997). The intensity (Table 15.3) and social context (Table 15.4), or apparent proximate "cause" of every agonistic interaction was recorded.

15.2.3. Hypothesis Testing

Mason (1993) suggested no criteria with which to evaluate the hypothesis of status striving; but de Waal (1996) did, highlighting the sudden energy expenditure and voluntary exposure to risk in mounting reversal attempts, followed by quick reduction of aggression. Indeed, these features and others should characterize any initiative adapted to force dominants to become subordinates while minimizing

TABLE 15.3. Low and High extremes were evaluated among five interaction intensities scored.

Intensity	Operational definition[a]	
	Aggression	Submission
Low		
Very low	No effort to contact or solely nosepush	< 2 m displacement; no spat
Low	Mild contact or limited or truncated effort to hit or push	Jump away ≥ 2 m from opponent > 2 m but < 3 m away
High		
Moderate	Charge or sharp, brief contact (< 1 s) (two-handed grab or bite)	Full spat-grimace display and/or flee < 3 m
High	Chase < 10 m or contact lasting > 1 s but < 2 s	Flee >3 m and < 10 m or deep spat and jump away
Very high	chase > 10 m or contact lasting > 2 s or grappling	Flee ≥ 10 m

[a] Interactions ascribed higher intensity whenever aggressive and submissive evidence contrasted.

TABLE 15.4. Ten operationally defined proximate causes for agonistic interaction.

Operational definition[a]	Social context[b]
"AGGRESSION OR SUBMISSION GIVEN OR RECEIVED..."	
...by/from member of other social group	Territoriality
...directly after gazing at groupmate [while][c] feeding on forest products	Feeding
...directly after gazing at groupmate [while] feeding on artificial provisions	Provisions
...during or < 5 s since groupmate's agonistic interaction with third party	Aggression (intervention)
...directly after gazing at groupmate [while] scent-marking	Scent-marking
...during or < 5 s since groupmate's nonagonistic interaction with third party	Interaction with third party
...during nonagonistic interaction with same groupmate	Affinitive interaction
...during nonaggressive approach to or by groupmate (within 10 m)	Proximity
...when 5 prior s of unobstructed observation satisfied no other criterion	None obvious
...when 5 prior s of behavior were obstructed or insufficiently well attended	Unknown

[a] Wordings are for subjects' initiations, whereas subject or opponent was identified as actual initiator.
[b] Context higher on list recorded whenever 2 definitions' criteria simultaneously fulfilled.
[c] Definition satisfied when word "while" (in brackets) is or is not included.

risk and loss of time or energy. To explore possible status striving in adolescent females, I evaluated of five specific predictions for behavioral change following abrupt reversals whose affirmation would reject Mason's null model of simple hedonism (Table 15.5). Success or failure of reversal attempts is clearly signaled by one ringtails' prominent unilateral use of the spat-grimace display as a formal signal of subordination (de Waal and Luttrell, 1985) following brief episodes of undecided dominance relations with another (Pereira and Kappeler, 1997). (1) Rate and (2) intensity of aggression and (3) intensity of preemptive submission (interactions lacking aggresion) were expected to decline directly after reversals, while proximate causes of aggression would revert to normative patterns of (4) greater diversity and (5) never lacking (i.e., no aggression seeming unprovoked).

15.2.4. Analyses

Five years of filing data at the conclusion of each calendar month guided the formulation and testing of the five predictions (Table 15.6). To distinguish putative dominance contests from relations in which aggression increased more indefinitely (some causing eviction), agonistic values from months of data filed

TABLE 15.5. Testing the hypothesis: Dominance is the goal of adolescent aggression.

	Predicted pattern of behavioral change across period of status change		
Agonistic dimension	Before	During	After
Rate of aggression	Low, then rising	High	Sharp decline, then low
Intensity of aggression	Low, then rising	High	Sharp decline, then low
Intensity of submission	Low, then rising	High	Sharp decline, then low
Diversity of causes	High	Low	Rising, then high
% aggression unexplained	Near zero	Higher	Near zero

TABLE 15.6. Evaluation of predictions: One example among 17 total cases.

Behavioral/ dimension	Variable	Predicted after pattern	Case 1 of 17 (1989, Lc1: KT-NM)			
			"Before"	"During"	After	Affirmed?
1. Rate	Aggressions/ hour	< Higher preceding value	4	2.7	4.8	No
2. Intensity	% aggressions = intense	< Higher preceding value	43	0	0	Yes
3. Intensity	% vol. Submits[a] = intense	< Higher preceding value	50	25	31	Yes
4. Proximate cause	No. of contexts	> Lower preceding value	3	2	3	Yes
5. Proximate cause	% unprovoked aggression	< Higher preceding value	60	13	7	Yes

[a]Voluntary submission: submissive signaling preceding or preempting aggression.

"before" and "during" each dominance reversal were contrasted with values from the earliest month of data known to have been collected soon (days to weeks) but entirely after the reversal. As predicted, reversals entailed sudden emergence of aggression of increased intensity from former subordinates that lasted only briefly. While some occurred relatively squarely amidst their monthly data files, others were completed largely between one month's sampling and that for the next, and most occurred relatively near the beginning or the end of their data file (by design, subjects' 10 to 12 monthly focal samples were spaced over at least three different weeks). Consequently, any extreme behavioral rates or intensities in months "during" reversals were, in fact, diluted by roughly five times as much data, on average, from normalized behavior preceding and/or following reversals. Increasing rate of post-reversal normalization after sudden outbreaks of aggression would progressively suppress chances to affirm that very pattern in present contrasts, gross monthly values providing *conservative* tests then. The overwhelming results from unusually many and diverse subjects studied longitudinally across a half-decade made conventional abuse of familiar statistical tests superfluous.

15.3. Results

Five years of weekly focal-animal sampling in two social groups detailed 17 successful reversal attempts by nine adolescent females overturning eight adult females in dominance. All reversals began and ended over periods of 1 to 36 hours, longer cases accumulating more fights of highest intensity levels, manifesting unprovoked targeted aggression for a period (Table 15.8), and more often generating sufficient data for testing all five predictions prior to resolution (Table 15.7).

Maturing females' targeted their older aunts/nieces (4 targeted by 3 adolescents) or cousins (1 targeted by 3 adolescents), an unrelated adult (3 targeted by 2 adolescents), and a grandmother (1 targeted by 4 adolescents). Only two

TABLE 15.7. All adolescent females (n = 9) reversing adults (n = 8) from 1989 to 1993 showed most behavioral changes predicted by hypothesis that equinox aggression is designed to subordinate specific adult female groupmates.

Subject	Opponent	Group	Date	Predictions Tested	Predictions Affirmed	Fraction affirmed/ subject
KT	NM older aunt	1	1989 APR	5	4	
KT	CN older aunt	1	1989 MAY	5	4	8/10
ER	TR older aunt	1	1989 OCT	5	4	
ER	LY grandmother	1	1989 OCT	5	4	
ER	NM older aunt	1	1990 MAR	4	3	
ER	KT older cousin	1	1990 APR	4	4	15/18
CH	AT unrelated adult	2	1990 MAR	3	3	
CH	ML unrelated adult	2	1990 MAR	4	4	7/7
NI	KT older cousin	1	1990 NOV	2	1	1/2
BI	LY grandmother	1	1990 NOV	5	4	4/5
SE	LY grandmother	1	1990 DEC	5	4	4/5
BU	LY grandmother	1	1990 OCT	4	4	
BU	KT older cousin	1	1990 DEC	1	1	
BU	KT older cousin	1	1991 FEB	2	0	
BU	LY grandmother	1	1991 MAY	4	3	8/11
AL	CH unrelated adult	2	1991 MAY	3	2	2/3
NS	KT older niece	1	1992 AP-MY	4	3	3/4
TOTAL				65	52	(80%)

relatedness classes typical as social partners were not represented among targets: matrilineal sisters, which also targeted one another infrequently among adult pairs (Pereira, 1993a; Pereira and Kappeler, 1997), and mothers and daughters, which never targeted one another at any ages throughout 15 consecutive years managing and observing these lineages.[1]

Every adolescent reversing an adult female soon exhibited all or most predicted behavioral adjustments (Table 15.7; median individual: 4 of 5 predicted changes observed), leading all five predicted behavioral changes for days and weeks

---

[1] The only two daughter–mother reversals in the 20-year DUPC record connected to my project are of unknown significance. LY, born 1976, was separated from her mother at weaning and returned to her as an adolescent, joining also a new juvenile sister and unfamiliar non-kin. Bearing twins within 15 minutes of one another proved too much for LY and her mother in their caging and the latter was judged to have suffered the worst fall-out (DUPC records). In Lc2 Group, AL, born 1989, reversed her mother, CL, after I had departed the DUPC. AL was hormonally odd growing up: uniquely nonreproductive for at least 7 years and one of only two females *of any age* I have ever seen behave as though scent-marking their tails with forearm glands of unknown activity. Throughout life, by contrast, male ringtailed lemurs characteristically use this signaling behavior during competitive interaction: first during play, as infants and juveniles, and next during adolescent and adult efforts to intimidate other males, especially in late summer and early fall, before the short annual mating season (Jolly, 1966; Evans, 1986; Pereira and Kappeler, 1997; Pereira, unpublished data).

15. Obsession with Agonistic Power   255

TABLE 15.8. Unprovoked aggression occurred only (a) "before" or "during" and (b) not even *soon* after reversals.[a]

| Subject | Opponent | Year | % proximate contexts = "none obvious" | | | Prediction affirmed? |
			Month "before"	Month "during"	Month after	
KT	NM	1989	60	13 (APR)	7	YES
KT	CN	1989	17	0 (MAY)	0	YES
ER	TR	1989	29	0 (OCT)	0	YES
ER	LY	1989	20	0 (OCT)	0	YES
ER	NM	1990	13	0 (MAR)	0	YES
ER	KT	1990	33	0 (APR)	0	YES
BI	LY	1990	0	25 (NOV)	0	YES
SE	LY	1990	0	33 (DEC)	0	YES
NS	KT	1992	0	25 / 0 (AP/MY)	0	YES

[a] Eight of 9 cases for five adolescents manifested unprovoked aggression during only one month and neither the preceding nor following month. *Ad libitum* data contained records of aggression lacking proximate cause for adolescents whose successful reversals (other 8 cases) contributed no such records to focal-animal data.

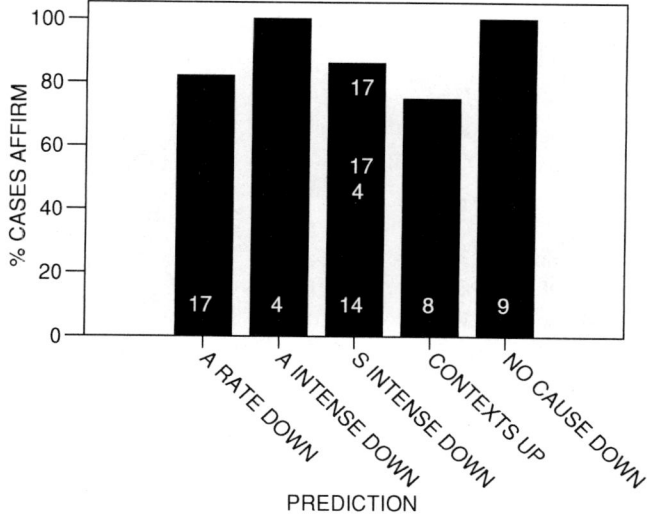

FIGURE 15.2. All five changes in agonistic behavior predicted for the first few days to weeks after adolescent reversals of adults were affirmed by the large majority of cases. Sample sizes varied due to various obstacles across the 17 cases precluding collection of requisite data. Comparison of Tables 15.7 and 15.8, for example, shows that cases failing to generate unprovoked aggression during focal samples tended to be reversals achieved so easily that insufficient data of that sort and sometimes others (e.g., aggression rated as "intense") were captured with which to test predictions. See Methods for more on analysis of very brief but important behavioral events.

following reversals to be affirmed by most cases (Figure 15.2; median: 86%). In hours or, more rarely, about one day after targeted aggression began, rates and intensities of aggression next plummeted to small fractions of prior values, and aggression that seemed unprovoked disappeared as suddenly as it had earlier appeared (Table 15.8). In more dangerous protracted cases, adolescents suspended their own formal signaling of subordination to dominants targeted for reversal (Pereira and Kappeler, 1997) before working harder than average to succeed in invoking targets' willing submission (see Table 15.8 footnote). Following reversals, social contexts of aggression soon diversified again and intensities of submission declined former opponents reinstalling normative social relations quickly.

15.4. Discussion

Serving a function like that of an artist, the scientist sees things in a way that no one has seen them before and finds a way to describe [it] so that other people can see it . . . widening and enriching the content of human consciousness . . . increasing the depth of the contact that human beings . . . can have with the world around them . . . arousing and satisfying a sense of wonder and curiosity about [its] riches

—D. S. Lehrman, 1971

There are many truths of which the full meaning cannot be realised until personal experience has brought it home.

—John Stuart Mill, 1859

Only after maturing and soon after equinoxes, nulliparous female ringtailed lemurs suddenly sought to over-turn particular adult females in dominance to whom they had previously been subordinate since birth (Pereira, 1993a). No parallel change of resource supply or demand in Madagascar or DUPC enclosures explains reliable manifestation of this aggression in these relationships at these times of year, nor the sudden reversion, following every reversal, to normative, relatively non-agonistic relations resembling those observed consistently between females of smaller groups (Table 15.9). Particularly by suddenly applying and extinguishing intense, unprovoked, targeted aggression, adolescent females provided the strongest quantitative evidence available that nonhuman primates have evolved some aggression specifically to promote economic over-turning of particular social superiors. Predicting 50% for all five outcomes evaluated so long as changes of no other over-riding factor, like photoperiod or nutritive supply, also influenced results, Mason's null model of simple hedonism was rejected by females maturing in expanding forest-living groups by restricting analyses to the first 10 weeks following each equinox (Oct.–Nov.–plus and April–May–plus; Table 15.7), photic phases throughout which females most commonly initiated targeted aggression over more than 20 years of research on naturalistic lineages [from indoor–outdoor caging to both forest enclosures and a controlled indoor

TABLE 15.9. "Good things" in life are most at risk for female ringtailed lemurs in summer.[a]

Season of dominance reversal	Supply/demand changes in:		Would simple hedonism likely change the intensity or predictability of aggression following the equinox?
	Madagascar	DUPC enclosures	
Spring equinox (warming)	Food supply expands during gradual resumption of hair growth and increased levels of daily activity, including sudden aggressive targeting of some new mothers	Ample provisioning enriched by spring forest foods (young leaves, flowers) during gradual metabolic increased (see left). Births induce sudden aggressive targeting of some new mothers	NO
Autumn[b] equinox (cooling)	Summer food boom declines slowly, as fat accumulates, sunning increases, weaning is completed and growth and activity levels decline	Food supply continued or increased (by oak fruits) while weaning was completed, sunning increased, and growth, activity, and metabolism declined (see left)	NO

[a] see also Pereira et al. (1999).
[b] Whereas spring pattern of aggression is well-known, evidence of fall aggression from wild yet comes only from Sauther (1993).

cage complex (Taylor, 1986; Pereira and Izard, 1989; Kappeler, 1993; Pereira, 1993a, 1993b; Pereira and Kappeler, 1997; Pereira et al., 1999; Cavigelli and Pereira, 2000; Nunn and Pereira, 2000)]. Results importantly complement varied evidence of status striving in other lemurs and anthropoids (e.g., Pereira, 1995, in prep; de Waal, 1996; Silk, 1993, 2002), including male common chimpanzees (*Pan troglodytes*; Wrangham and Peterson, 1997).

Generally speaking, gregarious lemurs are demonstrably agreeable mammals, assisting highly insulative fur, group huddling and sunning, and other energy-conserving traits to project their endearing "cuddliness" to animal lovers. Between distantly-related females in large groups, however, this species-typical mode is predictably punctuated, suddenly and briefly, by fiercely competitive *dominance contest* behavior, sharing mechanisms perhaps with similar campaigns of intimidation manifesting annually between some mature males [pre-mating: summer solstice to fall equinox (Pereira and Kappeler, 1997; Koyama, 1988; Sauther, 1991)]. Infant ringtails of both sexes also vie for dominance, midway through lactational support, directly after the summer solstice, before preparations for first over-wintering must be achieved independently and eight harsh, dry months begin [post-weaning and post-equinox; (Pereira, in prep)].

In sum, *Lemur catta* evolved developmental capacities enabling individuals to pass through successive life-history phases entailing adaptive cognitive preoccupation with social status relating to particular social partners. Competition for dominance itself evolved without any need for unusually large forebrains like those of lemurs' anthropoid relatives. Whether mechanisms are shared with non-primate status strivers, including perhaps even invertebrate taxa, are questions to be addressed throughout biology's ongoing "eco-evo-devo" paradigm shift (Plomin and Ho, 1991; Schlichting and Pigliucci, 1998; Laland et al., 2000; Wcislo, 2000; Ho, 2001; Moore, 2001; Oyama, 2003; West et al., 2003; Wake, 2004).

15.4.1. Behavioral Biology in the New Millennium: Proximate Causation

Research on proximate causes of social behavior illuminates the behavioral mechanisms that *actualize* life histories, importantly extending conjoint understanding of evolution and development (West et al., 2003). Wilson's (1975) *Sociobiology* helped field primatologists document adaptive values of species-typical adult behavior among long-lived subjects (ultimate causes), but largely that has continued among fieldworkers these past 30 years. Wilson's "new synthesis," like its first edition (Huxley, 1942), relegated investigators of proximate causes in whole organisms to the cluttered rear of the overall machine gaining steam (Pereira and Fairbanks, 1993). Those back there full- or part-time, however, understood biology's deepest richness to revolve around development. Huxley (1942) knew it, in emphasizing that developmental genetics needed at least as much attention as genetic mutation or selection processes. Mason, his forebears, and contemporaries in psychology, especially Schneirla (1957) and protégé Danny Lehrman (1970), devoted careers to the fact that function reveals nothing about how complex traits are installed.

Adaptive value teaches us nothing necessarily or much about social sign stimuli either, and respecting that is essential to focusing real attention on real species-typical behavior (cf. Huntingford and Turner, 1987; Klama, 1988; Moynihan, 1998). Behavior evolves to promote survival and reproduction, and Primates in diurnal niches evolved cohesive sociality to guard against predation (van Schaik and van Hooff, 1983). Where aerial and terrestrial predators presented divergent threats, distinctive calls evolved to enable allies to adjust blindly against "inverse" dangers (upward vs. downward; Pereira and Macedonia, 1990). In such contexts of risk, vital resources of moderate sizes distributed in ways favoring dominance behavior promoted its evolution (van Schaik, 1989; Sterck et al., 2000). And, *certain primates operate in such important daily relation to their social status that, opportunistically and systematically, species-typical behavior has them strive for the resource of dominance itself*, several demographic classes of ringtailed lemurs providing unrivaled examples (cf. Pereira et al., 1999; this paper; Pereira, in prep.).

15.4.2. *Characteristic Development, Metabolism, and Cognitive Ethology Manifest Social Systems*

Ringtailed lemur groups grow in rich gallery forest because all maturing daughters reliably continue to adore their mothers, remaining subordinate and working to stay with her (Jolly, 1966; Pereira, 1993a; Nakamichi and Koyama, 2000; Takahata et al., 2005). Under some circumstances ringtails actually overproduce daughters (Nunn and Pereira, 2001; see below). Moreover, matriarchs never evict any daughter and favor none over any other, beyond normative under-respect for oldest and leniency toward youngest (Vick and Pereira, 1989; Pereira, 1993a; cf Trivers, 1974).

Only certain adult female competitors target one another for post-equinox aggression (Jolly, 1966; Taylor and Sussman, 1985; Vick and Pereira, 1989; Pereira, 1993a; Hood and Jolly, 1995; Pereira and Kappeler, 1997; Nakamichi and Koyama, 1997, 2000; Jolly and Pride, 1999; Sussman et al., 1999; Koyama et al., 2001; Gould et al., 2003; Takahata et al., 2005). Adult half-sisters are huffy but generally not hostile with one another until female group size might surpass four, whereupon one might select another for unprovoked seasonal aggression. More often, sisters target and get targeted by certain of others' maturing daughters (different-aged aunt-nieces: Hamilton's r: 0.125 [see below]). Over 20 years, adolescents in DUPC groups routinely selected *older* aunts, nieces, or cousins (r: 0.0625) for post-equinox aggression while almost never targeting birth peers representing those same matrilineal classes of female relative (Pereira and Kappeler, 1997). Granddaughters seemed universally willing to target their grandmothers, evoking no aggressive vigilance or systematic punishment from own mothers or nonadversarial aunts for doing so (grandmothers' otherwise devoted daughters). Alongside inter-related indicators (Pereira, 1991; Pereira and Weiss, 1991), patterns predict that one male typically reproduces with most females in a group each year, population density notwithstanding, while females switch mates at intervals of less than four years, the average span between births of surviving sisters (Koyama et al., 2001; Gould et al., 2003).

Since 1980, DUPC-based research has quantified inter-related lemur physiology and behavior much of which was later confirmed in Madagascar, with most things ringtailed lemur seeming to reflect a polyphenism integrating information on population density, individual growth rate, success with seasonal metabolic allocations (Pereira et al., 1999), and adaptive modulation of female toleration of female partners. Low to high set-points for intensity of aggression exhibited in neutral-to-adversarial relations (Pereira and Kappeler, 1997) are displayed by females in DUPC enclosures and fast-growing tourist-fed Berenty groups, as well as Berenty groups on rich gallery territories or nourished in part by introduced exotic plants (e.g., T Group, which I studied before and after births in 1992 and a principal longterm study group of Koyama, Nakamichi, and colleagues). The highest known tolerance levels among females occur in smaller groups living sparsely in the southwest's Beza-Mahafaly Special Reserve and Berenty's driest areas.

Ringtails in real desert have yet to be studied, but primary cues coming to these primates wherever one finds them relate photoperiod to developmental rate, demography, fatness, and success with ongoing lean-fat allocations via modulation of circulating IGF-1 and thyroxine (Pereira et al., 1999), and presumably leptin, insulin, glucagon, other growth-factor hormones, and environmental signals yet undescribed (e.g., Suter et al., 2000). This physiological system presumably contributes mechanistically to adaptive vigilance regarding local demography (e.g., female group size, degree of territoriality, size of maturing cohort, opponent ages/conditions, etc.) and personal issues of status (e.g., steepness of dominance relations; targeted/not) to turn unprovoked aggression "on" and "off" against particular adversaries by season (Vick and Pereira, 1989; Pereira, 1993a, 1995; Pereira and Kappeler, 1997) differently easily and intensely according to dyads' other circumstances.

Many important links remain to be identified among regulatory factors, including any relating within- to between-group patterns of female aggression (Hood and Jolly, 1995; Takahata et al., 2005; cf. Pride, this volume). Clearly, though, we have succeeded over the past 20 years in describing the fundamental developmental system comprising *Lemur catta* to depths exceeding those known for most other primates studied in lab and field settings. Life-history modulations determine the population patterns that, in turn, modulate competition within each sex. Increased population density among ringtailed lemurs, for example, ramps up prospects for male- and female-perpetrated infanticide (Vick and Pereira, 1989 Pereira, 1996; Cavigelli and Pereira, 2000; Jolly et al., 2000) and female biasing of offspring sexes (Nunn and Pereira, 2000), while ages at first reproduction (female) and natal dispersal decrease [both sexes (Pereira, 1993a; Pereira and Leigh, 2003)] and secondary male dispersal shifts more predominantly to the birth season (Pereira and Weiss, 1991).

Identifying what animals think they are fighting about can be vital to efforts to prevent extinction. And, that's true for this precious primate, no matter what anyone might judge more broadly about efforts to know primate or other animal minds. A specialist of very tough environments, this clear, "simple" primate exhibits all the "plasticity" and individual variability needed to please any who wish to document same (cf. Marler and Peters, 1982; Nelson et al., 1995). But, my specific suggestion is that much variation needs be described yet as types within types of individuals, characterization of which could only make ringtailed lemurs still more valuable as model organisms. Every fractal organism relies on it physiologicogenetic systems to facilitate characteristically *variable responses*, for their next developmental phase, next season, next month, next week, next day, next night, and next moment, all in accord with effects of lineage, sex, social status, personal history, and current circumstance. While coarse on any printed page, this full concept of *species-typical behavior* enhances all research as well as every more casual appreciation of natural history and adaptive complexity (cf. Tinbergen, 1953; de Waal, 2003). The reaction-norm concept, synonymous with population ecology's "polyphenism," describes the basic variations of over-riding importance that emerge from differential interaction among genes, organs, nutrients, actual development, and key experiences in every species (Schmaulhausen, 1946; Schlichting and Pigliucci, 1998). To my knowledge, no one has yet encouraged biologists to apply the concept to *every time scale relevant to organisms under study*, while that is my own prescription (Pereira et al., 1999; Cavigelli and Pereira, 2000; Nunn and Pereira, 2000; Pereira and Leigh, 2003; Pereira in prep.).

Important hypotheses become testable only after initial understanding has been achieved, and much more integrative research can now be achieved with ringtailed lemurs, and a very few other primates (e.g., Suter et al., 2000; Buchan et al., 2003), than with representatives of almost any other mammalian kind (Stearns et al., 2003). As such, efforts should begin immediately to establish a research grid for permanent monitoring of development and behavior across this primates' entire geographic range. Ringtails have not yet been studied in their sparsest populations, where matriarchs travel with just one or two young adult daughters (personal observation). Given a third, one pair will probably begin not to appreciate togetherness as much as the others do. Males hang around females in comparably small numbers (van Schaik and Kappeler, 1993), each striving to dominate the others, if necessary, at the right time, and meantime probably work to curry female favor by effectively staying out of their ways and shielding their feeding and youngsters from threats (cf Overdorff, 1998). Survival of infants from any active, healthy, helping male should favor that sire to receive rare female choice for two or three consecutive years (cf. Smuts, 1985; Pereira and Weiss, 1991).

Integrative longterm collaborations are needed to address such predictions, but also insights to be garnered would augment every investigator's independent projects. Genetic testing via fecal analysis, made almost routine in recent field primatology, for example, provides methods with which to evaluate data for decades across a grid covering ringtails' geographic range, each collection point 100 km or so from the next in each cardinal direction. Ensuing population analyses could monitor genetic, developmental, and social evolution conjointly, integrating known socioecology and cognitive ethology, each generation of students digging deeper into networks of causal factors over however many additional decades with wild ringtails these same efforts simultaneously assist us to arrange.

Judicious and strategic transplantations would likely be useful soon in a collaboration between researchers from Berenty and Beza-Mahafaly Reserves. Such will be needed to secure healthy genetic futures for all vertebrate populations isolated by further human expansion, and the 500-plus ringtails on "Berenty island" might not best be made to wait much longer for injections of new genes. Most important, reciprocal transplants between ringtails' only two major field sites (leaving other animals aside initially) would secure the exquisitely rare opportunity before us to investigate directly the reaction norm of a nonhuman primate (Stearns et al., 2003; cf. Ricklefs, 2001). Many integrative advances would inevitably result, particularly for conservation biology.

Photoperiod regulates basic *metabolic mechanisms* in lemurs (e.g., Genin, 2000; Genin and Perret, 2000; Schmid, 2001) and seasonal issues indicate different likely functions for unprovoked aggression potentially following each equinox. Spring aggression threatens infants directly (Vick and Pereira, 1989; Jolly et al., 2000), often actually triggered by victims' parturition (Pereira, unpubl. data), when females and their infants are easily attacked (Vick and Pereira, 1989; Pereira, 1993a). By contrast, fall aggression threatens to disrupt victims' estrous cycling and fat accumulation prior to conception, followed by eight months of cold dry weather and scant high-quality forage. Delaying victims' conception relative to other females determines bottom rank for victims' infants (Pereira 1993a, 1995, in prep.). Late-born infants surviving mothers' sub-maximal fat storage, targeting for aggression, and possibly attacks on self from mothers' adversaries or immigrating males (Pereira and Weiss, 1991; Jolly et al., 2000) must next survive their first harsh southern Malagasy winter and spring while subordinate to every groupmate, enjoying only last (least) access to rare and valuable winter foods.

In nature, targets' options are not good (Nakamichi and Koyama, 1997; Jolly and Pride, 1999; Sussman et al., 1999; Takahata et al., 2005). Occurring only at high densities, evictions jettison departing females into minimal interstices between unrelated matriarchies' fully occupied, harshly defended, slightly overlapping territories, or to marginal habitat farther from local rivers. Sustaining one's victimization by striving not to disperse seems stressful anthropomorphically; but initial evidence suggests that females are well-adapted to their harsh social world: subordinates exhibit *lower* circulating levels of cortisol ("stress" hormone) than dominants in large groups (Cavigelli, 1999; Cavigelli et al., 2003; Pride, 2005).

Older evictees and *young* fall targets (just starting reproductive careers) are vulnerable to lacking partners altogether with whom to guard against predators and unrelated females come spring, when very harsh aggression breaks out, and then convert into daughter makers (Nunn and Pereira, 2000). Ringtails' hardiness and productivity in captivity make such functional sub-systems possible to explore thoroughly in the lab, where offspring sex ratios should be examined under controlled but realistic variation of stressors and other factors known and suspected to impact females of various ages and social statuses (cf Pride, 2005), while census data from the wild contrasts life-history matrices for low-versus high-density populations (cf. Alberts and Altmann, 2003). Nonintrusive fecal analysis throughout ringtails' geographic range would reveal whatever groups of various histories have yet to teach us about normative ringtailed-lemur hormonal function.

15.5. Conclusions

Sharing Mason's (1993) general concerns about work on primate aggression, I disagree with his characterization of theory and today's full literature. Few consistent contributors report either that competition causes most primate aggression directly or that much aggression reflects status striving in any primate (cf. Silk, 1993, 2002). Evenly or sparsely distributed resources may make only "scramble competition" possible, apparently accounting for *lack* of dominance behavior among relatively folivorous females [e.g., langurs, colobines, and gorillas (van Schaik, 1989; Sterck et al., 1997; Watts, 1994)]. Dominance is understood to be irrelevant also between monogamous adults, like the Hylobatids and Callitrichines, and the Aotines Mason studied most. Some dominant monkeys seem to *attempt* to coerce subordinate grooming partners' to *conduct* rather than *receive* more work. But, whether they *can* identifies a range of empirical projects that Mason and I would both be quick to suggest has hardly been considered, let alone attempted. Ultimately, most of us could better limit how often we write as though we already know answers to questions whose empirical investigation has, in fact, not yet been undertaken.

Today's researchers seek to understand the development of intra-adapted systems entailing variable dyadic toleration, aggression, subordination, third-party aggression, reconciliation, consolation, retribution and other behavioral *regulation* within groups, species, and clades (Moynihan 1998). Unhelpful, by contrast, are one-dimensional contrasts of the "significance" of aggression versus amicability, as the two social demeanors coexist experientially, one comprises behavioral events, the other behavioral states (Altmann, 1974), and the two domains are mechanistically opposed in neither physiological function (Leshner, 1975) nor behavioral sequences (Pereira and Kappeler, 1997). *De rigor*, serious discussion among seasoned practitioners systematically revisits conceptual, semantic, and methodological issues (Fedigan, 1982; Altmann, 1984), while possibilities are considered in relation to new perspectives (e.g. reproductive skew theory), trait by-products (Alcock, 2002; Gould, 2002), interannual and interdecadal ecological effects, and more. Dr. Mason joins others of us who ask for greater evidence of our thorough-going reflections in *published* discussions. Fair consideration of all possibilities, accounting for all detail, is how science is done, to avoid explaining things only according to the day's favored flavor (Klama, 1988; Oyama, 2003). Discussions lacking evidence of awareness indicate that someone – author, reviewer, and/or editor – failed to complete their job, and panicked perspective on putative pressure to publish promises primarily the perishing of probing.

Mill (1859) saw our "tendency . . . to leave off thinking about a thing when it is no longer doubtful" causing half of our errors. While that reflects one adaptive boundary for certain real-time decision-making, for science it prescribes deliberate, open-minded reflection, and interpretation of all available data in the full context of complexity. In the end, hypotheses are meant not for fussing over in club literature but for testing, with results communicated clearly and broadly, and the only empirical truths worth seeking are those that ultimately look much the same to most of us. So, while some pine yet to discover in animals evidence of a noble savage they also hope still lurks inside of us (Pinker, 2002), all known about vertebrates and mammals in particular today, including data from diverse primates, does not generate that particular hypothesis.

Evolutionary theory does, however, explain why diverse primates' adaptively strive to dominate particular social partners of particular types, while anticipating equally surely our inability to detect any meaningful dominance competition in many other primates, including New World groups of Mason's valued expertise. Similarly, while much of my research has been devoted to primates needing little help to generate data reflecting obsessions with agonistic power itself, the female squirrel monkeys (*Saimiri sciureus*) and all sifakas (*Propithecus* spp.) that I examined comprised unrelated primates demonstrably little invested in dyadic dominance, and for reasons yet obscure, redfronted lemurs (*Eulemur fulvus rufus*), sharing size and most gross traits with their close, ringtailed relatives, including metabolic and reproductive seasonality, hormonal patterns, and both rates and intensities of aggression, evince few adult relationships in which agonistic dominance plays important roles (Sussman, 1974; Pereira and Kappeler, 1997; Pereira and McGlynn, 1997; Overdorff, 1998; cf Ostner and Kappeler, 2004). Without discounting competition, researchers have repeatedly described similar *in*significance of agonistic dominance among females in several arboreal African monkeys (Cords, 2002) and the great apes, including humans (Watts, 1994; Wittig and Boesch, 2002).

Primates occupy divergent species-typical social systems and related cognitive ethologies within every one of which competition, aggression, and social dominance incontrovertibly remain different but related phenomena, variably expressed among life-history phases, dyadic social classes, and individual relationships. We help Mason to communicate an important complaint when we insist that every sociobiologist also must keep *all* that well in mind. Ethology is materially cheap, requiring only pens, paper, a watch, and a good diet. But, if it were easy, everyone would do it. It is not, because behavior is vexingly intangible, an emergent property of immensely complex physical systems of immensely complex histories. It does not tumble willy-nilly from differentially adapted anatomical machines, but constitutes major phenotypes simultaneously expressing instantaneous hedonic preferences, tactics of various modest time frames, and strategies for entire developmental phases, timing, achieving, and managing all necessary anatomical adjustments at every physical scale. Behavior also vitally includes various nonresponses, contextual and conditional every time.

Most strikingly, *there is no meaningful line to draw, in fact, between behavior and development*, identifying behavior as life itself, evolving even while affecting its own evolutionary trajectory (Polly, 1998; Oyama, 2003; Weber and Depew, 2003; West et al., 2003; West, 2005). Most experts of allied sub-disciplines complex enough on their own, like genetics, neurophysiology, and paleontology, thus best avoid publishing personal ruminations about behavior (cf. Eldredge, 2004; Hamer, 2004), while more ethologists need to begin taking more seriously their feedback-to-society portion of job responsibility.

The best thing here, though, is that dominance striving and innumerable other processes of species-typical cognitive development await exploration among scores of primates and thousands of other animals yet surviving on Earth. Each moment of every day, all around us, real data await expert collection, resourceful analysis, deliberate interpretation, and enthusiastic, well-reasoned communication.

References

Alberts, S., and Altmann, J. (2003). Matrix models for primate life history analysis. In: Kappeler, P. M., and Pereira, M. E. (eds.), *Primate Life Histories and Socioecology*. University of Chicago Press, Chicago, pp. 66–102.

Alberts, S., Sapolsky, R., and Altmann, J. (1992). Behavioral, endocrine, and immunological correlates of immigration by an aggressive male into a natural primate group. *Horm Behav* 26:167–178.

Alcock, J. (2001). *The Triumph of Sociobiology*. Oxford University Press, New York.

Altmann, J. (1974). Observational study of behavior: sampling methods. *Behaviour* 49:227–267.

Bentham, J. (1780). *An Introduction to the Principles of Morals and Legislation* (Clarendon Press, Oxford, 1907).

Bentham, J. (1824). *The Book of Fallacies* (unpublished papers).

Bernstein, I. S. (1981). Dominance: the baby and the bathwater. *Behav Brain Sci* 4:419–458.

Bogart, M. H., Kumamoto, A. T., and Lasley, B. L. (1977). A comparison of the reproductive cycle of three species of *Lemur. Folia Primatol* 28:134–143.

Buchan, J. C., Alberts, S. C., Silk, J. B., and Altmann, J. (2003). True paternal care in a multi-male primate society. *Nature* 425:179–181.

Buskirk, J., van McCollum, S. A., and Werner, E. F. (1997). Natural selection for environmentally induced phenotypes in tadpoles. *Evolution* 51:1983–1992.

Cavigelli, S. A. (1999). Behavioural patterns associated with faecal cortisol levels in free-ranging female ring-tailed lemurs (*Lemur catta*). *Anim Behav* 57:935–944.

Cavigelli, S. A., and Pereira, M. E. (2000). Mating season aggression and fecal testosterone levels in male ringtailed lemurs (*Lemur catta*). *Horm Behav* 37:246–255.

Cavigelli, S. A., Dubovick, T., Levash, W., Jolly, A., and Pitts, A. (2003). Female dominance status and fecal corticoids in a cooperative breeder with low reproductive skew: ring–tailed lemurs (*Lemur catta*). *Horm Behav* 43:166–179.

Cords, M. (2002). Friendship among adult female blue monkeys (*Cercopithecus mitis*). *Behaviour* 139:291–314.

de Waal, F. B. M. (1996). *Good Natured: The Origins of Right and Wrong in Humans and Other Animals*. Harvard University Press, Cambridge, Mass.

de Waal, F. B. M. (2003). Darwin's legacy and the study of primate visual communication. *Ann NY Acad Sci* 1000:7–31.

de Waal, F. B. M., and Luttrell, L. M. (1985). The formal hierarchy of rhesus monkeys: an investigation of the bared–teeth display. *Am J Primatol* 9:73–85.

Dunlap, J. C., Loros, J. J., and DeCoursey, P. J. (eds.). (2004). *Chronobiology: Biological Timekeeping*. Sinauer, Sunderland, Mass.

Eldredge, N. (2004). *Why We Do It: Rethinking Sex and the Selfish Gene*. W. W. Norton, New York.

Fedigan, L. (1982/1992). *Primate Paradigms: Sex Roles and Social Bonds*. Eden Press, Montreal.

Ganzhorn, J. U. (1986). Feeding behavior of *Lemur catta* and *Lemur fulvus. Int J Primatol* 7:17–30.

Ganzhorn, J. U., Fietz, J., Rakotovao, E., Schwab, D., and Zinner, D. (1999). Lemurs and the regeneration of dry deciduous forest in Madagascar. *Conservation Biol* 13:794–804.

Genin, F. (2000). Food restriction enhances deep torpor bouts in the grey mouse lemur. *Folia Primatol* 71:258–259.

Genin, F., and Perret, M. (2000). Photoperiod-induced changes in energy balance in gray mouse lemurs. *Physiol Behav* 71:315–321.

Glander, K. E., and Rabin, D. P. (1983). Food choice from endemic North Carolina tree species by captive prosimians (*Lemur fulvus*). *Am J Primatol* 5:221–229.

Gould, L., Sussman, R. W., and Sauther, M. L. (2003). Demographic and life-history patterns in a population of ring–tailed lemurs (*Lemur catta*) at Beza Mahafaly Reserve, Madagascar: A 15–year perspective. *Am J Phys Anthropol* 120:182–94.

Gould, S. J. (2002). *The Structure of Evolutionary Theory*. Harvard University Press, Cambridge, Mass.

Grober, M. S. (1998). Socially controlled sex change: Integrating ultimate and proximate levels of analysis. *Acta Ethol* 1:3–17.

Hamer, D. (2004). *The God Gene: How Faith Is Hardwired into Our Genes*. Doubleday, New York.

Hinde, R. A. (1981). Animal signals: Ethological and game–theory approaches are not incompatible. *Anim Behav* 29:535–542.

Hinde, R. A. (1987). *Individuals, relationships and Culture: Links Between Ethology and the Social Sciences*. Cambridge University Press, New York.

Ho, M.-W. (2001). The human genome map, the death of genetic determinism and beyond. *Institute of Science in Society Report*, 14 Feb. 2001. Available at http://www.ratical.org/co–globalize/maewanho/.

Huntingford, F., and Turner, A. (1987). *Animal Conflict*. Chapman and Hall, New York.

Huxley, J. (1942). *Evolution: The Modern Synthesis*. Harper, New York.

Hood, L. C., and Jolly, A. (1995). Troop fission in female *Lemur catta* at Berenty Reserve, Madagascar. *Int J Primatol* 16:997–1015.

Jolly, A. (1966). *Lemur Behavior: A Madagascar Field Study*. University of Chicago Press, Chicago.

Jolly, A., and Pride, R. E. (1999). Troop histories and range inertia of *Lemur catta* at Berenty, Madagascar: A thirty year perspective. *Int J Primatol* 20:359–373.

Jolly, A., Caless, S., Cavigelli, S., Gould, L., Pitts, A., Pereira, M. E., Pride, R. E., Rabenandrasana, H. D., Walker, J. D., and Zafison, T. (2000). Infant killing, wounding, and predation in *Eulemur* and *Lemur. Int J Primatol* 21:21–40.

Kappeler, P. M. (1993). Variation in social structure: The effects of sex and kinship on social interactions in three lemur species. *Ethology* 93:125–145.

Kappeler, P. M., Pereira, M. E., and van Schaik, C. P. (2003). Primate life histories and socioecology. In: Kappeler, P. M. and Pereira, M. E. (eds.), *Primate Life Histories and Socioecology*. University of Chicago Press, Chicago, pp. 1–20.

Klama, J. (1988). *Aggression: Conflict in Animals and Humans Reconsidered*. Longman Group UK, Essex, England.

Koyama, N. (1988). Mating behavior of ring–tailed lemurs (*Lemur catta*) at Berenty, Madagascar. *Primates* 29:163–175.

Koyama, N., Nakamichi, M., Oda, R., Miyamoto, N., Ichino, S., and Takahata, Y. (2001). A ten-year summary of reproductive parameters for ring–tailed lemurs at Berenty, Madagascar. *Primates* 42:1–14.

Lehrman, D. S. (1971). Conceptual and semantic issues in the nature–nurture problem. In: Aronson, L. R., Tobach, E., Lehrman, D.S., and Rosenblatt, J. S. (eds.), *Development and the Evolution of Behaviour: Essays in the Memory of T. C. Schneirla*. WH Freeman, San Francisco, pp. 17–52.

Laland, K. N., Odling-Smee, J., and Feldman, M. W. (2000). Niche construction, biological evolution and cultural change. *Behav Brain Sci* 23:131–175.

Leshner, A. (1978). *An Introduction to Behavioral Endocrinology*. Oxford University Press, New York.

Lorenz, K. Z. (1965). *Evolution and Modification of Behavior.* University of Chicago Press, Chicago.
Margulis, L., and Sagan, D. (1995). *What Is Life?* University of California Press, Berkeley.
Marler, P., and Peters, S. (1982). Developmental overproduction and selective attrition: new processes in the epigenesis of bird song. *Develop Psychobiol* 15:369–378.
Mason, W. A., and Mendoza, S. P. (eds.). (1993). *Primate Social Conflict.* SUNY Press, Albany.
Mason, W. A. (1993). The nature of social conflict: a psycho–ethological perspective. In: Mason, W. A., and Mendoza, S. P. (eds.), *Primate Social Conflict.* SUNY Press, Albany, pp. 13–47.
Moore, D. S. (2001). *The Dependent Gene: the Fallacy of Nature vs.Nurture.* Times Books, New York.
Mousseau, T. A., and Fox, C. W. (1998). *Maternal Effects as Adaptations.* Oxford University Press, New York.
Moynihan, M. H. (1998). *The Social Regulation of Aggression Among Animals.* Smithsonian Institution Press, Washington, DC.
Nakamichi, M., and Koyama, N. (1997). Social relationships among ring-tailed lemurs (*Lemur catta*) in two free–ranging troops at Berenty Reserve, Madagascar. *Int J Primatol* 18(1):73–93.
Nakamichi, M., and Koyama, N. (2000). Intra-troop affiliative relationships of females with newborn infants in wild ring-tailed lemurs (*Lemur catta*). *Am J Primatol* 50(3):187–203.
Nelson, D. A., Marler, P., and Palleroni, A. (1995). A comparative approach to vocal learning: intraspecific variation in the learning process. *Anim Behav* 50:83–97.
Nunn, C. L., and Pereira, M. E. (2000). Group histories and offspring sex ratios in ring-tailed lemurs (*Lemur catta*). *Behav Ecol Sociobiol* 48:18–28.
Overdorff, D. J. (1998). Are *Eulemur* species pair-bonded? Social organization and mating strategies in *Eulemur fulvus rufus* from 1988-1995 in Southeastern Madagascar. *Am J Phys Anthropol* 105:153–166.
Oyama, S. (2003). On having a hammer. In: Weber, B. H., and Depew, D. J. (eds.), *Evolution and Learning: The Baldwin Effect Reconsidered.* MIT Press, Cambridge, mass., pp. 169–191.
Ostner, J., and Kappeler, P. M. (2004). Male life history and unusual sex ratio of redfronted lemur (*Eulemur fulvus rufus*) groups. *Anim Behav* 67:249–259.
Pereira, M. E. (1989). Agonistic interactions of juvenile savanna baboons. II. Agonistic support and rank acquisition. *Ethology* 80:152–171.
Pereira, M. E. (1991). Asynchrony within estrous synchrony among ringtailed lemurs (Primates: Lemuridae). *Physiol Behav* 49(1):47–52.
Pereira, M. E. (1993a). Agonistic interaction, dominance relation, and ontogenetic trajectories in ringtailed lemurs. In: Pereira, M. E., and Fairbanks, L. A. (eds.), *Juvenile Primates: Life History, Development, and Behavior.* Oxford University Press, New York, pp. 285–305.
Pereira, M. E. (1993b). Seasonal adjustment of growth rate and adult body weight in ring-tailed lemurs. In: Kappeler, P. M., and Ganzhorn, J. U. (eds.), *Lemur Social Systems and Their Ecological Basis.* Plenum Press, New York, pp. 205–221.
Pereira, M. E. (1995). Development and social dominance among group-living primates. *Am J Primatol* 37:143–175.
Pereira, M. E., and Kappeler, P. M. (1997). Divergent systems of agonistic behaviour in lemurid primates. *Behaviour* 134:225–274.

Pereira, M. E., and Fairbanks, L. A. (1993). What are juvenile primates all about? In: Pereira, M. E., and Fairbanks, L. A. (eds.), *Juvenile Primates: Life History, Development, and Behavior.* Oxford University Press, New York, pp. 3–12.

Pereira, M. E., and Izard, M. K. (1989). Lactation and care for unrelated infants in forest-living ring-tailed lemurs. *Am J Primatol* 18(2):101–108.

Pereira, M. E., and Leigh, S. R. (2003). Modes of primate development. In: Kappeler, P. M., and Pereira, M. E. (eds.), *Primate Life Histories and Socioecology.* University of Chicago Press, Chicago.

Pereira, M. E., and Macedonia, J. M. (1991). Ring-tailed lemur antipredator calls denote predator class, not response urgency. *Anim Behav* 41:543–544.

Pereira, M. E., and Weiss, M. L. (1991). Female mate choice, male migration, and the threat of infanticide in ring-tailed lemurs. *Behav Ecol Sociobiol* 28:141–152.

Pereira, M. E., Strohecker, R., Cavigelli, S. A., Hughes, C., and Pearson, D. (1999). Metabolic strategy and social behavior in Lemuridae. In: Rasamimanana, H., Rakotosamimanana, B., Ganzhorn, J., and Goodman, S. (eds.), *New Directions in Lemur Studies*, Plenum Press, New York, pp. 93–118.

Pinker, S. (2002). *The Blank Slate: The Modern Denial of Human Nature.* Viking Press, New York.

Plomin, R., and Ho, M-W. (1991). Brain, behavior, and developmental genetics. In: Gibson, K. R., and Petersen, A. C. (eds.), *Brain Maturation and Cognitive Development: Comparative and Cross–Cultural Perspectives.* Aldine de Gruyter, Hawthorn, New York.

Polly, P. D. (1998). Variability, selection, and constraints: development and evolution in viverravid (Carnivora, Mammalia) molar morphology. *Paleobiology* 24:409–429.

Pride, R. E. (2005). Optimal group size and seasonal stress in ring-tailed lemurs (*Lemur catta*) *Behav Ecol* 16:550–560.

Ricklefs, R. E. (2001). *The Economy of Nature*, 5th ed. Freeman, New York.

Ridley, M. (2001). *The Cooperative Gene: How Mendel's Demon Explains the Evolution of Complex Beings.* Free Press (Simon and Schuster), New York.

Rowell, T. E. (1974). The concept of social dominance. *Behav Biol* 11:131–154.

Sade, D. S. (1967). Determinants of dominance in a group of free-ranging rhesus monkeys. In: Altmann, S. A. (ed.), *Social Communication Among Primates.* University of Chicago Press, Chicago, pp. 99–114.

Sauther, M. L. (1991). Reproductive behavior of free-ranging *Lemur catta* at Beza Mahafaly Special Reserve, Madagascar. *Am J Phys Anthropol* 84:463–477.

Schlichting, C. D., and Pigliucci, M. (1998). *Phenotypic Evolution: A Reaction Norm Perspective.* Sinauer, Sunderland, Mass.

Schmalhausen, J. J. (1946). *The Factors of Evolution*, Moscow–Leningrad (reprinted 1986: U. Chicago Press).

Schmid, J. (2001). Daily torpor in free-ranging gray mouse lemurs (*Microcebus murinus*) in Madagascar. *Int J Primatol* 22:1021–1031.

Schneirla, T. C. (1957). The concept of development in comparative psychology. In: Harris DB (ed.), *The Concept of Development.* University of Minnesota Press, Minneapolis, pp. 78–108.

Silk, J. B. (1993). The evolution of social conflict among female primates. In: Mason, W. A., and Mendoza, S. P. (eds.), *Primate Social Conflict.* SUNY Press, Albany, pp. 49–83.

Silk, J. B. (2002). Practice random acts of aggression and senseless acts of intimidation: The logic of status contests in social groups. *Evol Anthropol* 11:221–225.

Singer, P. (1999). *A Darwinian Left: Politics, Evolution and Cooperation*, Yale University Press, New Haven.

Smuts, B. B. (1985). *Sex and Friendship in Baboons*. Harvard University Press, Cambridge, Mass.

Stearns, S. C. (1992). *The Evolution of Life Histories*. Oxford University Press, Oxford.

Stearns, S. C., Pereira, M. E., and Kappeler, P. M. (2003). Primate life histories and future research. In: Kappeler, P. M., and Pereira, M. E. (eds.), *Primate Life Histories and Socioecology*. University of Chicago Press, Chicago.

Sterck, E. H. M., Watts, D. P., and van Schaik, C. P. (1997). The evolution of female social relationships in nonhuman primates. *Behav Ecol Sociobiol*. 41:291–309.

Strum, S. C. (1982). Agonistic dominance in male baboons: an alternative view. *Int J Primatol* 3:175–202.

Sussman, R. W. (1974). Ecological distinctions in sympatric species of lemur. In: Martin, R. D., Doyle, G. A., and Walker, A. C. (eds.), *Prosimian Biology*. University of Pittsburgh Press, Pittsburgh, pp. 75–108.

Sussman, R. W., Gould, L., and Sauther, M. (1999). The socioecology of the ringtailed lemur: Thirty–five years of research. *Evol Anthropol* 8:120–132.

Suter, K. J., Pohl, C. R., and Wilson, M. E. (2000). Circulating concentrations of nocturnal leptin, growth hormone, and insulin–like growth factor I increase before the onset of puberty in agonadal male monkeys: potential signals for the initiation of puberty. *J Clin Endocrinol Metab* 85:808–814.

Takahata, Y., Koyama, N., Ichino, S., and Miyamoto, N. (2005). Inter- and within-troop competition of female ring-tailed lemurs: a preliminary report. *African Study Monogr* 26:1–14.

Taylor, L. L. (1986). Kinship, dominance and social organization in a semi-free-ranging group of ringtailed lemurs (*Lemur catta*). PhD diss., University of Washington, St. Louis.

Taylor, L. L., and Sussman, R. W. (1985). A preliminary study of kinship and social organization in a semi–free–ranging group of *Lemur catta*. *Int J Primatol* 6:601–614.

Tinbergen, N. (1953). *Social Behaviour in Animals*. TJ Press Ltd, Padstow, Cornwall, UK.

Trivers, R. L. (1974). Parent-offspring conflict. *Am Zool* 14:249–264.

van Hooff, J. A. R. A. M. (1970). A component analysis of the structure of the social behaviour of a semi-captive chimpanzee group. *Experientia* 26:549–550.

van Schaik, C. P. (1989). The ecology of social relationships amongst female primates. In: Standen, V., and Foley, R. A. (eds.), *Comparative Socioecology: The Behavioral Ecology of Humans and Other Animals*. Blackwell, Oxford, pp. 195–218.

van Schaik, C. P., and Kappeler, P. M. (1993). Life history and activity period as determinants of lemur social systems. In: Kappeler, P. M., and Ganzhorn, J. U. (eds.), *Lemur Social Systems and Their Ecological Basis*. Plenum Press, New York, pp. 241–260.

van Schaik, C. P., and van Hooff, J. A. R. A. M. (1983). On the ultimate causes of primate social systems. *Behaviour* 85:91–117.

Vick, L. G., and Pereira, M. E. (1989). Episodic targeting aggression and the histories of *Lemur* social groups. *Behav Ecol Sociobiol* 25:3–12.

Wake, M. H. (2004). Integrative biology: the nexus of development, ecology, and evolution. Plenary Lecture: *Int. Union Biol Sci, General Assembly, Jan. 18, 2004*, Cairo Egypt.

Walters, J. (1980). Interventions and the development of dominance relationships in female baboons. *Folia Primatol* 35:61–89.

Watts, D. P. (1994). Agonistic relationships between female mountain gorillas (*Gorilla gorilla beringei*). *Behav Ecol Sociobiol* 34:347–358.

Wcislo, W. T. (2000). Environmental hierarchy, behavioral contexts, and social evolution in insects. In: Martins, R. P., Lewinsohn, T. M., and Barbeitos, M. S. (eds). *Ecologia e comportamento de Insectos*. Série Oecologia Brasiliensis, vol. VIII. PPGE–UFRJ. Rio de Janeiro, Brasil, pp. 49–84.

Weber, B. H., and Depew, D. J. (eds.). (2003). *Evolution and Learning: the Baldwin Effect Reconsidered*. MIT Press, Cambridge, Mass.

West, M. J., King, A. P., and White, D. J. (2003). The case for developmental ecology. *Anim Behav* 66:617–622.

West, M. J. (2005). *Developmental Plasticity and Evolution*. Oxford University Press, New York.

Wilson, E. O. (1975). *Sociobiology: The New Synthesis*. Harvard University Press, Cambridge, Mass.

Wittig, R. M., and Boesch, C. (2003). Food competition and linear dominance hierarchy among female chimpanzees of the Taï National Park. *Int J Primatol* 24:847–867.

Wrangham, R., and Peterson, D. (1996). *Demonic Males: Apes and the Origins of Human Violence*. Houghton Mifflin, New York.

16
Male and Female Ringtailed Lemurs' Energetic Strategy Does Not Explain Female Dominance

HANTANIRINA RASAMIMANANA, VONJY N. ANDRIANOME, HAJARIMANITRA RAMBELOARIVONY, AND PATRICK PASQUET

16.1. Introduction

Female dominance in terms of female feeding priority and female mate choice is characteristic of many lemur species. (Pollock, 1979; Jolly, 1984; Richard 1987; Kappeler, 1993; Meyers and Wright, 1993; Radespiel and Zimmermann, 2001). Female dominance is also found in *Pan paniscus* (Stanford, 1998). Malagasy Lemurs do not show any sexual dimorphism (Kappeler, 1991), while male bonobo weigh more than females. It seems that the dominance of one sex within a primate social group does not always depend on the weight.

Jolly (1984), Young et al. (1990), Pereira (1999), and Wright (1999) suggest as a hypothesis that female dominance is determined by important energy constraints during seasonal reproductive periods. Available energy is also seasonal and may be insufficient to satisfy the increased needs of the organism due to a very high rate of growth of the embryo and the infant.

In contrast, Kappeler (1996) observed that captive lemur mothers do not undergo any higher energy constraints than that of prosimians with no female dominance. In the same vein, Tilden et al. (1997) and von Engelhardt et al. (2000) have not found any obvious arguments to support a high rate of maternal investment during reproduction in lemur species. Embryo growth rate is not high during the gestation period, and the milk is neither richer nor of greater quantity than that of other prosimians, even in conditions of intense feeding competition. Finally, Sauther (1992) and Hemingway (1999) found no sexual differences in the activity budget and the feeding duration of *Lemur catta* and *Propithecus diadema edwardsi*. However, the latter showed a difference in the dietary composition of males and females, because females chose some differing plant parts during the lactation period.

Thus, according to Kappeler (1996), Malagasy primate female dominance would not directly depend on their physiological state (estrus, gestation, and lactation) given the results obtained from captive animals. This author concludes that there is probably no reason to tie the physiology of reproduction to the social behavior, but that this interpretation should still be restudied in natural conditions with limited food. Therefore, I undertook to study energy budgets of wild male and female ringtailed lemurs, to sort out the relationships between female dominance, social behavior, and the physiological state of lemurs.

I compared male and female ringtailed lemur behavior, diet quality, and energy expenditure, taking into account social status and interactions within the troop. The questions areas follow: Does male and female energy expenditure reflect intrasexual dominance hierarchies, and/or degree of female aggression toward males? And does male and female energy expenditure explain female dominance? Finally, what is the physical activity level, that is, the ratio of total energy expenditure to basal metabolism of male and female ringtailed lemurs as compared to that of other primates?

16.2. Methods

I compared 10 adult males and 10 adult females living in two different troops of *Lemur catta* in the Berenty Private Reserve, 25°05′ east and 46°18,5′ south (see Jolly *et al.* in this volume for a detailed description of the site). One of the two studied troops (troop D1A) was located in natural gallery forest and the second troop (troop G3) in the tourist area with introduced plants and garbage from the kitchen of the restaurant (see Rasamimanana and Rafidinarivo, 1993).

Focal sampling allowed us to observe each member of each group and to analyze individual characteristics that could influence the interaction between animals and that in turn defined the whole group behavior. Troop composition is shown in Tables 16.1 and 16.2.

In troop D1A during the 2002 lactation period, there were two adult males, and between 2002 lactation and 2003 mating periods two more immigrated. Then one of the two original males in the troop died before mating.

One of the dominant females, Diqua, mated on the same day as a subordinate one, Dana, but did not succeed in giving birth. During the gestation period, Diqua regressed in rank to become the most subordinate at the end of the study during 2003 lactation period when she was always beaten by the other females, even the subordinate one. Despite this fact, Diqua was one of the most dominant over males, as much so as Dido the new alpha female.

Within troop G3, no rank changes were seen, but the most dominant female, Antitra, died at the end of the study being about 15 years old. She was the eldest among the troop members and the mother of all the more dominant females. She had lost 3 offspring for two successive years, two infants and one subadult. She was also one of the females most dominant over males; the second one being the most subordinate female, Bobo, who lost her infant at the same period as Antitra. Male immigration was seen only during the mating period, when a new immigrant succeeded in being first to mate a subordinate primiparous female.

TABLE 16.1. Characteristics of adult animals of D1A troop.

		Diqua Female Dominant	Dido Female Dominant	Doso Female Subordinate	Dana Female Subordinate	Deba Male Dominant	Scar Male Dominant	Star	Doma
2002	Lactation period	Lactating	Lactating	Lactating	Lactating				
2003	Mating	Dominant one of the latest to mate	Dominant the second to mate	Dominant the first to mate	Subordinate one of the latest to mate	Dominant The latest to mate with Dido, Diqua, and Dana	Dead 04/21/03 in the morning	Subordinate Male second to mate with Doso took off the mating cork	Subordinate Male first to mate with Doso
	Gestation	Subordinate nongestating	Alpha dominante Gestating	Dominant Gestating	Subordinate Gestating	Dominant	Dead	Subordinate	Subordinate
	Birth	Subordinate	Dominant Gave birth	Dominant Gave birth	Subordinate Gave birth	Dominant	Dead	Subordinate	Subordinate
	Lactation	Subordinate	Dominant Lactating	Dominant Lactating	Subordinate Lactating	Dominant	Dead	Subordinate	Subordinate

TABLE 16.2. Characteristics of adult animals of G3 troop.

	Antitra	Cœur	Tata	Mavo	Kelilo	Bobo	RE	Triangle	Rabitro	Point noir	LE	Telosof	Vakitampo
2002 Lactation period	Female	Subadulte Antitra's daughter	Female Antitra's daughter	Female	Female	Female	Male	Male	Male	Male	Male	Male	Male
	Equivocal rank No lactating	Not followed No lactating	Dominant alpha lactating	Subordinate lactating	Not followed No lactating	Subordinate lactating	Dominant	Not yet in troop	Subordin.	Subordin.	Subordin.	Subadult not followed	Subordin.
2003 Mating	Dominant not observed to mate	Primipare Dominant first to mate	Dominant not observed to mate	Subordinate ordinate observed to mate	Subordinate not to mate Peripheral	Sub-first second observed to mate	Dominant to not mate with Cœur took off the mating cork. Second to mate with Kelilo	Immigrant Subordin. First to mate with Kelilo. Chased by RE	Subordin. peripheral	Subordin. peripheral	Subordin.	Subordin.	Subordin. third to mate with Kelilo
Gestation	Dominant gestating	Dominant alpha Gestating	Dominant Gestating	Subordinate Gestating	Subordinate Gestating	Subordinate Gestating	Dominant	Subordin.	Subordin. Peripheral	Subordin. Peripheral	Subordin.	Subordin.	Subordin.
Birth	Dominant last to give birth, lost its infant at the second day	Dominant First to give birth	Dominant Second to give birth	Subordinate Fourth to give birth	Subordinate Third to give birth	Subordinate fifth to give birth, lost its infant at the fifth day	Dominant	Subordin.	Subordin. Peripheral	Subordin.	Subordin.	Subordin.	Subordin.
Lactation	Died 10/25/03 late morning, 3 days after its subadult son	Dominant lactating	Dominant lactatordin.	Subordinate lactating	Subordin.	Subordin.	Dominant	Subordin.	Subordin. Peripheral	Subordin. Peripheral	Subordin.	Subordin.	Subordin.

16.2.1. Behavioral Observations

Individual troops were followed continuously through out the day from 0600 to 1800 h, with a break from 1200 to 1400 h, because the animals mostly sleep in that interval of time (Rasamimanana and Rafidinarivo, 1993; Ramasiarisoa, 2000), during 1502 hours of observation spread across the four reproductive periods (mating, gestation, birth, and lactation) between October 2002 and November 2003. More details on the divisions of these periods are presented in Rasamimanana and Rafidinarivo (1993).

The activity of the animals was recorded using instantaneous scan sampling and focal animal sampling (Altmann, 1974). Data collected with instantaneous scan sampling every five minutes included troop activities such as sleeping, sunning, resting, foraging, grooming, moving and traveling, type of food consumed (plant species and its part), and location of the troop within 25×25 m quadrats on a map of the study area (Williams, 1998). Scan sampling was conducted once a week in both troops (i.e., for 180 hours each troop). Data collected with focal animal sampling included the preceding activities, the duration of feeding of each individual on a particular plant food and its parts, the number of steps, jumps and leaps in order to calculate the distance covered during moving. This last was also estimated by GPS during traveling over a relatively long distance. Focal samples were 5 minutes long, but with three samples taken consecutively on each individual to minimize lost data. On average each individual was focal sampled for 45 hours valid for analysis in D1A troop and for 29 hours in G3 troop.

Using the duration of feeding it was possible to quantify the proportion of each food item that made up the ringtailed lemur diet. Activities were standard as in other primate studies, except that moving was defined as displacement over less than three meters, (thus usually within a single food patch.) Traveling was defined as displacement beyond 3 m. For calculation of total activity versus inactivity, sleeping, resting, and sunbathing were classed as inactivity, and feeding, moving, traveling, and grooming classed as activity.

Dominance within each sex was calculated by direction of aggression and submission, as is usual in primate studies. This does not describe individual relations between sexes, because in ringtailed lemurs all adult females dominate all adult males. Dominance between sexes was therefore defined by the frequency of aggression shown by each individual female toward all the males. So females were categorized in two groups: those more aggressive toward males and those less.

Binoculars, 8×30 with $7.5\,°$ field, a GPS device and a compass, pen and paper were used to collect these data.

16.2.2. Energy Expenditure

Coelho (1974) uses the term socio-bioenergetics to indicate the study of the energy expenditure of an animal as a group member interacting within that group. The study rests on the basis of combination of physiological principles and ethological techniques. By means of a "factorial" approach, the metabolic cost of an activity is related to the time each individual spent on it. This time is known from continuous focal observation of the animal. Coelho et al. (1976) established some indexes "K" of energy expenditure for each main activity (Table 16.3) of the animal which are used in allometric equation to calculate the energy expenditure of male and female *Lemur catta*.

Total energy expenditure in multiples of basal metabolism is calculated:

$$\text{Total energy expenditure TEE} = \sum_{i}^{n} C_i$$

- C_i = K_i BMT_i = energy expended for an activity "i" by a ringtailed lemur individual within a 10-h observation day.
- K_i = index of energy expenditure for an activity "i" (Table 16.3), other than traveling. Traveling is calculated by another allometric equation below so does not have an index "K."
- T_i = time (hours) spent by a ringtailed individual for an activity "i" within a 10 h observation day.
- BM = basal metabolism of *Lemur catta* predicted from Kleiber's formula (1961) MB = $70W^{0.75}$ adjusted for lemur species. Kleiber's calculated basal metabolism was reduced to 65% of its value according to the results obtained by Daniels (1984), Richard and Nicoll (1987), and Dracks et al. (1999), respectively, on *Eulemur fulvus* ssp., *Propithecus verreauxi verreauxi*, and *Lepilemur ruficaudatus*, who showed that prosimians have a much lower basal metabolism than other primates. Furthermore, it is known that the metabolism of a gestating and lactating mammal is respectively 1.25 times and 1.5 times higher than that of nongestating and nonlactating mammal (Crampton and Lloyd, 1959; Portman, 1970, in Coelho, 1974) so the theoretical values of basal metabolism of females were raised depending on their reproductive state.
- W = weight. I attributed to every male the same average weight (W= 2.6 Kg) obtained from other individuals of adjacent troops weighed in March 2003 (Crawford et al., pers. comm.). The same process was done with the females (W = 2.3 kg) except for one individual we could weight (W = 1.750 kg), who died during the study.

TABLE 16.3. Indexes of energy expenditure for some ring-tailed activities.[a]

Activities	Index of energy expenditure (K)
Sleeping	1.00
Sunning and resting	1.25
Feeding	1.38
Grooming	2.35

[a] Values taken from Leonard and Robertson (1997).

The allometric equation calculating the energy spent during traveling is following:

$C_{travel} = (0.041\ W^{0.6})\ DC + (0.029 W^{0.75}) T_{travel}$ = energy expended by a ringtailed individual for traveling within a 10-h observation day (Leonard and Robertson, 1997).

- W = weight (g) from Crawford et al. in this volume, as above.
- T_{travel} = time spent for traveling by a lemur individual within a 10-h observation day (hours).
- DC = distance covered by a lemur individual within a 10-h observation day (km). This was estimated by GPS or calculated by the following formula:

$$DC = (23.10^{-5}\ \Sigma\ S) + (50.10^{-5}\ \Sigma\ J) + (10^{-3}\ \Sigma\ L)$$

where S = number of steps during traveling, J = number of jumps during traveling, and L = number of leaps during traveling.

The physical activity level (PAL) is calculated as TEE/BM. This allows a comparison of activity alone without the influence of the animal's weight or reproductive state (Leonard and Robertson 1997).

16.2.3. Statistical Analyses

We performed all statistical tests via Statistica 6.0 (Statsoft). As samples were not large, we mostly used non-parametric tests including chi-square test to estimate the dependence between the distributions of 10-h-daily inactivity and the physiological periods, the intrasexual hierarchy, the female dominance over males, and the sexes.

The parametric tests Student t-test was used to test the differences between energy expenditure and physical activity level of males and females and ANOVA to test the differences of energy expenditure and physical activity level within reproduction periods.

16.3. Results

16.3.1. Male and Female Activities and Inactivity

In order to sort out whether male and female energy expenditure could explain the female dominance in ringtailed lemurs, we caclulated both sexes' daily distribution of activities at each reproductive period, mating, gestation, lactation and birth.

There was no difference between troops, so results from both troops are combined.

Lemur catta as a species spends most of its time at rest which matches with the fact it has a low basal metabolic rate (Daniels, 1984; Richard and Nicoll, 1987). Feeding, moving, traveling, and grooming could be gathered in one category called activity, and the remaining three activities: sleeping, sunbathing and resting in another one called inactivity.

Lemur catta at every reproductive period more was inactive than active (Figure 16.1). However, during the mating and gestation periods the animals were more active than during birth and lactation. This due to the fact that males and females are mixed here, and feeding is a component of activity. While feeding during those two periods, they did more moving than during birth and lactation and more grooming while resting.

During mating and lactation periods, females were significantly more active than males (respectively mating: $\chi^2 = 6.8$; df $=1$; p < 0.05 and lactation: $\chi^2 = 5.7$; df $= 1$; p < 0.05) (Figures 16.2 and 16.3). The big differences drawn in Figure

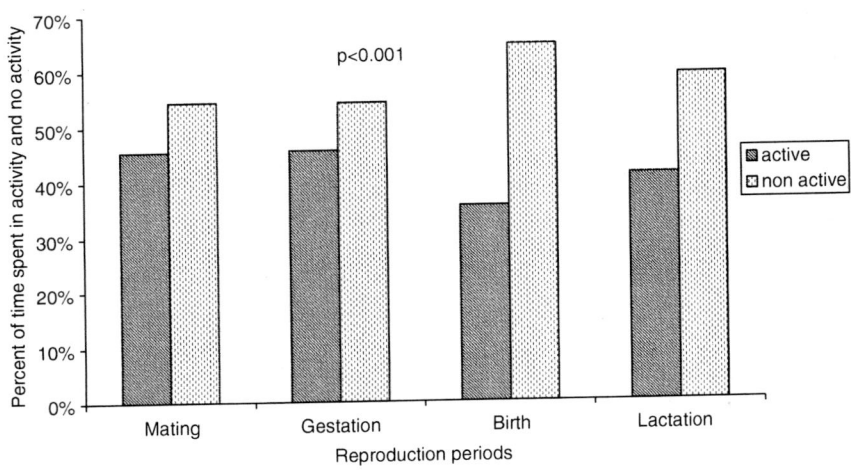

FIGURE 16.1. Inactivity of the animals during each reproductive period.

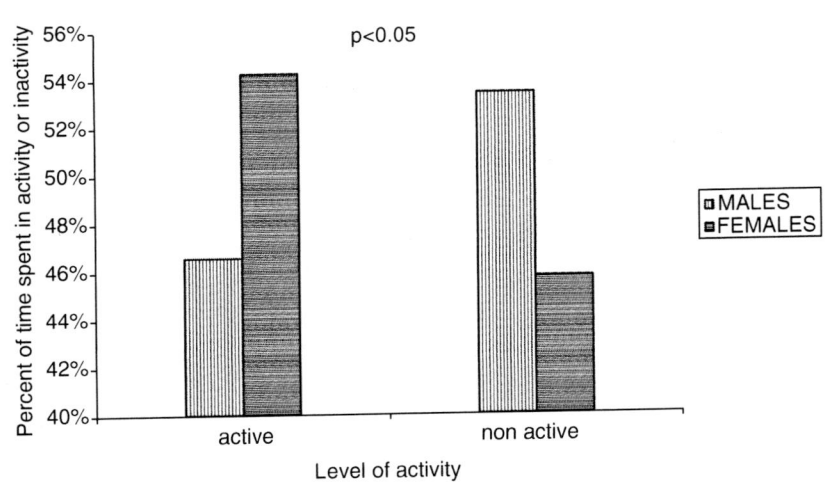

FIGURE 16.2. Level of activity of male and female *Lemur catta* during the mating period.

16. Energetic Strategy Does Not Explain Female Dominance   279

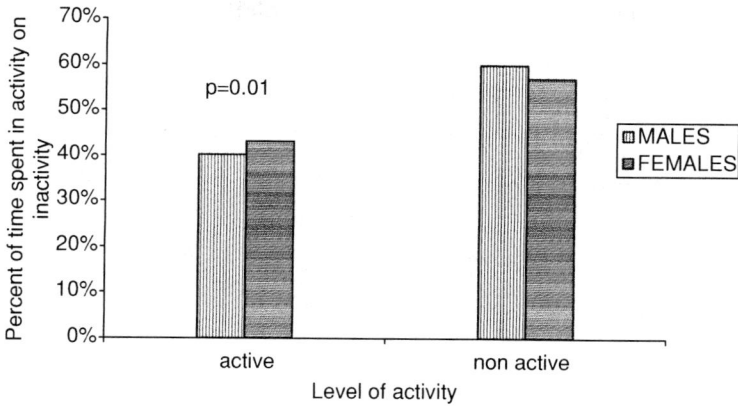

FIGURE 16.3. Level of activity of male and female *Lemur catta* during the lactation period.

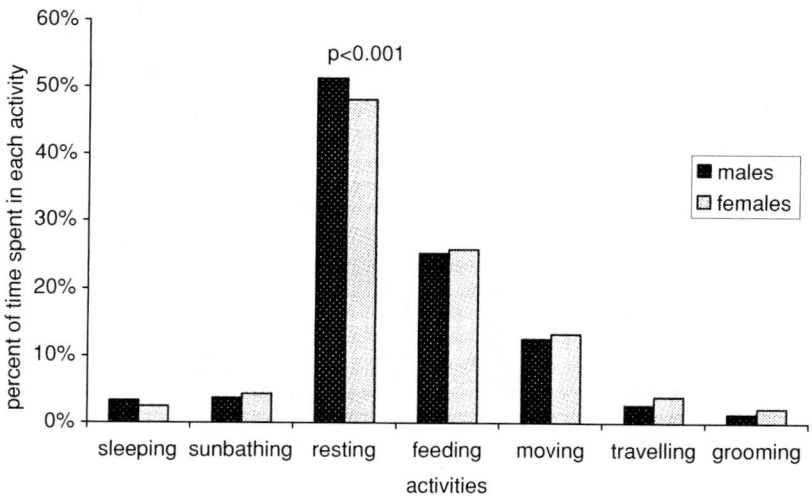

FIGURE 16.4. Percent of time animals spent in each activity according to their sex.

16.2 despite the fact that the difference is only significant at $p < 0.05$ is probably due to the different number of samples obtained during the two periods. The numbers of observation days are less in the mating period than in lactation period due to the different length of those two periods.

This greater activity of females also appears as an overall difference between males and females in percent of time they spent in resting. Males rested statistically longer than females, and females moved, traveled and groomed more than males ($\chi^2 = 11.17$; df = 6; $p < 0.001$) (Figure 16.4).

The components of activity differed according to the reproductive periods. Figure 16.5 shows that females spent more time in feeding during the gestation

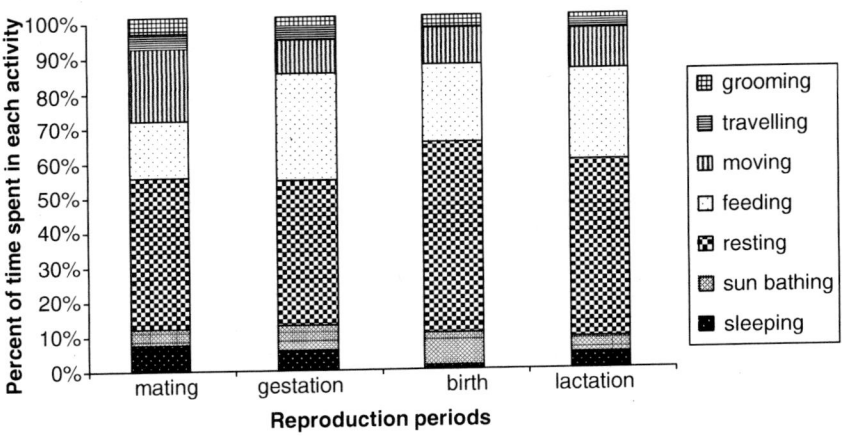

FIGURE 16.5. Distribution of female activities according to reproductive periods.

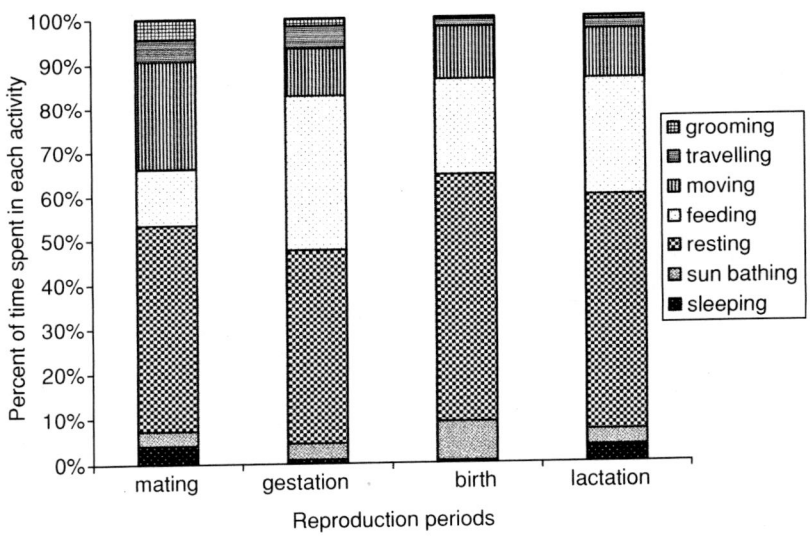

FIGURE 16.6. Distribution of male activities according to reproductive periods.

period and less time during mating period, while they were moved and travelled less during gestation and more during mating. Males displayed the same tendencies (Figure 16.6) during the gestation period, but during the birth period they traveled more than females.

16.3.2. Level of Females' Activity in Relation to Their Dominance

All females, whether dominant or subordinate in the female hierarchy, are dominant over males. In each study troop, the alpha and the most subordinate female

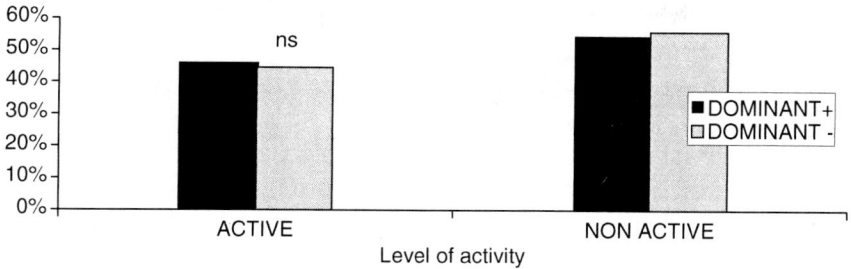

FIGURE 16.7. Percent of time females spent in inactivity in regard to their dominance over males.

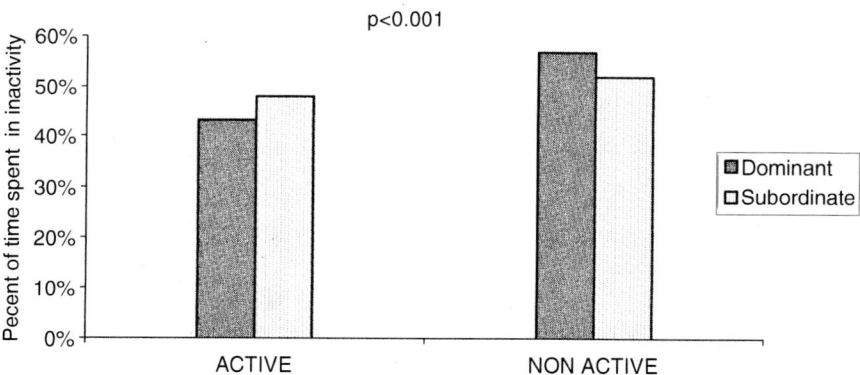

FIGURE 16.8. Percent of time spent in inactivity by females in regard to their intrasexual hierarchy.

were the most dominant over males as measured by frequency of aggression toward males. Curiously, those occupying the intermediate rank were less aggressive toward males. Female dominance toward males did not correlate with time inactive ($\chi^2 = 0.66$; df = 1; p = 0.42) (Figure 16.7).

When they were analyzed in regard to the female hierarchy, it appeared that subordinate females were significantly more active than dominants ($\chi^2 = 12.4$; df = 1; p < 0.001) (Figure 16.8).

On the other hand, the difference between dominants and subordinates in the male hierarchy was not significant ($\chi^2 = 2.7$; df = 1; p ≥ 0.05).

16.3.3. *Male and Female Energy Expenditure*

As mentioned above, energy expenditure depends on the animal's weight as well as its activity. The energy expenditure during 10-h observation active day was calculated on the basis of Leonard and Robertson's (1997) formula taking in account the activities, the distance covered, and the basal metabolism of each individual.

No troop difference was observed in terms of energy expenditure of male and female *Lemur catta* of Berenty, so the data could be combined. The average distances covered by members of both troops during the focal observations were respectively 0.450 km during the mating period; 0.270 km during gestation period; 0.140 km during birth period, and 0.221 km during 2003 lactation period. This showed that the animals covered a longer distance during mating and gestation periods than during birth and lactation periods, both sexes combined.

Males' and females' energy expenditure varied from one reproductive period to another F (4.39) = 15.01; $p < 0.001$. On average, females' energy expenditure showed a minimum of 82 kcal during the gestation period and a maximum of 104 kcal during the 2002 lactation period. In contrast, males had a minimum of 71 kcal during the 2003 lactation period and a maximum of 108 kcal during mating period. (Figure 16.9).

Activities counted as activity were feeding, moving, traveling, and grooming. The animals might spend a high percentage of time in those activities but covered a much shorter distance, so they might spend less energy, because total energy expended was significantly correlated with the distance covered ($r = 0.71$ $p < 0.001$). Figure 16.10 showed that ringtailed females of both troops covered the longest distances during copulation and lactation periods. The maximum energy expended by females during the lactation period is explained by the long distance covered in short time and also by the fact that we multiply the basal metabolic rate of lactating females by 1.5, following Crampton and Lloyd (1959) and Portman (1970) in Coelho (1974).

Males and females expended their energy differently each from other during the mating ($p = 0.02$) and lactation periods ($p < 0.001$) (Figure 16.9). That could mean a high need by one sex or the other during those periods. Generally, the lac-

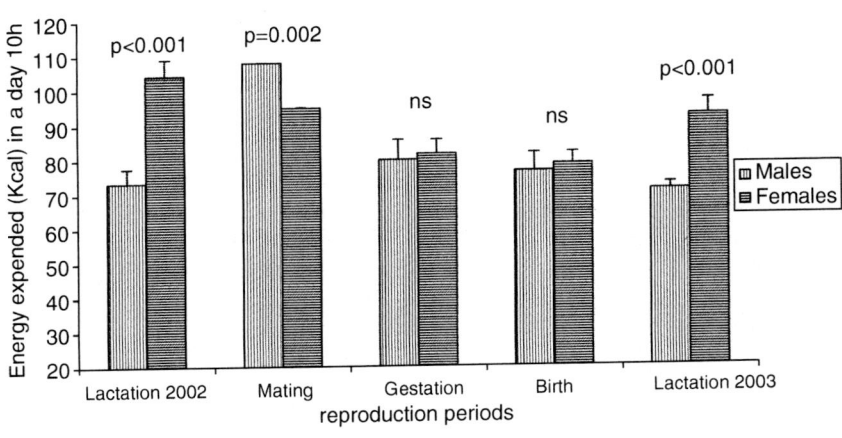

FIGURE 16.9. Energy expended on average in a 10-h-observation day by males and females during different reproductive periods.

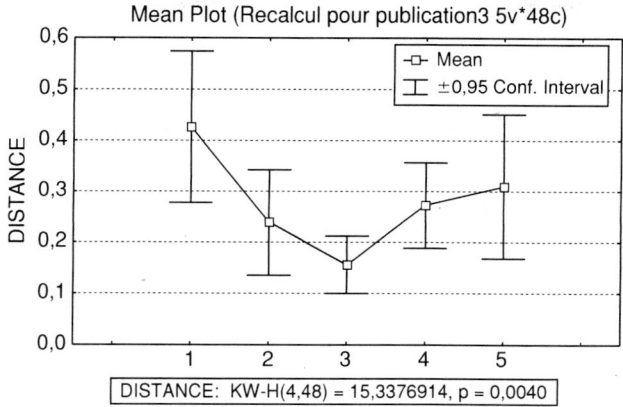

FIGURE 16.10. Mean distance covered by ring-tailed females during each reproductive period (1 = copulation period; 2 = gestation period; 3 = birth period; 4 and 5 = lactation periods in 2003 and 2002).

tation period is that of the highest energy expenditure for females and that of the least for males. On the contrary, the end of gestation to birth periods corresponded to the least energy expenditure for females (Figure 16.9) because of the short distance covered.

Sex and period influence on the energy expenditure were seen above, but there seemed not to be a significant intra-sexual hierarchy influence on this variable ($p \geq 0.05$) (Table 16.4) due to the fact that the hierarchy status of the females was not steady in D1A troop during our observation study. A dominant female did not succeed in giving birth, which suggests she had already lost her position during the gestation period before parturition. Subordinate females expended more energy than dominants but that was not significant. If there is a difference between males and females but none within males or within females in regard to their hierarchy, what about female energy expenditure in regard to their dominance over males? There was no relationship between female dominance over

TABLE 16.4. Analysis of variance of the average energy expenditure (10-h observation) of ringtailed lemurs depending on the period, sex, and hierarchy intra-sexual.

	SS	Degrees of freedom	MS	F	p
Intercept	616589.9	1	616589.9	2912.511	0.000000
Period	7780.5	4	1945.1	9.188	0.000003
Hierarchy	128.0	1	128.0	0.604	0.439052
Sex	1263.4	1	1263.4	5.968	0.016639
Standard error	17994.8	85	211.7		

males and female energy expenditure even when the reproductive period and troop were taken in account (F (1, 41) = 0.16; p = 0.69 NS).

16.3.4. *Male and Female Physical Activity Level*

The differences noted between males and females' energy expenditure was due either to their weight or to their activity differences. In order to compare individuals not taking their weight into account, one calculates the physical activity level (PAL), which is the ratio of the total energy expenditure to the basal metabolism.

As shown in Table 16.5, reproductive period influenced the PAL while troops, sex, and intrasexual hierarchy did not. The tendency effect of hierarchy is to be noted but was not significant. Although there was a significant difference between males' and females' energy expenditure, this was not the case for their physical activity level (F (1, 84) = 0.35; p = 0.55). Males and females were physically active at the same level during every reproductive period with no difference between troops.

The PAL of females more dominant toward males was statistically the same as that of females less dominant toward males.

When the females were analyzed by their intrasexual hierarchy, a difference in PAL was not statistically significant between dominant and subordinate females even though subordinate females had higher physical activity level than dominants (2.12 vs. 1.97). According to these results, it seemed that social group organization due to interactions within individuals did not have any direct relationship with the physical activity level of animals as Kappeler (1999) suggested. The females' PAL seemed neither to be linked to their social status nor to their reproductive state, for there was no significant difference between males' and females' PAL during gestation and lactation periods.

16.3.5. *Individual Variation*

The variation between individuals was striking. Although many of the comparisons between males and females or between dominants and subordinates are not significant, individuals within each category could differ sharply. The variation of

TABLE 16.5. Analysis of variance of PAL according to the period, troop, sex, and intrasexual hierarchy.

	SS	Degrees of freedom	MS	F	p
Intercept	335.64	1	335.64	3963.46	<0.001
Period	10.81	4	2.70	31.94	<0.001
Group	0.014	1	0.014	0.16	0.68
Hierarchy	0.29	1	0.28	3.387	0.069
Sex	0.03	1	0.03	0.348	0.556
Standard error	7.11	84	0.08		

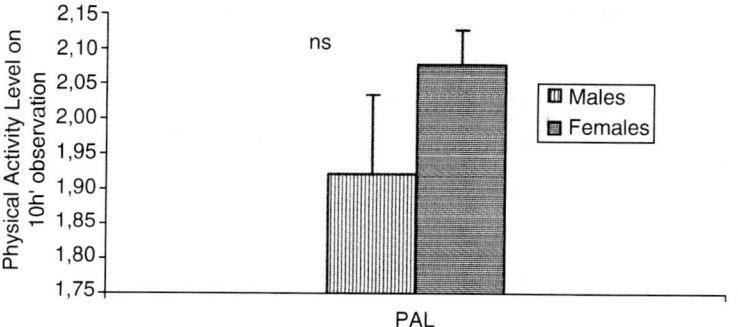

FIGURE 16.11. Comparison of the males and females' PAL in a 10-h day.

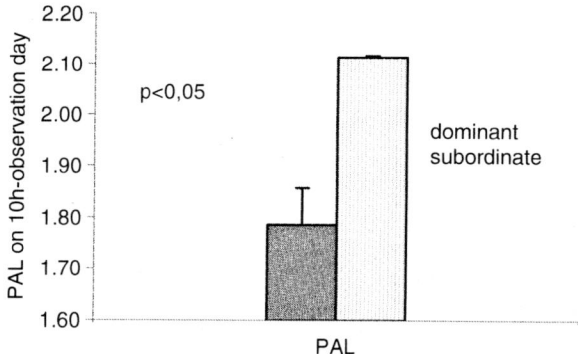

FIGURE 16.12. G3 troop females' PAL in respect to intrasexual hierarchy in a 10-h day during lactation, 2002.

the physical activity level between females was 2 times lower than that observed between males (Figure 16.11). During the 2002 lactation period, the variation of physical activity level among dominant troop G3 females was 7 times higher than among subordinate females (Figure 16.12). Individuality will be considered further in the discussion.

16.4. Discussion

At the present time, individual differences in non-human primate behavior, as well as that of other mammals, are being recognized and studied much more than in the recent past.

From this point of view, Lestel (2001) pointed out that ethological observations allow considering each animal as a subject having its own life history rather than just as an object. This life history along with its individual genetic characteristics determines the gamut of its reactions to its group-mates. Recent work on anthropoids (Mitani et al., 2002) follows this course of analyzing the behavior of individuals, to show that they are aware of their relationships with other individuals in their troop.

Could it be possible to consider such a cognitive level in the lemurs of Madagascar, and up to what point, when there are clearly big differences in their learning ability from that of monkeys or apes (Wilkerson and Rumbaugh, 1979)? The study of primate social organization has progressed enormously, mainly with the study of Cercopithecidae (Kappeler and van Schaik, 2002), since Jolly (1966) first observed the *Lemur catta* troops, aspects of whose interindividual relationships are studied in the current work.

Before discussing energy expenditure and the physical activity level, one example among our observations on this lemur species will be enough to illustrate that among ringtailed lemurs, one individual may be totally different from another one.

That example concerns the parental investment of different individuals. One primiparous female of troop G3, called Coeur, daughter of a dominant female and dominant herself, gave birth to an infant in which she seemed to not invest much. Another female, Tata, Coeur's sister, apparently kidnapped her infant. Tata had already her own infant born some hours before that of Coeur. Both infants were hanging onto Tata's belly, and that of Coeur was suckling more often. An hour later, Coeur approached Tata, but Tata threatened her by staring and chased her away. Coeur ran away but came back several times trying unsuccessfully to pick up her infant. This process lasted 3 hours. Finally the infant was restored to Coeur with no aggression from Tata.

A few days later as the troop rested in the trees, each individual in contact with another, Coeur's infant jumped from one individual to the next. Coeur then rose up and left the troop. At any moment it seemed the infant might fall down. Tata and her mother Antitra both ran to retain it. Twenty minutes later Coeur was back, and the adult females made a contact call that could be interpreted as calling the "flighty mother."

Coeur's deficient parental investment continued over several more incidents. On one occasion the troop was feeding on garbage in an excavation and Coeur's infant was trying to climb down from its mother's back to explore its surroundings. A few minutes later, Coeur left the troop with other troop members without taking the baby on her back. The infant was not yet able to walk by itself so it was not able to get out of the excavation and it screamed. Its calls made an adult male come to it, threatening. A subadult male staying near the infant faced the adult male and made him leave. Twenty minutes later, the subadult took the screaming infant on its back and went toward the troop and the mother, but Coeur was high in the trees feeding and did not approach the new arrivals.

Tata and Antitra met them. The infant jumped right away to Antitra to suckle her. Twenty more minutes later Coeur climbed down the tree and came to the trio, but Tata rose up and cuffed her making her scream a submission call. During the next 6 weeks of its life, the infant suckled alternatively its mother and Antitra whose own infant had died a few days after birth.

This kind of major difference between individuals' behaviors and between parental investments could partly explain the differences within males and females in regard to intrasexual hierarchy, daily distribution of activities, and physical activity level.

16.4.1. Interindividual Differences in Physical Activity Level

Table 16.6 shows each individual PAL in both troops. The value of the average PAL was 2.09. A PAL above this value characterized the animal as more active and under it as less active. Only two males of the 11 observed were less active. Both belonged to G3 troop, one dominant and the second subordinate. Among females, 4 of 10 observed were less active and 3 of these were lactating and dominant. The difference between the lowest PAL of females (1.85) and males (1.96) during a 10-h day was not significant. That might be explained by the higher basal metabolism of the pregnant and lactating females we took in account during

TABLE 16.6. Individual classification by the average PAL according to the social status and the females reproductive state.

Subject	Sex	PAL average	Position in regard to the average	Social status	Females reproductive state
Deba	M	2.15	+	Dominant	
Star	M	2.12	+	Subordinate	
Doma	M	2.43	+	Subordinate	
Scar	M	2.15	+	Dominant	
RE	M	2.03	−	Dominant	
TR	M	2.1	+	Subordinate	
Rabitro	M	2.1	+	Subordinate	
Point noir	M	1.96	−	Subordinate	
LE	M	2.1	+	Subordinate	
Sofina	M	2.1	+	Subordinate	
Tatape	M	2.3	+	Subordinate	
Dido	F	2	+	Dominant	P.L
Doso	F	1.86	−	Dominant	P.L
Dana	F	1.99	−	Subordinate	P.L
Diqua	F	2.13	+	Subordinate	NP.NL.
Antitra	F	1.93	−	Dominant	P.L
Cœur	F	2.13	+	Dominant	P.L
Tata	F	1.85	−	Dominant	P.L
Mavo	F	2.1	+	Subordinate	P.L.
Kelilo	F	2.13	+	Subordinate	P.L
Bobo	F	2.34	+	Subordinate	P.NL.

P.L., pregnant and lactating; P.NL., pregnant and nonlactating; NP.NL., nonpregnant and nonlactating; +, more active; −, less active.

calculation of the PAL. Neither was there any significant difference between PAL of dominant and subordinate within sex, although one could think dominance status should be advantageous to obtain a high diet quality cheaply.

The difference between individual PALS did not seem to be related to female dominance over males. Indeed, Diqua and Dido of D1A troop had high PAL and were both the most dominant over males, while Antitra and Bobo, which were the most dominant over males in G3 troop, had the two extreme PAL values. But on the other hand, that also could suggest an absence of relationship between female dominance and reproductive state, for neither Diqua nor Bobo were lactating during the observation period.

16.4.2. Evolution and Energetic Strategy

The most appropriate variable used to compare different species with respect to energetic strategy is the physical activity level, which does not take weight into account.

Data on *Lemur catta* displayed in Table 16.7 are those obtained by the current work using Leonard and Robertson's (1997) formula and doing the calculation on the basis of 24-h day to take in account the time spent sleeping. The PAL in 24h is 1.47 for males and 1.43 for females. *Lemur catta* is known as an animal with low basal metabolism, so it should be below Kleiber's regression line. To compare them with the other primate families that lie on Kleiber's regression line, we should make calculations that will elevate their basal metabolism. But if their Basal Metabolism is higher, it should enter both the total energy expenditure as higher and also the PAL, which is TEE/BM, so using a different BM should make little difference. Therefore they can be compared with those of other primate species, deduced from field work by different authors and reported in Leonard and Robertson (1997), in Warren and Crompton (1998) for other prosimian species, and in Dracks et al. (1999) for *Lepilemur ruficaudatus*.

Notably, among prosimians *Lemur catta* had the lowest PAL, with the highest weight. It had a less specialized diet than that of folivore *Lepilemur* and *Avahi*, or than that of insectivore *Tarsius* and *Galago* and the gummivore *Otolemur*. The high percent of time *Lemur catta* spends in resting could be explained by an energy saving strategy and folivore survival strategy. Indeed, leaves need a long time (around 5h in *Lemur catta*; Cabre-Vert and Feistner, 1996) to degrade the fibers almost entirely and to assimilate energy from that degradation.

Table 16.7 shows the high PAL of most prosimians correlated with their small size and thus their low total metabolism. Only the lorisidae, tarsiidae, and indriidae, with low basal metabolism, have PAL higher than 2. Thus, prosimians on the lowest level of the primate phylogenical scale and which had the best diet quality spent more energy for body size than other primates. From this

TABLE 16.7. Comparison of ringtailed lemur PAL with that of species from other families.

Species	Sex	Weight (kg)	PAL
LORISIDAE			
Galago moholi	M/F	0.182	4.9
Otolemur crassicaudatus	M/F	1.384	5.07
TARSIIDAE			
Tarsius bancanus	M/F	0.123	4.95
INDRIIDAE			
Avahi laniger	M/F	0.708	6.2
Lepilemur eawardsi	M/F	0.819	6.2
Lepilemur ruficaudatus	M	0.744	3.22
	F	0.747	2.8
LEMURIDAE			
Lemur catta	M	2.6	1.47
	F	2.3	1.43
CALLITRICHIDAE			
Saguinus fuscicollis	M/F	0.3	1.36
Saguinus imperator	M/F	0.4	1.29
CEBIDEA			
Cebus apella	M/F	2.6	1.29
Cebus. albifrons	M/F	2.4	1.27
Saimiri. sciureus	M/F	0.8	1.27
Aotus trivirgatus	M/F	0.85	1.50
Callicebus mcloch	M/F	0.7	1.22
Allouatta palliata	M	8.5	1.18
	F	6.4	1.17
Ateles geoffroyi	M/F	8.41	1.20
CERCOPITHECIDAE			
Cercocebus albigena	M/F	7.9	1.31
Macaca fascicularis	M/F	5.5	1.19
Papio anubis	M	29.3	1.34
	F	13	1.34
Colobus guereza	M/F	7	1.24
HYLOBATIDAE			
Hylobates lar	M/F	6	1.17
Siamea syndactylus	M/F	10.5	1.23
PONGIDAE			
Pan troglodytes	M	39.5	1.46
	F	29.8	1.36
Pongo pygmaeus	M	83.6	1.33
	F	37.8	1.40
HOMINIDAE			
Homo sapiens			
!Kung	M	46	1.68
	F	41	1.56
Ache	M	59.6	2.00

viewpoint, they might have reached the upper limit of their physiological adaptability and might have an energy-limited way of life (Warren and Crompton, 1998).

If one compares species with female dominance or with no female dominance, it is noticeable that males' and females' PAL varies from 1.17 to 1.46 and that of males could be higher than that of females and vice versa. There is no relation of PAL to female dominance.

The other primates' species have a PAL between 1.68 and 1.17, similar to *Lemur catta*. There is no correlation between diet quality within this group and PAL ($p = 0.2$).

In short, *Lemur catta* seems to be an exception as a prosimian. It has a folivorous–frugivorous diet, and its PAL matches with that of folivorous–frugivorous simians rather than the prosimians.

By my calculations, *Lemur catta* then seems to be at the limit allowed for energy expended in locomotion, which could explain the high percentage of time spent in inactivity during a 10-h day. Elsewhere its activity such as the sunbathing on waking allowed it to diminish the thermoregulation cost (Martin, 1974; Daniels, 1984; Peters, 1989). Its body temperature is also regulated by behaviors such as grouping together, one against another, in a big ball when the ambient temperature is cool, or on the contrary one away from another with spread limbs when the ambient temperature is hot. All these behaviors display a specific adaptational response to metabolic constraints.

16.5. Conclusion

Natural selection has presumably shaped the mechanism that ordered the social competition from which the dominance structure evolved. This competition would favor individuals with high degree of adaptation to their surroundings and with high degree of efficient reproduction of their genetic heritage.

Genes responsible for lemur female dominance may have come from a monogamous nocturnal lemur ancestor (van Schaik and Kappeler, 1996) especially because this behavior is observed in extant nocturnal lemurs when two different sexes encounter each other (Radespiel and Zimmermann, 2001, Dammhahn and Kappeler, 2005). Alternatively, they might come from pair-bonded—not exclusively nocturnal—ancestor (Jolly, 1998), whose male subordination was a paternal investment (Pollock, 1979).

The structures of prosimian and simian troops are amazingly alike in regard to interindividual relationship and the intersexual hierarchy. The main difference is the lemur female dominance. In simians, males are dominant with feeding priority. This male dominance, including that observed in humans, could be problematic in that female feeding priority might improve the diet quality during lactation period, which could in turn be advantageous for reproduction.

However, our results comparing the most and the least dominant females' energy expenditure during each reproduction period seems to assert the absence of correlation between the three parameters of dominance within sexes, dominance between sexes, and energy expenditure.

Thus, the systems involving female or male dominance or males may result mainly from a remnant of the evolution history of prosimians and simians, respectively, even though both systems are efficient and contribute to the animals' adaptability.

The example cited at the beginning of the "Discussion" section shows an example of apparent mutual aid between next of kin individuals of the troop. Jolly (1999) argued that among evolutionary mechanisms, cooperation as much as competition has progressively modeled the life forms on the planet Earth. This idea keeps recurring, although it is relatively recent, in interpreting selection pressures (Leigh, 1999). Neither sexual selection nor individual competition is sufficient to explain group structures for which one should actually take much more account of kin selection. Genes responsible for subordination behavior might be selected because such behavior might raise the reproductive possibilities of kin and dominant individuals.

Although social behavior is not directly linked to energy expenditure, we could conclude that adaptation mechanisms to energetic resources are based on the troop relationship with its environment, but that this relationship is more complex than simple dominance hypotheses can explain.

Acknowledgments. We are grateful to Marcel Hladik, Patrick Pasquet, Bruno Simmen, and Alison Jolly who recognized the difficulty of comparing social behavior with physiological state and gave very helpful and relevant advice to improve this work. We also thank the de Heaulme family for permission to conduct the study in their reserve. This research received support from WWF-US Russell E.Train Educational for Nature.

References

Alexander, R. (1974). The evolution of social behavior. *Annu Rev Ecol System* 4:325–383.
Altmann, J. (1974). Observational study of behavior: Sampling methods. *Behaviour* 49(3–4):227–267.
Altmann, S. A. (1974). Baboons, space, time and energy. *Am Zool* 14:221–248.
Archer, J. (1988). *The Behavioural Biology of Aggression.* Cambridge University Press, Cambridge.
Beaugrand, J. P. (1983). Modèles de dominance et théorie de l'evolution. In: *Darwin après Darwin,* (eds.), Lévy, J., and Cohen, H. E. PUQ, Québec, pp. 110–137.
Buckley, J. S. (1983). *The Feeding Behavior, Social Behavior, and Ecology of the White-Faced Monkey, Cebus capucinus at Trujillo, Northern Honduras, Central America.* PhD diss., University of Texas at Austin, Austin.
Chapman, C. A. (1990). Ecological constraints on group size in three species of neotropical primates. *Folia Primatol* 55:1–9.
Charles-Dominique, P., and Hladik, C. M. (1971). Le *lepilemur* du sud de madagascar: Ecologie, alimentation et vie sociale. *la terre et la vie* 1:3–66.

Coelho, A. M. (1974). Socio bioenergetics and sexual dimorphism in primates. *Primates* 15(2–3):263–269.

Coelho, A. M., Bramblett, C. A., Quick, L. B., and Bramblett, S. S. (1976). Resource avalaibility and population density in primates: a socio-bioenergetic analysis of the energy budgets of guatemalan howler and spider monkeys. *Primates* 17(1):63–80.

Cordain, L., Gotshall, R. W., Boyd Eaton, S., and Boyd Eaton, III, S. (1998). Physical activity, energy expenditure and fitness: An evolutionary perspective. Physiology and biochemistry. *Int J Sports Med* 19:328–335.

Dammhahn, M., and Kappeler, P. M. (2005). Social system of microcebus berthae, the world's smallest primate. *Int J Primatol* 26(2):407–436.

Daniels, H. L. (1984). Oxygen consumption in *Lemur fulvus*: Deviation from the ideal model. *J Mammal* 65(4):584–592.

Drack, S., Ortmann, S., Bürhrmann, N., Schmid, J., Warren, R. D., Heldmaier, G., and Ganzhorn, J. (1999). Field metabolic rate and the cost of ranging of the red-tailed sportive lemur (*Lepilemur ruficaudatus*). In: Rakotosamimanana, B., Rasamimanana, H., Ganzhorn, J., and Goodman, S. M. (eds.), *New Directions in Lemur Studies*, Kluwer Academic/Plenum, New York, pp. 83–91.

Ehrlich, S., and Flament, C. (1966). *Précis de statistique*. Presses universitaires de France.

Engen, S., and Stenseth, N. C. (1984). An ecological paradox: A food type may become more rare in the diet as a consequence of being more abundant. *Am Nat* 124:3.

Feen, M. D. (2003). The spiny forest eco-region. In: Goodman, S. M., and Benstead, J. P. (eds.), *The Natural History of Madagascar*, University of Chicago Press, Chicago and London, pp. 1525–1530.

Fleagle, J. G. (1988). *Primate Adaptation and Evolution*. Academic Press, New York.

Ganzhorn, J. U., Klaus, S., Ortmann, S., and Schmid, J. (2003). Adaptations to Seasonality: Some Primate and Non Primate Examples. In: Kappeler, P. M., and Pereira, M. E. (eds.), *Primate Life History and Sociobiology*. University of Chicago Press, Chicago, pp. 132–149.

Goodall, J. (1986). *The Chimpanzees of Gombe*. Harvard University Press, Cambridge, Mass.

Gresse, M., Gandon, B., Tarnaud, L., Simmen, B., Labat, J-N., and Hladik, C. M. (2002). Conservation et Introduction de Lémuriens sur l'Îlot Mbouzi (Mayotte). *Revue d' Ecologie (Terre Vie)* 57:75–82.

Gursky, S. (2000). Effect of Seasonality on the Behavior of an Insectivorous Primate, *Tarsus spectrum*. *Int J Primatol* 31(3):477–496.

Hemingway, C. (1999). Time Budgets and Foraging in a Malagasy Primate: Do Sex Differences Reflect Reproductive Condition and Female Dominance? *Behav Ecol Sociobiol* 45:311–322.

Hill, W. C. O. (1953). *Primates: Comparative Anatomy and Taxonomy. I Strepsirhini*. The University Press, Edinburgh.

Jenkins, M. D. (1987). Madagascar an environmental profile international union for conservation of nature and natural resources. United Nations Environment Programme. Prepared by the IUCN Conservation Monitoring Center, Cambridge UK.

Jolly, A. (1966). *Lemur Bebavior*. University of Chicago Press, Chicago.

Jolly, A. (1985). *The Evolution of Primate Behavior*. Macmillan, New York.

Jolly, A. (1999). *Lucy's Legacy: Sex and Intelligence in Human Evolution*. Harvard University Press.

Jolly, A., Oliver, W. L. R., and O'Connor, S. M. (1982). Population and troop ranges of *Lemur catta* and *Lemur fulvus* at Berenty Madagascar. *Folia Primatol* 31:106.

Kappeler, P. M. (1990). Female Dominance in *Lemur catta*: More than Just Female Feeding Priority. *Folia Primatol* 55:92–95.
Kappeler, P. M. (1991). Patterns of sexual dimorphism in body weight among prosimian primates. *Folia Primatol* 57:132–146.
Kappeler, P. M. (1993). Female dominance in primate and other mammals. In: Bateson, P., Klopfer, P., and Thompson, N. (eds.), *Perspectives in Ethology*, Vol. 10, Behavior and Evolution. Plenum, New York, pp. 143–158.
Kappeler, P. M. (1996). Causes and Consequences of Life-History Variation among Strepsirhine Primates. *Am Nat* 148(5):868–891.
Kappeler, P. M., and Ganzhorn, J. U. (1999). The evolution of primate communities and societies in madagascar. *Evol Anthropol* 159–171.
Koyama, N. (1988). Mating behavior of ringtailed lemurs (*Lemur catta*) at Berenty, Madagascar. *Primates* 29:163–174.
Leonard, W. R., and Robertson, M. I. (1997). Comparative primate energetics and hominid evolution. *Am J Phys Anthropol* 102:265–281.
Lestel, D. (2001). *Les origines animales de la culture*. Flammarion.
Mcnab, B. K. (1992). Energy expenditure: A short history. In: Tomasi, T. E., and Horton, T. H. (eds.), *Mammalian Energetics: Interdisciplinary Views of Metabolism and Reproduction*, Comstock Publishing, pp. 1–15.
Mertl-Millhollen, A. S., Gustafson, H. L., Budnitz, N., Dainis, K., and Jolly, A. (1979). Population and territory stability of the *Lemur catta* at Berenty, Madagascar. *Folia Primatol* 39:115.
Milton, K. (1980). *The Foraging Strategy of Howler Monkeys*. Columbia University Press, New York.
Mitani, J. C., Watts, D. P., and Muller, M. N. (2002). Recent developments in study of wild chimpazee behavior. *Evol Anthropol* 11:9–25.
Mowry, C. B., and Campbell, J. L. (2001). Nutrition. In *Ring-tailed Lemur (Lemur catta) Husbandry Manual*. American Association of Zoos and Aquariums.
Müller, E. F. (1985). Basal metabolic rates in primates—the possible role of phylogenetic and ecological factors. *Compar Biochem Physiol* 81A(4):707–711.
Nakamichi, M., and Koyama, N. (1997). Social relationships among ring-tailed lemurs (*Lemur catta*) in two free-ranging troops at Berenty reserve, Madagascar. *Int J Primatol* 18:1:73–93.
Pereira, M. E., and Kappeler, P. M. (1997). Divergent systems of dominance in lemuroid primates. *Behaviour* 132:225–274.
Pereira, M. E., Strohecker, R. A., Cavigelli, S. A., Hughes, C. L., and Pearson, D. D. (1999). Metabolic strategy and social behavior in lemuridae. In: Rakotosamimanana, B., Rasamimanana, H., Ganzhorn, J. U., and Goodman, S. M. (eds.), *New Directions in Lemur Studies*, Kluwer Academic/Plenum, New York, pp. 93–118.
Peters, R. H. (1989). *The Ecological Implications of Body Size*. Cambridge Studies in Ecology. Cambridge University Press, Cambridge.
Pollock, J. I. (1979). Female Dominance in *Indri indri*. *Folia Primatol* 31:143–164.
Rabenandrasana, D. (1996). Hiérarchie des mâles et connaissance générale sur *lemur catta* et leurs intérêts éducationnels au niveau des agents d'information touristique. Mémoire de CAPEN.
Ramanantsoa, H. (1997). Etude de la hiérarchie des mâles chez *lemur catta* et l'enseignement de cette espèce aux lycéens. Mémoire de CAPEN.
Ramasiarisoa, P. L. (2000). Contribution à l'étude de l'exploitation de l'espace à travers l'activité alimentaire de trois lémuriens *propithecus verreauxi verreauxi*,

lemur catta, eulemur fulvus rufus dans la réserve de berenty. DEA d'Ecologie Environnement.

Rasamimanana, H. R., and Rafidinarivo, E. (1993). Feeding behavior of *lemur catta* females in relation to their physiological state. In: Kappeler, P. M., and Ganzhorn, J. U. (eds.), *Lemur Social Systems and Their Ecological Basis.* Plenum, New York, pp. 123–133.

Ratsirarson, J. (1986). Contribution à l'étude comparative de l'éco-éthologie de *Lemur catta* dans deux domaines vitaux différents de la réserve spéciale de beza-mahafaly. Mémoire d'ingéniorat.

Reiss, M. J. (1985). The allometry of reproduction: Why larger species invest less in their offspring. *J Theor Biol* 113:529.

Richard, A. F. (1985). *Primates in Nature.* WH Freeman, New York.

Richard, A. F., and Nicoll, M. E. (1987). Female social dominance and basal metabolism in a malagasy primates, *Propithecus verreauxi verreauxi. Am J Primatol* 12:309–314.

Roeder, J-J., and Anderson, J. R. (1990). *Primates: Recherches Actuelles.* Masson.

Ruffie, J. (1982). *Traité du vivant.* 2ème volume. Champs Flammarion.

Sauther, M. L. (1991). Reproductive behavior in free-ranging *Lemur catta* at beza-mahafaly special reserve, madagascar. *Am J Phys Anthropol* 844–863.

Sauther, M. L. (1992). *The Effect of Reproductive State, Social Rank, and Group Size on Resource Use Among Free-Ranging Ring-Tailed Lemurs (Lemur catta) of Madagascar.* PhD diss., Washington University, St. Louis.

Sauther, M. L. (1993). Resource competition in wild populations of ring-tailed lemurs (*Lemur catta*): Implications for female dominance. In: Kappeler, P. M., and Ganzhorn, J. U. (eds.), *Lemur Social Systems and Their Ecological Basis.* Plenum, New York, pp. 135–152.

Sauther, M. L., Sussman, R. W., and Gould, L. (1999). The socio-ecology of the ring-tailed lemur: Thirty five years of research. *Evol Anthropol* 8(48):120–132.

Schmid, J., and Ganzhorn, J. U. (1996). Resting metabolic rates of *Lepilemur mustelinus ruficaudatus. Am J Primatol* 38:169–174.

Siegel, S. (1956). *Nonparametric Statistics for the Behavioral Sciences.* Series in Psychology. McGraw-Hill, New York, Toronto, London.

Simpson, G. G., Roe, A., and Lewontin, R. C. (1960). *Quantitative Zoology.* Harcourt Brace Jovanovich, New York.

Stanford, C. B. (1998). The Social Behaviour of Chimpanzees and Bonobos. *Curr Anthropol* 39:399–420.

Sussman, R. W. (1991). Demography and social organization of free-ranging *Lemur catta* in the beza-mahafaly reserve, madagascar. *Am J Phys Anthropol* 84:43.

Thornhill, R. (1986). Relative parental contribution of the sexes to their offspring and the operation of sexual selection. In: Nitecki, M. H., and Kitchell, J. A. (eds.), *Evolution of Animal Behavior: Paleontological and Field Approaches.* Oxford University Press, New York, Oxford, pp. 113–136.

Tilden, C. D., and Oftedal, O. T. (1997). Milk composition reflects pattern of maternal care in prosimian primates. *Am J Primatol* 41:(3)195–211.

Trivers, R. L. (1972). Parental investment and sexual selection. In: Campbell, B. Aldine, (ed), *Sexual Selection and the Descent of Man, 1871–1971*, Chicago, pp. 136–179.

van Schaik, C. P., and van Hoof, J. A. R. A. M. (1983). On the ultimate causes of primates social systems. *Behaviour* 85:91–117.

van Schaik, C. P. (1989). The ecology of social relationships amongst female primates. In: Standen, V., and Foley, R. A. (eds.), *Comparative Socio-ecology: The Behavioral Ecology of Humans and Other Mammals*, Blackwell Scientific, Oxford, pp. 195–218.

Von Engelhardt, N., Kappeler, P. M., and Heistermann, M. (2000). Androgen levels and female social dominance in *Lemur catta. Proc R Soc London B Biol Sci* 267(1452): 1533–1539.

Walker, A. (1979). Prosimian locomotor behavior. In: Doyle, G. A., and Martin, R. D. (eds.), *The Study of Prosimian Behavior*. Academic Press, New York, pp. 543–565.

Warren, R. D., and Crompton, R. H. (1998). Diet, body size and energy costs of locomotion in saltatory primates. *Folia Primatol* 69(1):86–100.

Watt, D. (1985). Relations between group size and composition and feeding competition in mountain gorilla groups. *Anim Behav* 33:72–85.

Wilkerson, B. J., and Rumbaugh, D. M. (1979). Learning and intelligence in prosimians. In: Doyle, G. A., and Martin, R. D. (eds.), *The Study of Prosimian Behavior*, Academic Press, New York, pp. 207–246.

Wright, P. C. (1999), Lemur traits and Madagascar ecology: Coping with an island environment. *Yrbk Phys Anthropol* 42:31–72.

Young, A. L., Richard, A. F., and Aiello, L. C. (1990). Female dominance and maternal investment in strepsirhine primates. *Am Nat* 135:473.

17
Male Sociality and Integration During the Dispersal Process in *Lemur catta*: A Case Study

LISA GOULD

17.1. Introduction

The ultimate reason for male dispersal in multimale, female-bonded non-human primate groups is suggested to be the reduction of inbreeding and enhancement of reproductive success (Itani, 1972; Packer 1979, Cheney, 1983; Moore and Ali, 1984; Melnick and Pearl, 1987; Clutton-Brock, 1989), whereas sexual attraction to unfamiliar, unrelated females is proposed as the proximate cause (Enomoto, 1974; Sugiyama, 1976; Packer, 1979; Pusey and Packer, 1987). The possibility of rank improvement is also cited as a proximate reason for dispersal, as a male who becomes high-ranking in his new group may experience an increase in mating opportunities (Altmann and Altmann, 1970; Henzi and Lucas, 1980; Cheney and Seyfarth, 1983; van Noordwijk and van Schaik, 1985; Sprague, 1992; Borries, 2000).

How does a dispersing male primate choose a new group? Pusey and Packer (1987) suggested that several factors, including proximity, the presence of individuals from the natal group, and the number and quality of mates can affect and influence a migrating animal's choice of group. Dispersing males likely gain information on the composition of neighboring social groups from intergroup encounters, thus, most migrants might be expected to transfer to adjacent groups. Packer (1979) found that most male baboons in his study groups transferred into neighboring groups, and in vervets, Cheney and Seyfarth (1983) found that young males often affiliate with individuals in neighboring groups during intergroup encounters and visit the home ranges of other groups before emigration from the natal group. Such predispersal activity may allow them to become familiar with resource locations, behavior of predators, and competitive abilities of residents before they make an immigration attempt (Isbell et al., 1993).

Costs to dispersal are numerous, and include predation, possible attack by extragroup males, starvation, and even death (Cheney, 1983; Cheney and Seyfarth, 1983; Pusey and Packer, 1987; Isbell et al., 1993; Kumar et al., 2001). One way to offset such costs, especially for a natal male, is to disperse with at least one other male. Japanese macaque, vervet, long-tailed macaque, white-face capuchin, and ringtailed lemur natal males will often emigrate from their group with age-mates, brothers, or older males and join groups in which familiar males already reside (Sugiyama, 1976; Cheney and Seyfarth, 1983; van Noordwijk and van Schaik, 1985; Sussman, 1992; Jack and Fedigan, 2004a). Such a strategy benefits both natal and more experienced males in terms of enhanced predator protection, protection from attacks by resident males in other groups, and assistance/support in entering a new group (Cheney and Seyfarth, 1983; van Noordwijk and van Schaik, 1985; van Hooff 2000).

17.1.1. Male Migration Patterns in Lemur catta

Patterns of male migration in *Lemur catta* have been studied at two field sites in Madagascar: Berenty Reserve (Jones, 1983; Koyama et al., 2002) and Beza Mahafaly Reserve (Sussman, 1991, 1992; Gould, 1994). Sussman (1992) suggested that sexual competition and mate choice are the proximate causes for male migration in this species. For a natal male, attraction to non-natal females would be the primary impetus.

Adult male *Lemur catta* migrate approximately once every 3.5 years, and young natal males begin to disperse at 3 or 4 years of age (Sussman, 1992), although at Berenty, some natal males dispersed at 2 years of age (Koyama et al., 2002). Adult males tend to disperse several times during their lives (Sussman, 1990, 1992; Gould, 1994; Gould et al., 2003; Koyama et al., 2002). Sussman (1991) noted that male *Lemur catta* leave a social group of their own accord, but when attempting to enter a new group, they are subjected to aggression by resident adult males and occasionally adult females. Jones (1983) observed resident males continuously on guard against migrants, chasing and scent marking whenever they approached. At both Beza Mahafaly and Berenty reserves, males tend to migrate in pairs or threesomes (Jones, 1983; Sussman, 1991, 1992; Gould, 1994, 1997).

Whereas Jones, Sussman and Koyama largely focused on the demographic patterns of male migration, in this chapter I will present detailed information on the social process of dispersal. In 1992 and 1993, I closely examined male affiliative behavior over an annual cycle in nine adult males residing in three social groups at Beza Mahafaly Reserve. Of those nine focal males, two (males 118 and 121) from the same group, Red group, successfully immigrated into an adjacent study group (another male, no. 119, from the same group began to disperse, but returned to Red group after 4 weeks). Thus, I was able to quantitatively document affiliative and agonistic interactions of the two males during both emigration and immigration. Furthermore, as one male was natal and dispersing for the first time, and the other had migrated at least once previously, I was able to examine differences in dispersal behavior between a young male and a more experienced one.

The following questions will be addressed:

- When the two focal males emigrated from their group, did a "peripheralization process" occur? For example, did they affiliate less frequently with other group members for a period of time before leaving the group, and did they increase

their spatial distance from other group members? Were either forced out of the group by other group members?
- Upon entering a new group, with which group members did they affiliate initially? How did their affiliative patterns change as they became integrated into the new group? What were their positions in the male dominance hierarchy in the new group? Were there individual differences in group integration?

17.2. Methods

17.2.1. Study Animals/Study Groups

The information presented here was part of a larger study of adult male affiliative patterns over an annual cycle at Beza Mahafaly Reserve, particularly focusing on reproductive seasons (mating, gestation, lactation, migration and post-migration). The two males discussed in this chapter were a subset of nine focal males from three neighboring groups (Red, Green, and Blue groups) on which data were collected from mid-March of 1992 to mid-March of 1993. The periods covered in this chapter, and referred to as premigration, migration and postmigration periods occurred between September 29, 1992, and March 13, 1993.

The two males pertinent to this chapter resided in Red group during the first half of the 12-month study. Group composition before and after dispersal is presented in Table 17.1. The home ranges of the three neighboring groups were situated in the eastern, riverine forest part of the Beza Mahafaly reserve, and Red and Green group home ranges overlapped extensively. Inter-group encounters were frequent, and sleeping/siesta trees were adjacent to each other, therefore animals from each group saw each other on a daily basis.

17.2.2. Data Collection

The order of focal animal samples was equally rotated and determined at the beginning of each data collection week. Focal animal sampling (Altmann, 1974)

TABLE 17.1. Group composition before and after dispersal of the two focal males.

Red group	
Before dispersal	After dispersal
Adult males = 6 (including one natal male)	4
Adult females = 4	4
Subadults = 3	3

Green group	
Before dispersal	After dispersal
Adult males = 2 (no natal males)	4
Adult females = 5	5
Subadults = 4	4

was the primary method of data collection. Each male was followed for a one-hour period, and behavioral categories were noted on-the-minute, so that there were 60 behavioral records for each male per focal animal session. If the behavior was social (affiliative or agonistic), the initiator and direction of the behavior was also recorded. To determine true frequency of behavioral states, the onset of a "bout" of behavior was coded differently from the same behavior if it continued for a duration of more than one minute. For example, if a focal animal was engaged in a mutual grooming session of 5 minutes, the first 1-minute interval was recorded as the "onset" of mutual grooming between the focal male and his partner, and the remaining four intervals were recorded as simply "mutual grooming" with partner X. Rates of affiliative and agonistic behavior were calculated for each reproductive season by dividing all instances of each type of affiliative and agonistic behavior by the number of hour-long focal animal sessions collected on the focal male during that season. The natal male discussed in this chapter, identified as "118" was followed for 39 hours during the 20 weeks of the dispersal process, and his migration partner "121" was followed for 37 hours this period.

Affiliative behaviors included allogrooming, sitting and resting in contact, and sitting or resting near (<0.5 m). Behaviors considered agonistic included mild agonism, such as supplantations, submissive chattering, and tail-waving, while chasing, cuffing 'stink-fighting' and lunging was considered more severe.

17.2.3. Age-Class and Migration Status of the Two Males

Most of the adult *Lemur catta* living in the reserve population in 1992–1993 had been fitted with nylon collars and plastic numbered tags as part of a long-term demographic study (Sussman, 1991; Gould et. al, 2003). The last round of collaring before I arrived at Beza Mahafaly in 1992 had occurred in 1990, therefore not all adult animals were collared when my study began. Neither of the two males discussed in this chapter had identification collars at the onset of my study, as one was 2½ in March of 1992, and the other had immigrated into the reserve population from the continuous forest outside of the reserve in 1991. I captured both in early September, 1992 and they were fitted with collars and tags before the onset of the migration season.

Relative age class of each focal male was determined by a set of dental wear criteria developed by Bob Sussman and Michelle Sauther (in Sussman, 1991, 1992). A rough estimate of age was also be made by observing scrotum length: young adult males have small testicles, and testicles in this species tend to continue to descend until past middle age.

Male 118 was a young natal male approximately three years of age at the beginning of migration season in 1992. Male 121 was considered a prime-aged male, likely between 5 and 8 years of age, and he had immigrated into Red group from outside of the reserve during the migration season of 1991/92 (Gould, 1994).

17.3. Results

17.3.1. Affiliative and Agonistic Behavior During the Premigration Period

The two males began to display occasional premigration behavior during the first week of September 1992. "Premigration behavior" consisted of spending short periods (usually part of the morning or afternoon) away from Red group. They often spent time at the periphery of neighboring groups. Time away from Red group gradually increased until the males spent the majority of the day, and occasionally the entire night, away from Red group.

I determined the actual premigration period to be September 29 to November 3. The two males were successful in immigrating together into the adjacent Green group in mid-to-late November, but they were observed on the periphery of Blue, Green, and Black groups prior to their their successful immigration.

I wanted to examine whether these two males affiliated less frequently with Red group members (other males, females, and subadults) during their last four weeks as residents. Figure 17.1 represents all affiliative behavior between natal male 118 and other group members during the 20-week dispersal process. Weeks 1–4 on the graph represent his affiliative behavior with his natal group members during the four-week pre-migration period. In Figure 17.2, affiliation rates for male 121 are presented over the 20-week period. Natal male 118 did not affiliate less often with Red group members during the pre-migration period, in fact, he affiliated more frequently with females and juveniles during weeks 3 and 4 of premigration compared with the previous two weeks (Friedman 2-way analysis of variance: $p = 0.55$, $df = 3$, Figure 17.1). Male 121 affiliated more with his migration partners (both 118 and 119, as 119 had initially begun to disperse with 118 and 121) and with adult females (particularly in week 3) than with Red group males and subadults, though the difference is not quite significant (Friedman two-way analysis of variance: $p = 0.07$, $df = 3$, Figure 17.2).

Very little agonism was directed toward the two males over the 4-week period by Red group members. Male 121 was chased once by a resident male, and he was displaced by females four times. No agonistic behavior was observed between natal male 118 and other group members in the four-week period. Because rates of agonism are extremely low (less than 2 instances per week), it was not possible to calculate a chi-square value to test for significance.

Agonistic behavior was also noted on an *ad libitum* basis during the migration process (i.e., outside of focal animal sampling). During weeks three and four of the pre-migration period, the resident Green group males (G10 and G23) were observed chasing or exhibiting tail-waving behavior towards the dispersing Red group males on four occasions when the dispersing males were spending time on the periphery of Green group.

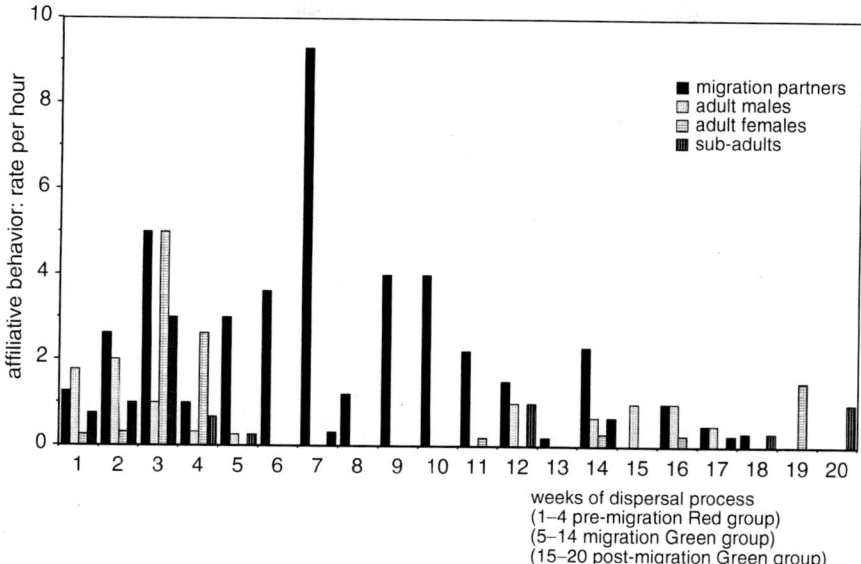

FIGURE 17.1. Rates of affiliative behavior between the natal male 118 and other group members during the dispersal process. Note that weeks 1–4 represent affiliative behavior in the natal group (Red group) during the premigration period, while weeks 5–14 represent affiliation during immigration/integration into Green group, and weeks 15–20 represent his rates of affiliative behavior with Green group members during the postmigration period.

17.3.2. Affiliative Behavior During Migration and Integration into Green Group

During the subsequent 10-week period (November 4, 1992, to January 25, 1993), males 118 and 121 spent increasing amounts of time on the periphery of Green group. During this period, they began to affiliate with Green group members, though their rates of affiliative behavior with each other are significantly higher than with Green group members (Friedman two-way analysis of variance: $p = 0.000$, $df = 3$ for male 118, $p = 0.002$, $df = 3$ for male 121, see weeks 5–14 on Figures 17.1 and 17.2). Only during the final 2 weeks did the immigrants affiliate with the other Green group males on a consistent basis, and even then, the rates of affiliative behavior at this time with these males were low. In only 4 of the 10 migration weeks did any affiliation with females occur. The frequency of affiliative behaviors with subadults was similar to the frequencies found between the immigrants and resident males.

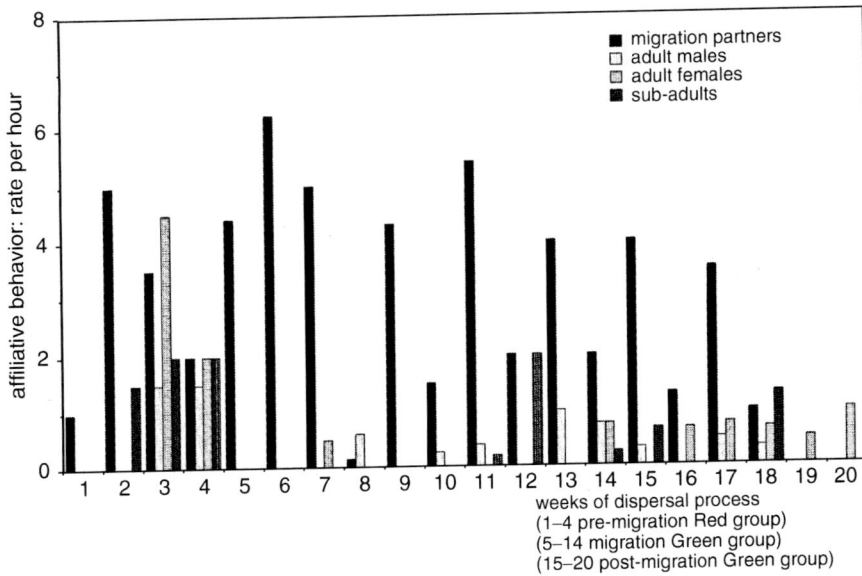

FIGURE 17.2. Rates of affiliative behavior between the older male 121 and other group members during the dispersal process. Weeks 1–4 represent affiliative behavior in Red group during the premigration period, while weeks 5–14 represent affiliation during the immigration/integration process into Green group, and weeks 15–20 represent the rates of affiliative behavior between 121 and Green group members during the postmigration period.

17.3.3. Agonistic Behavior During Migration/Integration

The rate of agonism between the two males and members of their new group were highest in the first migration week, fell between weeks 2 and 5, and then increased again weeks six through ten (weeks 1–10, Figure 17.3).

In weeks 4 and 5 of the migration period, resident male G10 began exhibiting submissive behaviors toward male 121 and in one instance, the two immigrants jumped on G10 while he was passing under a tree. From that point onward, G10 was subordinate to new male 121.

17.3.4. The Postmigration Period

17.3.4.1. Affiliative Behavior During the Postmigration Period

Postmigration occurred between January 26, 1993, and the end of the study period, March 13, 1993. Rates of affiliative behavior between the two immigrants fell during this period, and they began to affiliate more and more with

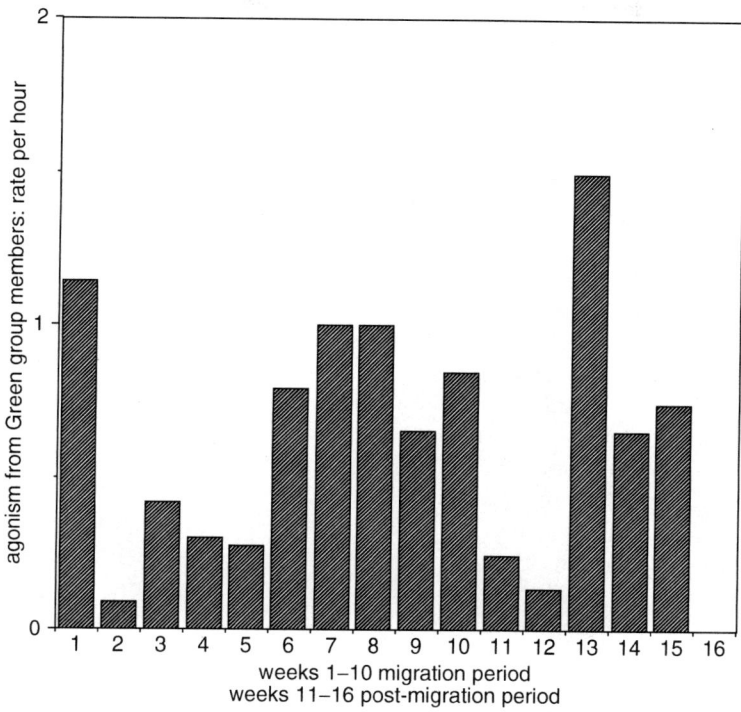

FIGURE 17.3. Rates of agonistic behavior between the two dispersing males and Green group members during the 16-week migration/postmigration period in Green group.

Green group members (weeks 15–20 in Figures 17.1 and 17.2). In fact, no significant differences were found for either male in terms of their rates of affiliation with each other compared with affiliation with adult males, adult females, or subadults in their new group (Friedman analysis of variance, $p = 0.68$, $df = 3$ for male 118 and $p = 0.172$, $df = 3$ for male 121). Male 118 affiliated more frequently with the resident males than did male 121 and male 121 affiliated more consistently with the Green group females. During the last 2 weeks of the postmigration period no affiliative behavior was observed between the two migration partners.

17.3.4.2. Agonistic Behavior During the Postmigration Period

Rates of agonism between the two immigrants and Green group members are low throughout the 6-week postmigration period, and in the final week of the study, no agonism was observed (weeks 11–16, Figure 17.3).

17.4. Discussion

17.4.1. Migration Partners and Choice of New Group

Dispersing ringtailed lemur males have been observed migrating alone, in pairs, or in threesomes (Jones, 1983; Sussman, 1991, 1992; Sauther, 1991, Gould, 1994; 1997; Koyama et al., 2002). The two males described here spent several months in close proximity, often as the only affiliative partner of the other. Aside from social benefits, they experienced better predator protection when together, which was demonstrated on two occasions when the older 121 was traveling alone during the early dispersal period, and he was targetted by a raptor. When the two males traveled together, such targetting was not seen.

At Beza Mahafaly, males have been observed transferring into adjacent groups, non-neighboring groups in the reserve and groups outside the reserve, sometimes several kilometers away (Sussman, 1991, 1992; Gould et al., 2003). Similarly, at Berenty reserve, Koyama et al. (2002) found that about one-third of the males followed during their 10-year study moved to neighboring groups upon first migration, and the remainder immigrated into groups much further away.

Immigration into an adjacent group can be particularly advantageous for a natal male, as he would have previous information on group composition and familiarity with group members via the frequent inter-group encounters that occur at this field site, and prior familiarity with adult females could eventually confer a mating advantage (Pusey and Packer, 1987). Frequent interaction with neighboring groups has been suggested as an important influence on natal immigration in Japanese macaques (Sugiyama, 1976); vervets (Cheney and Seyfarth, 1983) white-faced capuchins (Jack and Fedigan, 2004a). In addition to the natal male described in this chapter, I have observed natal male immigration into neighboring groups in subsequent studies in 1994; 2001 and 2003 (Gould, unpublished data).

17.4.2. Agonism and Number of Males in the New Group

Sussman (1991) suggested that male *Lemur catta* migrate from a social group of their own accord, and my observations support this suggestion: the males in this study were not "evicted" in any way by other group members, in fact, no agonism whatsoever was directed toward the natal male in the weeks prior to his departure.

Clearly, patterns of agonistic behavior directed towards potential immigrants is varied in this species, as Sussman (1991) and Sauther (1991) reported that all observed transfers at Beza Mahafaly Reserve in the 1988–1989 seasons were met with resistance by all male members and some female members of new groups into which the migrating males were attempting to enter. Number of males in a new group may strongly affect such resistance. There were only two resident males in Green group when the two dispersing males began spending time in and around the group, and very little agonism between resident and immigrant males was observed. In groups with more resident males, resistance is likely to be much

FIGURE 17.4. Three *L. catta* males who all immigrated into the same group the year the photo was taken rest together amicably after the rigors of the mating season. All animals in the study troops at Beza Mahafaly wear distinguishing collars and tags, which allow them to be traced from group to group and year to year. (Photo: L. Gould)

stronger, as mating competition would increase for resident males if immigration was successful. In six of ten of the migrations that I documented while censusing the Beza Mahafaly population in 1992 and 1993, males moved to groups with fewer males than the group from which they dispersed, and in four cases, males transferred to groups with more females. Sussman (1992) also found a greater tendency for males at Beza Mahafaly to leave groups with high male-to-female ratios.

During the early weeks of the migration period, the two immigrants were chased and lunged at by the females in Green group; however, agonistic behavior from the females changed in nature in the middle of the migration period, and from that point on, consisted primarily of milder supplantations, which may be seen as simply an aspect of group living.

17.4.3. Group Defense: Which Side Are We On?

The matter of group defense during intergroup encounters may be a time of ambivalence for males transferring to adjacent groups. In the first month of the dispersal process, the immigrants participated in defending Green group during Red/Green encounters; but in one instance, they changed sides three times, tail-waving to both Green and Red group males. In the fifth week of migration, the two immigrants engaged in a stink-fight with two Red group males, including 119 who had begun to disperse with them only weeks before, but had returned to Red group.

17.4.4. Affiliative Behavior of Immigrant Males

Henzi and Lucas (1980) and Smuts (1984) found that immigrating male vervets and olive baboons affiliated with adult females and subadults upon entering a new group, and Japanese macaque immigrants will first establish affiliative relationships with peripheral juvenile and subadult males before associating with other group members (Sugiyama, 1976). Drickamer and Vessey (1973) noted that rhesus macaque immigrants associated initially with only migration partners, but affiliation with resident males increased as the amount of time in the group increased. Similarly, the two males in this study affiliated primarily with each other during the 10-week migration period, although their rates of mutual affiliation decreased in the latter three weeks, as they became more integrated into Green group. Male 118, who was 3 years of age, affiliated somewhat with the subadults in Green group in the early weeks of immigration, and not at all with the females. Male 121, older and more experienced, affiliated more with the two resident males during the 10-week period than with females or subadults. During postmigration, male 121 affiliated less often with adult males and consistently with adult females. Because females are dominant to males in *Lemur catta*, acceptance by resident females is key to successful integration. Young male 118 rarely affiliated with the females and spent most of the time on the periphery of the group, while 121 approached females early on, exhibiting submissive gestures (squealing, tail-waving, and chattering), and was accepted as a grooming partner. Females and subadults are spatially central in *Lemur catta* (Jolly, 1966), and affiliating with adult females outside of mating season is important with respect to female mate choice (Koyama, 1988; Sauther, 1991; Gould, 1994, 1996). Thus, it is important for a new immigrant to risk agonism by females and attempt to affiliate with them.

17.4.5. Immigrant Males, Rank, Nontransitive Dominance Relationships, and a Possible Instance of Coaltion/Alliance Behavior

Rank improvement after dispersal has been noted in Japanese macaques (Norikoshi and Koyama, 1975; Sprague, 1992; Suzuki et al., 1998), vervets (Cheney and Seyfarth, 1983) long-tailed macques (van Noordwijk and van Schaik, 1985) and white-faced capuchins (Jack and Fedigan, 2004b), and Sussman (1992) suggested that *Lemur catta* males might transfer to groups where the attainment of highest-ranking central male status is more likely than in the group from which they are migrating. For the older male 121, this appeared to be the case. In their former group, 121 had been a high-ranking male (1/6 then 2/6), and 118, as a natal male, had been the lowest-ranking of the six males. When they entered Green group, they were initially third- and fourth-ranking of the four males, respectively; however, during the fifth week of the migration period, 121 began displacing G10, the higher-ranking of the two resident Green group males. The following week, both immigrant males chased and jumped on G10. Similar types of agonistic encounters between 121 and G10 or both immigrants and G10 were observed on four more occasions during the remainder of the migration/postmigration periods. However, 121's dominance relationship in the group was of a nontransitive nature, as 121 quickly became dominant to G10, but remained subordinate to the other resident male, G23, and G23 was subordinate to G10. Male 118 was submissive to both resident males, and although he participated in agonistic encounters with 121 toward G10, he remained the lowest-ranking male in Green group until the end of the study.

The agonistic behaviors directed by both immigrants toward male G10 and the subsequent rank reversal between immigrant male 121 and resident male G10 might be seen as a form of coalition behavior by the two immigrants to aid in rank improvement. Alliances and coalitions are reported as rare in this species, but they have occasionally been observed, though not to the same degree or persistence as in some anthropoid species (Jolly, 1966; Pereira, 1993; Nakamichi et al., 1997; Nakamichi and Koyama, 1997; Sauther et al., 1999). That male 118 participated in agonistic events with 121, which resulted in 121 becoming dominant to G10, would imply that some sort of reciprocity may have been expected in future by 121 to aid 118 (*sensu* Trivers, 1971).

17.4.6. Behavioral Differences During Dispersal Between the Natal and Older Male

There were clear behavioral differences in migration behavior between natal male 118 and 121, who had experienced at least one previous dispersal. During the premigration season when they began to leave Red group for short periods, the natal male always followed 121 during sojourns to the periphery of other groups and behaved submissively toward him. On the initial journeys away from his natal group, 118 often contact called to the group while separated from them, while 121 did not. Furthermore, 118 would occasionally leave 121 when they were traveling to the periphery of other groups, and return to Red group. Initially, 118 appeared much more skittish and unsure when away from Red group than did 121, and he was far more spatially peripheral to Green group during the 10-week migration process.

17.4.7. Follow-up: A Case of Tenuous Tenure

Length of male tenure in ring-tailed lemur groups at Beza Mahafaly is extremely variable (Sussman, 1992; Gould, 1994) and furthermore, this population occasionally experiences severe droughts, which markedly affects population size and survival (Gould et al., 1999). Though both males appeared to be integrated into Green group by mid-March of 1993, in a subsequent study just 18 months later, none of the four Green group males discussed in this chapter were present. The years 1991 and 1992 were drought years that did have a serious effect on survival, and between 1993 and 1994 the entire adult population of ringtailed lemurs in the reserve had fallen by 27% (Gould et al., 1999, 2003). Only one of the nine focal males from the larger 1992–1993 study was still present in 1994. None of the remaining eight males were found during a census of all groups in the reserve in November 1994, nor were they spotted in a survey covering a 1-km radius outside of the reserve in the same month. Furthermore, in subsequent censuses between 1994 and 2003, these males were not seen. Therefore, it appears that the two immigrant males likely died not long after successful integration, possibly as a result of the 1991–1992 drought.

Acknowledgments. The primary research referred to in this chapter, occurring in 1992 and 1993, was funded by NSF Grant no. BNS-9119122, Wenner-Gren Predoctoral Research Grant no. 5401, National Geographic Research Grant no. 4734-92, and the Boise Fund of Oxford. Censuses conducted in 1994, 2001, and 2003 and referred to here were part of two larger projects funded by an I.W. Killam postdoctoral fellowship and Natural Sciences and Engineering Research Council of Canada. I am grateful to the following people for their help in facilitating my research at Beza Mahafaly over the years: Mme. Berthe Rakotosamimanana, Benjamin Andriamihaja, the late Pothin Rakotomanga, Joel Ratsirarson, Andrianansolo Ranaivoson, Joseph Andrianmampianina, Celestine Ravaoarinoromanga, the School of Agronomy at the University of Antananarivo, and Direction des Eaux et Fôret, Madagascar. I also thank Jonah Ratsimbazafy for his excellent research assistance in 1992, including his help collecting the data on the dispersing males referred to in this chapter.

References

Altmann, S. A., and Altmann, J. (1970). *Baboon Ecology.* Chicago: University of Chicago Press.

Altmann, J. (1974). Observational study of behaviour: Sampling methods. *Behaviour* 49:227–265.

Borries, C. (2000). Male dispersal and mating season influxes in Hanuman langurs living in multi-male troops. In: Kappeler, P. M. (ed.), *Primate Males: Causes and Consequences of Variation in Group Composition.* Cambridge: Cambridge Press, pp. 146–158.

Cheney, D. L. (1983). Proximate and Ultimate factors related to the distribution of male migration. In: Hinde, R. A. (ed.), *Primate Social Relationships: An Integrated Approach.* Blackwell: Oxford Press, pp. 241–249.

Cheney, D. L., and Seyfarth, R. M. (1977). Behavior of adult and immature male baboons during intergroup encounters. *Nature* 269: 4.4–406.

Drickamer, L. C., and Vessey, S. H. (1973). Group changing in free-ranging rhesus monkeys. *Primates* 14:359–368.

Enomoto, T. (1974). The sexual behavior of Japanese monkeys. *J. Human Evol.* 3:351–372.
Gould, L. (1994). Patterns of affiliative behavior in adult male ringtailed lemurs (*Lemur catta*) at the Beza-Mahafaly Reserve, Madagascar. Ph.D. diss., Washington University, St. Louis.
Gould, L. (1996) Male-female affiliative relationships in naturally occurring ringtailed lemurs (*Lemur catta*) at Beza-Mahafaly Reserve, Madagascar. *Am. J. Primatol.* 39:63–78.
Gould, L. (1997). Intermale affiliative relationships in ringtailed lemurs (*Lemur catta*) at the Beza-Mahafaly Reserve, Madagascar. *Primates* 38:15–30.
Gould, L., Sussman, R. W., and Sauther, M. L. (2003). Demographic and life-history patterns in a population of ringtailed Lemurs (*Lemur catta*) at Beza Mahafaly Reserve, Madagascar: A 15-year perspective. *Am. J. Phys. Anthropol.* 120:182–194.
Gould, L., Sussman, R. W., and Sauther, M. L. (1999). Natural disasters and primate populations: The effects of a two year drought on a naturally occurring population of ringtailed lemurs in southwestern Madagascar *Int. J. Primtol.* 20:69–84.
van Hooff, J. A. R. A. M. (2000). Relationships among non-human primate males: A deductive framework. In: Kappeler, P. M. (ed.), *Primate Males: Causes and Consequences of Variation in Group Composition.* Cambridge: Cambridge University Press, pp. 183–191.
Henzi, S. P., and Lucas, J. W. (1980). Observations on the inter-troop movement of adult vervet monkeys (*Cercopihtecus aethiops*). *Folia Primatol.* 33:220–235.
Isbell, L. A., Cheney, D. L. C., and Seyfarth, R. M. (1993). Are immigrant vervet monkeys, *Cercopithecus aethiops*, at greater risk of mortality than residents? *Anim. Behav.* 45:729–734.
Itani, J. (1972). A preliminary essay on the relationship between social organization and incest avoidance in non-human primates. In: Poirier, F. E. (ed.), *Primate Socialization.* New York: Random House, pp. 165–171.
Jack, K. M., and Fedigan, L. (2004a). Males dispersal patterns in white-faced capuchins, *Cebus capucinus* Part 1: Patterns and causes of natal emigration. *Anim. Behav.* 67:767–769.
Jack, K. M., and Fedigan, L. (2004b). Males dispersal patterns in white-faced capuchins, *Cebus capucinus* Part 2: Patterns and causes of secondary dispersal *Anim. Behav.* 67:771–782.
Jolly, A. (1966). *Lemur Behavior.* Chicago:University of Chicago Press.
Jones, K. C. (1983). Inter-troop transfer of *Lemur catta* males at Berenty, Madagascar. *Folia Primatol.* 40:145–160.
Koyama, N. (1988). Mating behavior of ring-tailed lemurs (*Lemur catta*) at Berenty, Madagascar. *Primates* 29:163–174.
Koyama, N., Nakamichi, M., Ichino, S., and Takahata, Y. (2002). Population and social dynamics changes in ring-tailed lemur troops at Berenty, Madagascar between 1989–1999. *Primates* 43:291–314.
Kumar, A., Singh, M., Kumara, H. N., Sharma, A. K., and Bertsch, C. (2001). Male migration in lion-tailed macaques. *Primate Report* 59:5–13.
Melnick, D. J., and Pearl, M. C. (1987) Cercopithecines in multimale groups: Genetic diversity and population structure. In: Smuts, B. B., Cheney, D. L., Seyfarth, R. M., Wrangham, R. W., and Struhsaker, T. T. (eds.), *Primate Societies.* Chicago: The University of Chicago Press, pp. 121–134.
Moore, J., and Ali, R. (1984). Are dispersal and inbreeding avoidance related? *Anim. Behav.* 32:94–112.
Nakamichi, M., and Koyama, N. (1997). Social relationships among ring-tailed lemurs (Lemur catta) in two free-ranging troops at Berenty Reserve, Madagascar. *Int. J. Primatol.* 18:73–93.

Nakamichi, M., Rakototiana, M. L. O., and Koyama, N. (1997). Effects of spatial proximity and alliances on dominance relations among female ring-tailed lemurs (*Lemur catta*) at Berenty Reserve, Madagascar. *Primates* 38:331–340.

van Noordwijk, M. A., and van Schaik, C. P. (1985). Male migration and rank acquisition in wild long-tailed macaques (*Macaca fascicularis*). *Anim. Behav.* 33:849–861.

Norikoshi, K., and Koyama, N. (1975). Group shifting and social organization among Japanese monkeys. Kondo, S., Kawai, M., Ehara, A., and Kawamura, S. (eds.), *Proc. 5th Congress of the Int. Prim. Soc.* Tokyo: Japan Science Press, pp. 43–61.

Packer, C. (1979). Inter-troop transfer and inbreeding avoidance in *Papio anubis*. *Anim. Behav.* 27:1–36.

Pereira, M. E. (1993). Agonistic interactions, dominance relation, and ontogenetic trajectories in ringtailed lemurs. In: Pereira, M. E., and Fairbanks, L. A. (eds.), *Juvenile Primates: Life History, Development and Behavior.* New York: Oxford University Press, pp. 285–305.

Pusey, A. E., and Packer, C. (1987). Dispersal and philopatry. In: Smuts, B. B., Cheney, D. L., Seyfarth, R. M., Wrangham, R. W., and Struhsaker, T. T., (eds.), *Primate Societies.* Chicago: University of Chicago Press, pp. 250–266.

Sauther, M. L. (1991). Reproductive behavior of free-ranging *Lemur catta* at Beza Mahafaly Special Reserve, Madagascar. *Am. J. Phys. Anthropol.* 84:463–477.

Sauther, M. L., Sussman, R. W., and Gould, L. (1999). The myth of the typical *Lemur catta*: 30 years of research on the ringtailed lemur. *Evol. Anthropol.* 8:120–132.

Smuts, B. B. (1985). *Sex and Friendship in Baboons.* New York: Aldine.

Sprague, D. S. (1992). Life history and intertroop mobility among Japanese Macaques (*Macaca fuscata*). *Int. J. Primatol.* 13:437–454.

Sprague, D. S., Suzuki, S., Takahashi, H., and Sato, S. (1998). Male life history in natural populations of Japanese macaques: Migration, dominance rank, and troop participation of males in two habitats. *Primates* 39:351–363.

Sugiyama, Y. (1976). Life history of male Japanese monkeys. In: Rosenblatt, J. S., Hinde, R. A., Shaw, E., and Beer, C. (eds.), *Advances in the study of behavior,* Vol. 7. New York: Academic Press, pp. 255–284.

Sussman, R. W. (1991). Demography and social organization of free-ranging *Lemur catta* in the Beza Mahafaly Reserve, Madgascar. *Am. J Phys. Anthropol.* 84:43–58.

Sussman, R. W. (1992). Male life histories and inter-group mobility among ring-tailed lemurs (*Lemur catta*). *Int. J. Primatol.* 13:395–413.

Trivers, R. L. (1971). The evolution of reciprocal altruism. *Q. Rev. Biol.* 46:35–57.

Part IV

Health and Disease

18
Patterns of Health, Disease, and Behavior Among Wild Ringtailed Lemurs, *Lemur catta*: Effects of Habitat and Sex

MICHELLE L. SAUTHER, KRISTA D. FISH, FRANK P. CUOZZO,
DAVID S. MILLER, MANDALA HUNTER-ISHIKAWA, AND HEATHER
CULBERTSON

18.1. Introduction

The Malagasy lemurs are ranked among the most endangered primates due to unprecedented levels of endemism, hunting, and habitat loss/fragmentation (Mittermeier et al., 1992). This is a consequence of a rapidly increasing human population forcing primates and humans into increased contact. Major conservation initiatives are in place to mediate this conflict, but none include assessing basic health and disease parameters in any wild extant lemur species (Ganzhorn et al., 2002). Published health evaluations on wild lemurs are important, but are rare, have been based on small sample sizes, and have focused on species living within parks and reserves (Garell and Meyers, 1995; Junge and Lewis, 2002; Dutton et al., 2003). Stressors derived from human–primate conflicts, combined with their negative effects on population parameters, make extinction events a very real possibility for extant lemurs, and developing an understanding of the disease and health ecology in lemur populations is key. Because Malagasy lemurs have evolved in the absence of many disease pathogens found on other continents, they may be especially susceptible to pathogen pollution from human-introduced species such as rats, domestic animals, and the human populations themselves. Such introductions have been implicated in the extinction of a number of animal species elsewhere (Daszak et al., 2000). Understanding health and disease patterns of animal populations is now recognized as an important component of their conservation and as critical as data on their biogeographical patterns, population dynamics and individual behavior (Wolfe et al., 1998; Daszak et al., 2000). Nevertheless, "few data sets exist to establish the 'normal' or expected range of values for most of the world's threatened or endangered wild species" (Wolfe et al., 1998, p. 1229). Information of this sort should be viewed as an essential component of long-term conservation management. The goal of the current project was to survey how habitat and sex are affecting the behavior, health, and disease ecology of wild ringtailed lemurs living within and around a protected reserve. We anticipated that habitat would affect behavior in ways that could be quantified. We expected that lemurs within the reserve would exhibit enhanced health when compared with those exploiting environments altered by human activities. As this is a female dominant society (Jolly, 1966; Sauther et al., 1999), we also expected females to exhibit better health indices when compared with males.

18.2. Materials and Methods

18.2.1. Research Site and Study Groups

The ringtailed lemurs of the Beza Mahafaly Special Reserve, Madagascar (23°30′S, 44°40′E), have been the focus of ecological, behavioral, and biological studies since 1987 (see summary in Sauther et al., 1999). Much more recently, research has been expanded outside of this protected reserve to focus on how anthropogenic factors are affecting the behavior and biology of this species. During May–August 2003, a survey on the disease ecology of wild ringtailed lemurs was begun. This is a seasonal habitat with both dry (June–September) and wet (October–May) seasons (Sauther et al., 1999). Study groups within three habitats were studied: Reserve, Degraded, and Marginal. The Reserve habitat is within the Beza Mahafaly Reserve and is an intact gallery forest that has not been affected by human disturbance for more than 18 years (Figure 18.1). The Degraded habitat includes the research camp where a number of Mahafaly families live on-site throughout the year. Researchers also live here when carrying out fieldwork. This site includes trash pits, traditional toilet areas, a deep-pit latrine used by visiting researchers, a number of dwellings, and a well. The adjoining forest has been highly impacted by villagers that live in the area, with a large portion of the riverine forest now removed and grazing by goats and cattle a common occurrence (Figure 18.2). Lemurs within this habitat commonly encounter humans and domestic animals and readily exploit human resources such as discarded food, water sources, and cattle forage (Figure 18.3). The Marginal habitat is located in a dry Didieracea forest approximately 3 km from the gallery forest reserve. Lemurs within this habitat are living within a habitat dramatically impacted by grazing and the destruction/removal of forest products (Figure 18.4). Both Degraded and Marginal habitats are characterized by heavy grazing and fecal contamination by domestic animals as well as tree cutting by humans when compared with the protected reserve (Whitelaw et al., 2005). It should be noted that the Mahafaly who live here do not hunt lemurs, and it is a cultural taboo to kill them (Ratsirarson, 2003).

18.2.2. Lemur Sampling Protocol

All methods and materials followed animal handling guidelines (IACUC-University of Colorado). A field laboratory was set up, with a centrifuge, microscope, and digital camera run by a gas generator. Samples were preserved in

FIGURE 18.1. The reserve habitat within Beza Mahafaly Special Reserve, Madagascar (see text).

portable liquid nitrogen containers. The research team included two members of the Beza Mahafaly Ecological Monitoring group, Mr. Enafa Efitroatamy (who has more than 15 years of darting experience) and Mr. Ehandidy Ellis (who has 5 years darting experience). The team darted and safely captured 70 adult and subadult individuals using a Tel-inject blow gun system using a mixture of ketamine hydrochloride (60 mg) and diazepam (2 mg) based on doses that have been worked out over the past 15 years of captures. Each ringtailed lemur received a collar and tag and a subcutaneous microchip (PIT tag), for long-term identification. All captures occurred in the morning to allow lemurs to recover and be released before nightfall on the same day. Captured individuals were transported to the camp's laboratory, and biological data were collected while they were under general anesthesia. After the biological data were collected, the lemurs were placed in covered dog kennels and kept in a quiet place for recovery. Once the anesthesia completely wore off, the lemurs were released in the same area they were captured. All individuals successfully integrated back into their groups

FIGURE 18.2. The degraded habitat near Beza Mahafaly Special Reserve, Madagascar (see text).

and maintained their previous social ranks. Behavioral studies were carried out on two camp groups both when in camp and when these same groups were in adjacent forests. Scan samples at 1-minute intervals were used, sample size was 26 individuals, and 70 hours of observation were carried out.

18.2.3. Biological Data

The collection of lemur biological data was based on protocols in the Prosimian Biomedical Survey Manual developed by Drs. Randy Junge, R. Eric Miller, and DVM of the St. Louis Zoo. All samples were collected using techniques that assure representative samples and the accuracy of results. A basic medical evaluation included heart rate, respiratory rate, temperature, and a physical examination of the whole body with specific dental, ocular, aural, and dermal protocols. Body weights were taken using a hanging scale, linear measurements were collected with a measuring tape using standard anatomical landmarks, and body fat was indirectly assessed with skin-fold calipers. External parasites (ectoparasites) were removed with a cotton swab or forceps and placed into vials containing 70% isopropyl alcohol for later identification. Ectoparasites were counted on the face and genitals where they would cluster during the lemur's examination. Feces were collected as they were voided or by inserting a

FIGURE 18.3. Use of cattle forage by ringtailed lemurs within the Degraded habitat (see text).

cotton-tipped applicator for fecal assays, which included fecal float and direct smear techniques. Feces were examined under the microscope and if helminth, ova, protozoa, or other parasites were identified, a digital photograph of the microscope image was made for future taxonomic identification. Fecal samples were also placed into containers with 10% formalin for molecular analyses of selected parasites. A small amount of feces was also placed in tubes and stored in liquid nitrogen for enteric pathogen bacterial culture in the United States. Blood was collected using standard venipuncture techniques and was replaced by an equal volume of Lactated ringer's solution (Abbot Laboratories, Chicago, Illinois, USA). No more than 1% body weight (1mL/100 g; the accepted safe volume) was collected from ringtailed lemurs and placed in appropriate serum or anticoagulant tubes for standard medical assays and on special isocode DNA paper for genetic studies. Both white and red blood cell counts and hematocrits (packed cell volume) were made for each lemur in the field within 4 hours of collection. Blood smears (2/individual) were made, fixed, stained, and coverslipped in the field, thereby permanently preserving them for hemoparasite examinations as well as complete blood counts (CBC), and platelet counts at a later date. Serum samples were processed within 2 hours of collection, aliquoted into 1-mL cryotubes to simplify dispersal for multiple tests, stored and transported in liquid nitrogen, and transferred to a −70 °C freezer until analyzed.

FIGURE 18.4. The marginal habitat 3 km from the Beza Mahafaly Special Reserve, Madagascar (see text).

We also collected dental health data that included information on the presence of canine abscesses, tooth damage (i.e., broken, cracked or chipped teeth), and antemortem tooth loss. Teeth were recorded as lost if either no trace of the tooth remained, or if worn down to or below the gum line with only worn and/or damaged roots remaining, hence being "functionally absent" (see Cuozzo and Sauther, 2004, in press, for details on scoring tooth loss).

18.2.4. Data Analysis

Patterns of health were compared across habitats and between the sexes. Linear measurements, skinfold data, and body weights are based on 64 adult individuals (≥3 years old; Sauther et al., 2002). Iron, albumin, Na (sodium), Cl (chloride), and osmolality values are based on a subset of individuals from the three habitats: Reserve n = 6, Degraded n = 10, Marginal n = 11. All other measurements come from 70 individuals. Adult male and female data were combined only when there were no significant sex differences for the measure or as noted below. Behavioral data were based on n = 26. Standard t-tests and X^2 analyses were used with significance set at $p = 0.05$.

18.3. Results

18.3.1. Hydration and Nutrition

Results indicate that habitat affects the health and nutrition of this population, and that groups in low-quality habitats and those coming into direct or indirect contact with humans and domestic animals are adversely affected in ways that can be measured.

Serum samples on a subset of individuals indicate that members of the Marginal group had higher Na, Cl, and osmolality values. These values, taken together, indicate that members of the Marginal group were less hydrated than individuals living within the other two habitats. It is important to note that none of these individuals were clinically dehydrated; they were simply less hydrated when compared to the other groups. The Degraded group had elevated levels of albumin, iron, and packed cell volume (hematocrit) values when compared with Marginal and Reserve individuals (Table 18.1). For hematocrit and albumin, values are within the range reported for wild ringtailed lemurs at Tsimanampetsotsa, although iron levels are outside this range (Dutton et al., 2003).

18.3.2. Body Weights, Skinfold Measurements, and Selected Linear Measurements

Degraded groups also had greater skinfold measurements than individuals from the other two habitats (Table 18.2). Marginal groups had smaller abdominal skinfold measurements than both Degraded and Reserve groups (Table 18.2). Comparing females only, Degraded group individuals were heavier compared with those living in marginal and reserve habitats (Table 18.2).

Limb lengths potentially reflect environmental effects during development. Females normally remain in their natal groups and so can demonstrate habitat effects more directly, in contrast with males who migrate in and out of other groups. We thus present data only on females. Degraded group females were larger than females of the reserve and marginal habitats as represented here by greater lower arm and muzzle lengths (Table 18.2).

18.3.3. Dental Health

When data from groups in habitats influenced by humans ("Degraded" and "Marginal" groups) are compared with those from the protected reserve, the former show a significantly greater frequency of both tooth loss and tooth damage (e.g., broken, cracked, or chipped) (Table 18.3). Of the 64 adults studied, four display obvious maxillary canine abscesses. Only one of these is a female, and all are older individuals, (with the female known to be 11 years old). Two of these abscesses (one male and the aforementioned female) are associated with broken and decayed canines. All four cases of maxillary canine abscesses occur in groups influenced by humans, with three of the four coming from individuals using

TABLE 18.1. Nutritional differences by habitat.

Nutritional/health measure	Reserve Mean ± SD	Degraded Mean ± SD	Marginal Mean ± SD
Percentage packed cell volume (Hematocrits)	40% ± 10.91	47% ± 4.36	42% ± 4.45
Iron (µg/dL)	142 ± 25.7	188 ± 41.3	134 ± 31.27
Albumin(g/dL)	4.33 ± 0.39	4.68 ± 0.35	4.05 ± 0.35
Na (sodium)	141.7 ± 2.3	142.8 ± 3.5	148.6 ± 8.7
Osmolality	295.8 ± 4.9	297.4 ± 7.1	305.9 ± 7.4
Cl (chloride)	101.0 ± 4.1	102.3 ± 4.6	105.6 ± 2.4
Nutritional/health measure by habitat	Mean diff.	t value	p value
Percentage packed cell volume			
Reserve, Degraded	6.862	2.68	0.009
Reserve, Marginal	2.475	0.942	0.36
Degraded, Marginal	4.39	3.11	0.004
Iron			
Reserve, Degraded	46.63	2.47	0.03
Reserve, Marginal	7.49	0.50	0.63
Degraded, Marginal	54.12	3.41	0.003
Albumin			
Reserve, Degraded	0.480	2.71	0.02
Reserve, Marginal	0.145	0.833	0.42
Degraded, Marginal	0.625	4.08	0.0006
Na			
Reserve, Degraded	1.11	0.69	0.50
Reserve, Marginal	6.97	1.91	0.05
Degraded, Marginal	5.85	1.90	0.05
Osmolality			
Reserve, Degraded	1.61	0.48	0.63
Reserve, Marginal	10.08	2.96	0.009
Degraded, Marginal	8.47	2.59	0.02
Cl			
Reserve, Degraded	1.33	0.57	0.58
Reserve, Marginal	4.54	2.90	0.01
Degraded, Marginal	3.21	2.02	0.04

Degraded habitat that includes exploiting human refuse. These abscesses, in addition to often displaying inflamed gingiva surrounding the base of the teeth, present as open or healed wounds on the maxillary area of the muzzle, due to infection and damage at the apex of the canine root (Figure 18.5).

18.3.4. *Behavioral Differences*

As noted before, we compared the behavior of the same groups when they used the camp area with their behavior within the adjacent forests (Figure 18.6). Our results indicate that there were no differences with regard to number of individuals feeding. However, more individuals spent time in social behaviors when in the forest. More individuals engaged in locomotion (moving) when in the camp habitat, and more lemurs rested in camp during scans than in the forest. Importantly,

TABLE 18.2. Body weights, skinfold measurements, and selected linear measurements by habitat.

Measure	Reserve Mean ± SD	Degraded Mean ± SD	Marginal Mean ± SD
Body weights (kg)	2.23 ± 0.16	2.40 ± 0.16	2.12 ± 0.21
Female abdominal skinfolds	1.42 ± 25.7	2.01 ± 0.82	1.29 ± 0.19
Subscapular skinfolds	2.17 ± 0.47	2.61 ± 0.60	2.05 ± 0.40
Biceps skinfolds	1.26 ± 0.26	1.56 ± 0.31	1.17 ± 0.16
Suprailiac skinfolds	1.93 ± 0.43	2.25 ± 0.61	1.90 ± 0.35
Female muzzle length (cm)	4.16 ± 0.36	4.56 ± 0.38	4.10 ± 0.37
Female lower arm length (cm)	10.19 ± 0.36	10.72 ± 0.68	10.20 ± 0.43

Measure	Mean diff.	t value	p value
Body weights			
Reserve, Degraded	0.17	2.47	0.02
Reserve, Marginal	0.12	1.58	0.13
Degraded, Marginal	0.29	3.47	0.003
Abdominal skinfolds			
Reserve, Degraded	0.52	2.16	0.04
Reserve, Marginal	0.20	2.10	0.04
Degraded, Marginal	0.72	2.80	0.01
Subscapular skinfolds			
Reserve, Degraded	0.45	2.00	0.02
Reserve, Marginal	0.12	0.67	0.51
Degraded, Marginal	0.57	2.57	0.02
Biceps skinfolds			
Reserve, Degraded	0.38	3.43	0.001
Reserve, Marginal	0.09	1.23	0.20
Degraded, Marginal	0.38	4.86	<0.0001
Suprailiac			
Reserve, Degraded	0.33	2.09	0.04
Reserve, Marginal	0.03	0.23	0.81
Degraded, Marginal	0.35	2.2	0.03
Muzzle length			
Reserve, Degraded	0.39	2.38	0.03
Reserve, Marginal	0.05	0.36	0.72
Degraded, Marginal	0.44	2.62	0.01
Lower arm length			
Reserve, Degraded	0.53	2.54	0.02
Reserve, Marginal	0.03	0.23	0.82
Degraded, Marginal	0.50	2.31	0.03

TABLE 18.3. Habitat differences in tooth loss and tooth damage.

	Reserve	Nonreserve	Chi-square	Chi-square p value
Tooth loss	Expected = 9 Observed = 3	Expected = 9 Observed = 15	6.0	< 0.05
Tooth damage	Expected = 12 Observed = 6	Expected = 12 Observed = 18	6.0	< 0.05

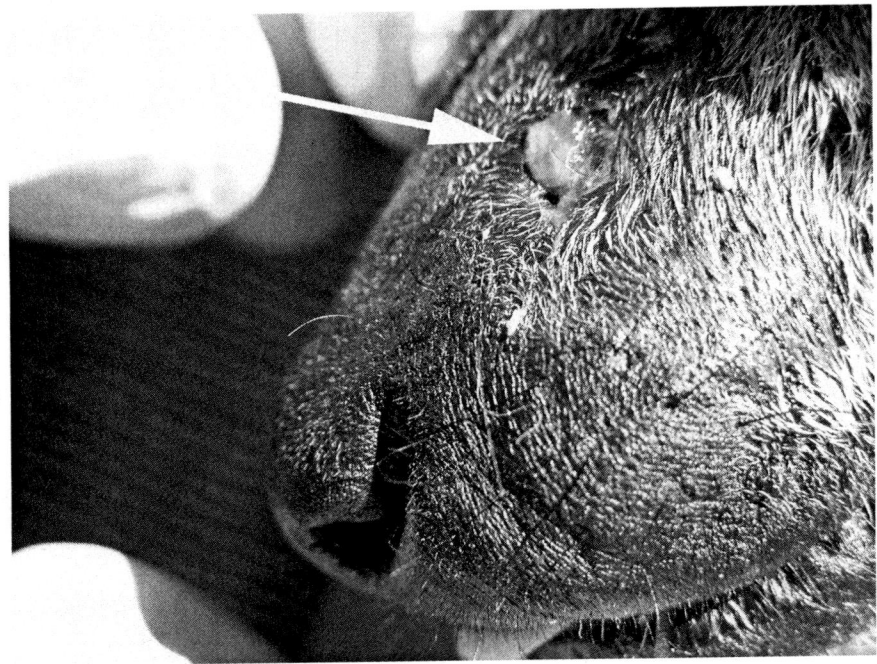

FIGURE 18.5. Open or healed wounds on the maxillary area of the muzzle of a ringtailed lemur, due to a canine dental abscess in an individual from the degraded habitat (see text).

their patterns of locomotion changed as more individuals engaged in terrestrial locomotion when in camp, but more used arboreal locomotion while in the forest. In addition, more lemurs engaged in a variety of atypical behaviors while in camp (noted as "other" in Figure 18.6). This category included self-grooming, or scent-marking, but also included a number of atypical behaviors such as exploiting human foods (even going into huts to steal food) and gazing at themselves in car windshields (Figure 18.7). The number of individuals who engaged in playing was greater in camp than in the forest. The number of lemurs involved in feeding agonism was higher in camp, but there were no differences in the number of individuals carrying out social agonism (Figure 18.8).

18.3.5. Sex Differences

Males exhibited a higher number of ectoparasites (mesostigmatid mites) than females, regardless of habitat. Approximately 22% of the sample population exhibited dermatitis, some with severe hair loss (Figure 18.9). We found this in individuals across all habitats and of varying severity. There were no differences by habitat regarding dermatitis. However, males primarily exhibited this condition, in fact only two females were found with dermatitis (Table 18.4). Males

18. Patterns of Health, Disease, and Behavior Among Wild *L. catta* 323

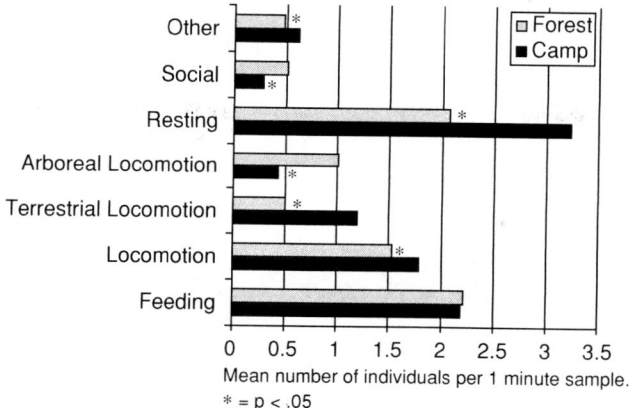

FIGURE 18.6. Behavioral differences in the same groups when in the camp and in the forest (see text).

FIGURE 18.7. Juvenile ringtailed lemur gazing at her reflection in a car window within the camp.

displayed a greater number of scars and wounds (Table 18.4). There were sex differences in abdominal and pectoral skinfolds, but this varied by habitat. In the Degraded habitat, there were no sex differences. In the Reserve habitat, both measures were greater in females. In the Marginal habitat, females had higher pectoral values only (Table 18.4).

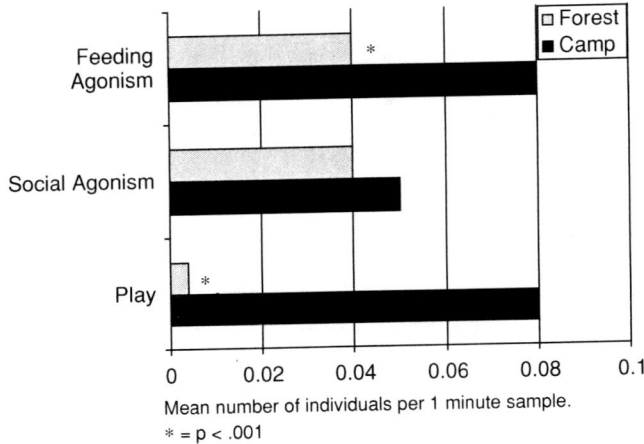

FIGURE 18.8. Behavioral differences in play and feeding and social agonism for the same groups when in the camp and in the forest (see text).

18.4. Discussion

18.4.1. Health and Growth and Development

Our results indicate that habitat and sex have measurable effects on health and nutritional measurements in this population. Skinfold measurements are an indirect measure of body fat. Dissections of white adipose tissue in captive ringtailed lemurs reveal that the skinfold measurements that differ between troops in our sample (e.g., abdominal, pectoral, etc.) are also locations of high amounts of adipose tissue in this species (Pereira and Pond, 1995). Among females, Marginal groups had lower abdominal skinfold measurements than both Reserve and Degraded habitats. The abdominal paunch is an important area for fat storage and thus an energy source in ringtailed lemurs (Pereira and Pond, 1995). As all lemurs were sampled during the period of reduced food availability, this indicates that individuals living in the Marginal habitat were using body fat selectively from this region at a time when they would be nutritionally stressed. In addition, hydration measures (higher Na, Cl, and osmolality) indicate that the members of Marginal groups were less hydrated. This may reflect differences in habitat. During our sample period, there were no direct sources of water within the Marginal habitat, which is also quite open, lacking any true closed forest canopy. The Reserve is a closed canopy forest, with a richer food source, and the lemurs are commonly observed licking morning dew from the leaves (Sauther, 1992). The Degraded habitat is a more open forest but also includes the camp and thus access to human water sources.

It is initially surprising that body weight, growth, and developmental measures (e.g., linear measurements) are not statistically different, when comparing individuals of the higher quality Reserve habitat with those of the Marginal habitat.

FIGURE 18.9. Dermatitis in a male ringtailed lemur at Beza Mahafaly Special Reserve.

However, although the group sizes are not significantly different, the reserve population density is much greater, at 3.2 groups/km, when compared with only 0.56 groups/km in the Marginal habitat (Whitelaw and Sauther, 2003). This suggests that while the reserve may support many more groups than the surrounding human altered landscapes, in reality this may create high levels of intergroup feeding competition year round as has been observed both at Beza (Sauther and Sussman, 1993) and at Berenty (Jolly et al., 1993), which is a gallery forest surrounded by sisal fields. Such competition may dampen some of the positive effects of a richer environment within the reserve and indicates that "island reserves," be they at Beza or Berenty (Crawford et al. this volume), may place pressures of their own on lemurs living within such habitats.

Having access to additional resources during the period of seasonal reduction in natural lemur foods appears to have both immediate and long-term

TABLE 18.4. Sex differences in scars/wounds, dermatitis, ectoparasite number, and skinfold measurements.

	Male Individuals	Female Individuals		χ^2	p
Scars/wounds: no. individuals	Observed = 18 Expected = 13	Observed = 12 Expected = 17		15.04	0.03
Dermatitis: no. individuals	Observed = 14 Expected = 7	Observed = 2 Expected = 8		15.45	0.0001
	Mean ± SD	Mean ± SD	Difference	t value	p
Ectoparasites	7.18 ± 5.34	4.65 ± 3.17	2.53	2.31	0.02
Reserve habitat					
Pectoral skinfolds	1.47 ± 0.44	2.07 ± 0.50	0.61	3.37	0.002
Abdominal skinfolds	1.20 ± 0.39	1.50 ± 0.07	0.29	2.28	0.03
Degraded habitat					
Pectoral skinfolds	1.78 ± 0.26	2.09 ± 0.46	0.31	1.61	0.13
Abdominal skinfolds	1.71 ± 0.34	2.01 ± 0.82	0.31	0.91	0.38
Marginal habitat					
Pectoral skinfolds	1.28 ± 0.20	1.68 ± 0.42	0.39	2.58	0.02
Abdominal skinfolds	1.17 ± 0.226	1.30 ± 0.19	0.12	1.24	0.23

developmental effects. Our current information indicates that individuals who come into camp have a more consistent resource base by relying on both natural forest products as well as domestic and human foods. These groups have been exploiting camp resources since at least 1987. Domestic animal resources were green forage brought in for cattle (Figure 18.5), and human resources included cooked sweet potatoes, a variety of melons, carrots, and mango peels (Sauther, pers. obs.). Being able to exploit these additional resources may buffer individuals especially during the dry season, when food availability in this area is dramatically reduced (Sauther, 1998). Groups coming into camp and exploiting human and domestic food were heavier, had greater skinfold measurements, and adult females were larger than adult females within groups living in both Marginal and Reserve habitats. Having such a food "buffer" may affect growth and developmental patterns, allowing these females to achieve a larger body size. Among Amboseli baboons, groups that use tourist lodge refuse also have greater body size and fatness (Altmann and Muruthi, 1988; Muruthi et al., 1991; Altmann et al., 1993). Access to human foods could account for the elevated values in the Degraded group, as a higher nutritional plane could account for increased albumin and packed cell values, and human foods are associated with elevated iron levels in lemurs (Junge, 2003). An alternative explanation, that the elevated albumin and PCV values reflect lower hydration in the Degraded group, is less likely, as lemurs in this habitat had access to and used water sources that were associated with human activities. As the data are preliminary, this needs to be directly tested during future studies.

There may also be costs to exploiting such an environment. Within camp, individuals competed for highly monopolizable resources such as human food remains, and behavioral results indicate individuals spent more time in energetically costly movement (Figure 18.7). There were also more intergroup fights when in camp (14) versus the forest (1) during sampling. These were dramatic fights that involved one group chasing the other out of the camp. In addition, all dental abscesses were found in individuals that lived in the Degraded habitat, and three were from individuals who daily exploited human resources in the camp. This has also been found among wild baboons, as those who spent more time exploiting human refuse were also the individuals with reduced periodontal health (Phillips-Conroy et al., 1993). In addition, Degraded group females exhibited higher measures of fluctuating asymmetry (Fish et al., 2004.). In a variety of mammals and birds, higher levels of fluctuating asymmetry have been associated with environmental stress (Lens et al., 1999). Given the clumped and easily monopolized nature of camp foods such as human trash pits and livestock feeding areas, the potential for contest competition is higher and is reflected in the greater feeding agonism and higher number of intergroup fights over foods in camp.

18.4.2. Behavior

Finding no differences in feeding was surprising given the vastly different nature of the camp as compared with the forest environment, as there were few trees and other natural sources of food within the camp. Most of the feeding within the camp was on non-natural food resources such as discarded root crops and other human food remains, and especially cattle forage. These were easily monopolizable resources, which relates to the higher number of individuals involved in feeding agonism when in camp. Habitat had direct effects on behavior with greater sociality exhibited while in the forest, and greater moving and resting in camp. It should be noted that social interactions are important, as greater levels can provide more opportunities for the transmission of pathogens with direct transmission routes (i.e., direct contact versus environmental). When in camp these same individuals had more time to focus on "leisure" behaviors such as playing and self-grooming, as well as unusual nonmaintenance behaviors. A similar pattern is seen among baboons exploiting human refuse at tourist lodges (Altmann and Muruthi, 1988).

18.4.3. Sex Differences

It is widely known that ringtailed lemurs are a female-dominant society. Females have social, spatial, and feeding priority over males in their groups (see Sauther et al., 1999 for a summary). This can lead to sex differences in both life history and patterns of disease and health (Sauther et al., 2002). As such, it was predicted that females would show more positive health and nutritional status than males. Within this female-dominant society, males suffer a greater number of health disorders that reflect both social and ecological factors. These include a greater number of ectoparasites, a higher frequency of males with dermatitis, and the greater presence of wounds and scars on males versus females. The exact cause of dermatitis in this population is not clear but it may be related to a nutritional disorder or ectoparasites (Randy Junge, pers. comm.). Analyses of skin biopsies suggest it is not caused by *Leucaena leucocephala* as is found at Berenty (see Crawford et al., this volume). Higher frequencies of scarring and wounds among males are primarily on the face and arms and may relate to males engaging in jump fights involving more direct physical contact, as argued elsewhere (Sauther et al., 2002; Sauther and Cuozzo, 2005).

Skinfold measurements show that females in both Reserve and Marginal habitats had markedly greater skinfold measures than did males. In the Reserve habitat this included both the pectoral and abdominal areas, but in the Marginal habitat it included only the pectoral area. As the abdominal paunch may play an important role in energy storage, it suggests that females in the Marginal habitat may have utilized these fat stores to the point that no sex differences could be detected, but that Reserve females still had greater stores. In the Degraded habitat there were no sex differences, signifying that access to camp resources may enhance male's fat stores to such a degree that sex differences are not in evidence. These results add to our previous data that specify how social structure can have strong effects on male versus female health (Sauther et al., 2002), but also indicate that characteristics of different habitats can modify such effects.

18.5. Conclusions

A better understanding of the factors contributing to environmental stress for endangered and vulnerable lemur species is an important avenue for their protection and for the development of feasible conservation initiatives. This multidisciplinary collaboration and comparative approach to health and disease assessment provides an opportunity to fully integrate conservation medicine with more traditional conservation-based research for the management of *in situ* and *ex situ* lemur populations and to maintain lemur biodiversity. A basic understanding of lemur disease patterns will facilitate predictions of the onset of health crises in populations, thus alleviating potential new threats to endangered populations. Field-based medical evaluations of wild populations are providing baseline normal health values (blood biochemical values, hematology) exhibited by lemurs within high-quality protected areas and how this may change with anthropogenic alterations. These results are valuable to zoo veterinarians by providing reference ranges of values from wild populations that will enhance health maintenance in captive populations. How well lemur groups do under both high-quality habitat and more challenging environments will inform the capacity for each species to cope under future habitat alteration.

The data presented here suggest that ringtailed lemurs living in areas influenced by human activity, either in terms of availability of human-derived resources (e.g., crops, food waste) or in areas degraded through pastoral activity and deforestation are affected both immediately and over time. These results are a first step toward understanding how lemur health can be affected by human induced changes. Given the already high levels of human impact on the Malagasy lemurs, and given that increasing contact between wildlife, humans, and domestic animals continues to increase, clarifying whether some species may be more at risk when faced with living within human-altered environments will directly inform developing conservation policies. The dynamic nature of this interaction between human and wildlife species in Madagascar requires similar studies that can provide an opportunity to understand the critical links between anthropogenic change, behavioral, and disease ecology.

Acknowledgments. We thank Enafa Efitroaromy, Ehandidy Ellis, Razanajafy Olivier, Emady Rigobert, and Elahavelo of the Beza Mahafaly Ecological Monitoring Team. We thank Robert W. Sussman and Jeff C. Kaufman for their help with collection of dental data in 1987 and 1988. We thank Robert W. Sussman, Randy Junge, Ingrid Porton, Joel Ratsirarson, Jo Ajimy, Randrianarisoa Jeannicq, and Ibrahim Jacky Youssouf and Dr. Rafidisoa Tsiory (ANGAP) for their strong support and facilitation of our project. Our appreciation also goes to the Département des Eaux et Forêts, Ecole Superieur des Sciences Agronomiques, Université d'Antananarivo, and ANGAP for allowing us to continue our research at Beza Mahafaly. Funding for this study came from the Lindbergh Fund, The Saint Louis Zoo, The John Ball Zoo Society, The National Geographic Society, and the University of Colorado, Boulder.

References

Altmann, J., and Muruthi, P. (1988). Differences in daily life between semiprovisioned and wild-feeding baboons. *Am. J. Primatol.* 15:213–221.

Altmann, J. D., Schoeller, S. A., Altmann, P., Muruthi, P., and Sapolsky, R. (1993). Body size and fatness of free-living baboons reflect food availability and activity levels. *Am. J. Primatol.* 30:149–161.

Cuozzo, F. P., and Sauther, M. L. (2004). Tooth loss, survival, and resource use in wild ring-tailed lemurs (*Lemur catta*): Implications for inferring conspecific care in fossil hominids. *J. Hum. Evol.* 46:623–631.

Cuozzo, F. P., and Sauther, M. L. (in press). Severe wear and tooth loss in wild ring-tailed lemurs (*Lemur catta*): a function of feeding ecology, dental structure, and individual life history. *J. Hum. Evol.*

Daszak, P., Cunningham, A. A., and Hyatt, A. D. (2000). Emerging infectious disease of wildlife: Threats to biodiversity and human health. *Science* 287:443–449.

Dutton, C. J., Junge, R. E., and Louis, E. E. (2003). Biomedical evaluation of free-ranging ring-tailed lemurs (*Lemur catta*) in Tsimanampetsotsa Strict Nature Reserve, Madagascar. *J. Zoo Wildlife Med.* 34:16–24.

Fish, K., Sauther, M. L., and Cuozzo, F. (2004). Effects of habitat on fluctuating asymmetry in a population of wild ring–tailed lemurs (*Lemur catta*). *Am. J. Phys. Anthropol. Suppl.* 38:94.

Ganzhorn, J. U., Lowry, P. P. II., Shatz, G. E., and Sommer, S. (2001). The biodiversity of Madagascar: One of the world's hottest hotspots on its way out. *Oryx* 35:346–348.

Garell, D. M., and Meyers, D. M. (1995). Hematology and serum chemistry for free-ranging golden crowned sifaka (*Propithecus tattersalli*). *J. Zoo Wildlife Med.* 26:382–386.

Gray, P. B., and Marlowe, F. (2002). Fluctuating asymmetry of a foraging population: The Hadza of Tanzania. *Ann. Hum. Biol.* 29:495–501.

Jolly, A. (1966). *Lemur Behavior*. University of Chicago Press, Chicago.

Jolly, A., Rasamimanana, H. R., Kinnaird, M. F., O'Brien, T. G., Crowley, H. M., Harcourt, C. S., Gardner, S., and Davidson, J. (1993). Territoriality in *Lemur catta* groups during the birth season at Berenty, Madagascar. In: Kappeler, P. M., and Ganzhorn, J. U. (eds.), *Lemur Social Systems and Their Ecological Basis*. Plenum Press, New York, pp. 85–109.

Junge, R. E., and Louis, E. E. (2002). Medical evaluation of free-ranging primates in Betampona Reserve, Madagascar. *Lemur News* 7:23–25.

Lens, L., van Dongen, S., Wilder, C. M., Brooks, T. M., and Matthysen, E. (1999). Fluctuating asymmetry increases with habitat disturbance in seven bird species of a fragmented afrotropical forest. *Proc. R. Soc. Lond. B* 266:1241–1246.

Mittermeier, R. A., Konstant, W. R., Nicoll, M. E., and Langrand, O. (1992). Lemurs of Madagascar: An Action Plan for their Conservation. IUCN/SSC Primate Specialist Group.

Muruthi, P. M., Altmann, J., and Altmann, S. (1991). Resource base, parity, and reproductive condition affect female's feeding time and nutrient intake within and between groups of a baboon population. *Oecologia* 87:467–472.

Pereira, M. E., and Pond, C. M. (1995). Organization of white adipose tissue in Lemuridae. *Am. J. Primatol.* 35:1–13.

Ratsirarson, J. (2003). Réserve Spéciale de Beza Mahafaly. In: Goodman, S. M., and Benstead, J. P. (eds.), *The Natural History of Madagascar*. University of Chicago Press, Chicago, pp. 1520–1525.

Sauther, M. L. (1992). *Effect of Reproductive State, Social Rank and Group Size on Resource Use Among Free-ranging Ringtailed Lemurs (Lemur catta) of Madagascar*, PhD thesis, Washington University, St. Louis.

Sauther, M. L. (1998). The interplay of phenology and reproduction in ringtailed lemurs: Implications for ringtailed lemur conservation. In: Harcourt, C. S., Crompton, R. H., and Feistner, A. T. C. (eds.), *Biology and Conservation of Prosimians. Folio Primatol.* 69:309–320.

Sauther, M. L., and Cuozzo, F. P. (2005). Patterns of mortality and trauma in a wild population of ring-tailed lemurs, *Lemur catta*. *Am. J. Phys. Anthropol. Suppl.* 40:181.

Sauther, M. L., and Sussman, R. W. (1993). A new interpretation of the organization and mating systems of the ringtailed lemur (*Lemur catta*). In: Kappeler, P. M., and Ganzhorn, J. (eds.), *Lemur Social Systems and Their Ecological Basis*. Plenum Press, New York, pp. 11–121.

Sauther, M. L., Sussman, R. W., and Gould, L. (1999). The socioecology of the ringtailed lemur: Thirty-five years of study. *Evol. Anthropol.* 8:120–132.

Sauther, M. L., Sussman, R. W., and Cuozzo, F. (2002). Dental and general health in a population of wild ring-tailed lemurs: A life history approach. *Am. J. Phys. Anthropol.* 117:122–132.

Whitelaw, D. C., and Sauther, M. L. (2003). A preliminary survey and GIS analysis of ring-tailed lemur habitat use in and around Beza-Mahafaly Reserve, Madagascar. *Am. J. Phys. Anthropol. Suppl.* 36:224.

Whitelaw, D., Sauther, M. L., Loudon, J. E., and Cuozzo, F. (2005). Anthropogenic change in and around Beza-Mahafaly Reserve: Methodology and results. *Am. J. Phys. Anthropol.* 40:222.

Wolfe, N. D., Escalante, A. A., Karesh, W. B., Kilbourn, A., and Lal, A. (1998). Wild primate populations in emerging infectious disease research: The missing link? *Emerg. Infect. Dis.* 4:149–158.

19
Bald Lemur Syndrome and the Miracle Tree: Alopecia Associated with *Leucaena leucocephala* at Berenty Reserve, Madagascar

GRAHAM C. CRAWFORD, LOUIS-EXPERT ANDRIAFANEVA,
KATHRYN BLUMENFELD-JONES, GARY CALABA, LINDA CLARKE,
LISA GRAY, SHINICHIRO ICHINO, ALISON JOLLY, NAOKI KOYAMA,
ANNE MERTL-MILLHOLLEN, SUSAN OSTPAK, R. ETHAN PRIDE,
HANTANIRINA RASAMIMANANA, BRUNO SIMMEN, TAKAYO SOMA,
LAURENT TARNAUD, ALISON TEW, AND GEORGE WILLIAMS

19.1. Introduction

Berenty reserve (25 °0.29′S, 46 °19.37′E) is a 2 km^2 forest fragment in southern Madagascar, surrounded on two sides by agriculture, on the third by the Mandrare River, and on the fourth by about 2 km^2 of more degraded forest. The reserve has been protected from grazing and hunting for almost 70 years. From north to south, Berenty forest consists of four adjacent habitats: Ankoba, a 50-year-old second-growth forest (500 ringtailed lemurs per km^2); tourist front (500 lemurs per km^2); gallery forest (250 lemurs per km^2); and scrub/spiny forest (100 lemurs per km^2) (Jolly et al., this volume).

Ringtailed lemurs (*Lemur catta*) in Berenty suffer during the dry season (May–September) from alopecia (fur loss) that appears to be associated with poor body condition and, in extreme cases, lethargy and death. The "bald lemur syndrome" first appeared in the late 1990s in some members of a troop at the northern edge of Ankoba, and then in 2001 in some individuals of troops ranging in the southern section of Ankoba and the tourist front, skipping troops between (Figures 19.1 and 19.2).

The alopecia syndrome varies in severity. In extreme cases, afflicted lemurs have a complete absence of fur on the trunk, legs, and tail, with sparse fur on the face and feet. Less severely affected cases have partial alopecia of the trunk or tail only. In Koyama and colleagues' individually known sample, 17 of 57 adults and 3-year-olds were classed as severely alopecic in October 2003. Four of the 17 died by 2004, or 24%. This compares with an annual death rate of 13% in the remaining 40, and a 13% annual death rate for all 3-year-old and adult females in the same troops in the period 1989–1999 (Koyama et al., 2002).

Affected animals that survive regrow their fur, beginning in October (Figure 19.3). They appear at first as though they have been spray-painted white. They

FIGURE 19.1. Feeding on green leucaena pods, May 2005. At the start of the leucaena-feeding season (May–October), most animals are fully furred. Photo, A. Jolly.

then progress through a short fuzzy coat to a coat of normal length. However, the regrown fur as late as March may still be dull and discolored red-orange and in some cases fur may be absent from the ventral trunk.

Alopecia may result from many causes, including dermal or systemic infectious or parasitic diseases, malnutrition, stress, or specific toxins. Thus, in the early stages of the investigation we looked very generally for all possible causes. Our studies then became focused with the realization that *Leucaena leucocephala* is a known cause of alopecia and wasting in livestock, and that some troops of *L. catta* were seasonally eating a very high percentage of *Leucaena* in their diet (Soma et al., this volume).

We initially considered the possibility of infectious disease. In animals evaluated in our March 2003 veterinary study, there was no evidence of commonly recognizable diseases, the skin appeared to be normal and there was no gross evidence of infectious or inflammatory disease or ectoparasites except

FIGURE 19.2. A bald lemur who has lost all tail fur and half of body fur, but retains fur on back and limbs (September 2001). Photo, S. Ichino.

FIGURE 19.3. A 2-year-old beginning to regrow very short fur and showing the wasting that accompanies extreme bald lemur syndrome. This animal recovered and is still alive, though with annual fur loss (September 2001). Photo, S. Ichino.

ticks. In March all lemurs were fully furred, but animals with the abnormal fur condition had lower red blood cell mass and plasma protein relative to animals with normal fur condition. The initial veterinary conclusion was that contagious disease was unlikely to be a factor in the syndrome but that a systemic disease or malnutrition might play a role.

A second possibility was malnutrition or stress associated with the high population densities, or specifically associated with tourism. A tornado in October 1999 damaged a quarter of Berenty's forest trees (Rasamimanana et al., 2000) and was followed by 3 years of drought. In 1999, banana-feeding of the tourist front troops was also banned, after a period when population in the tourist area had grown steeply (Jolly et al., 2002; Koyama et al., 2002; Jolly et al, this volume). The growing population of introduced brown lemurs (*Eulemur fulvus rufus x collaris*) competes directly for food with the ringtailed lemurs (Simmen et al., 2003; Pinkus et al, this volume). With the ageing and drying of the forest and with lemur population growth, the number of tamarind trees over 50 cm diameter at breast height (a keystone resource) in a well-studied 14.2 ha area has declined since 1990 from 2.8 per individual *Lemur catta* to 1.8 per individual, quite aside from the brown lemurs' use of the same trees (Koyama et al., this volume). Thus, the period since 1999 has been one of steep decline in the availability of food for the ringtails. If the food shortage differentially affected the densest population, it might be that we were seeing direct starvation. However, analysis of cortisol levels in troops in the tourist front, the gallery forest, and the scrub area of the forest in 1999–2000 suggested that at that time there was no difference in stress directly related to the fivefold population density difference between these zones (Pride, 2003; Pride, this volume).

Finally, there was the hypothesis of the effects of *Leucaena leucocephala*. Its seeds were planted in pilot stands in or near the northern edge of the forest in about 1990 as potential browse for cattle and ostriches, and the tree has spread since then by self-seeding (J. de Heaulme, pers. comm.). It is a fast-growing leguminous tree, originally from Mexico, Central America, and northern South America, one of a complex of species and varieties planted by indigenous farmers who consume the green pods as a vegetable (Figure 19.4). Its common names include leucaena, koa haole, jumbie bean, vai vai, ipil-ipil, lead tree, zarcilla, pipinac, and, in Madagascar, kantsa-kantsa. It is now pan-tropical. Leucaena seeds were taken to the Philippines in sailing ships as early as the 17th century, and since the 1970s forestry organizations have seen it as a "miracle tree" for fuelwood, polewood, erosion control, and high-protein livestock forage. Its rapid growth, self-seeding, and propagation by tillers are major benefits in the right circumstances, though it has also become a weed that invades cultivated fields from Ghana to Hawaii (Brewbaker, 1987; Shelton, 1994; Sethi and Kulkarmi, 1995; Hughes, 1998).

Leucaena leaves and seeds contain mimosine, a non-protein amino acid known to be toxic to mammals, particularly non-ruminants. The mechanisms of mimosine toxicity are not completely understood but include inhibition of DNA, RNA, and

FIGURE 19.4. *Leucaena leucocephala* leaves, flowers and mature seed pods. Photo, Cyril Ruoso.

protein synthesis; blockade of the metabolic pathways of aromatic amino acids and tryptophan; chelation of metals; and antagonizing the action of vitamin B_6. Commercially available mimosine is used in laboratories to synchronize cohorts of dividing cells because the effect is reversible when treatment stops. Mimosine toxicity has been reported to cause a number of symptoms in mammals, including alopecia, poor body condition, and poor weight gain, as well as infertility, fetal resorption, low birth weight, neurologic symptoms, mucosal ulceration, and cataracts. Signs of mimosine toxicity symptoms disappear after a short time and leave no residual effects when the plants are removed from the diet. In ruminants, mimosine is rapidly converted by the gut flora to another toxic compound,

3,4-dihydroxy pyridine (3,4-DHP). Although mimosine is directly toxic, 3,4-DHP is only indirectly so as a goitrogen. Thus, animals that convert mimosine to DHP can tolerate higher dietary levels of leucaena than other animals, and animals that can detoxify DHP can tolerate higher levels yet. Mimosine detoxification allows ruminants to tolerate levels of dietary leucaena up to 30% dry weight if they are unable to convert DHP and even higher if they can convert DHP. In contrast, mimosine toxicity has been reported in non-ruminant animals when leucaena is fed at levels above 5–10% dry weight (Shelton, 1994; Sethi and Kulkarmi, 1995).

19.2. Materials and Methods

19.2.1. Veterinary Evaluation

In March 2003, 60 adult lemurs (28 male, 32 female) were captured by a veterinary team headed by Crawford using tiletamine and zolazepam[1] delivered by blow dart.[2] Lemurs were given complete physical examinations. Thirty-one lemurs had normal fur and 29 lemurs had abnormal appearing fur that was brittle, short, and discolored red-orange. In September 2004, two male lemurs with severe alopecia were captured and evaluated. Blood was collected from the femoral vein, and collected blood was centrifuged in heparinized capillary blood tubes[3] at 2400 rpm. After calculating the percentage of red cell mass in the capillary tubes, plasma protein was measured in the remaining plasma with a refractometer. Skin biopsies were collected using a 6-mm biopsy punch[4] and immediately fixed in 10% buffered neutral formalin. The formalin-fixed tissues were embedded in paraffin, sectioned at 4 µm, and stained with hematoxylin and eosin, and the histopathologic sections were evaluated by a board-certified veterinary pathologist.[5] The capture protocol and population health evaluation was approved by the San Francisco Zoo Institutional Animal Care and Use Committee and was permitted by the Madagascar Ministry of Water and Forests (Eux et Foret). In accordance with CITES, CITES export permits were issued by Eux et Foret and CITES import permits were issued by the United States Fish and Wildlife Service for all biological samples.

19.2.2. Mapping of Leucaena and of Lemur Fur Condition

The location of all leucaena groves and individual trees were identified by walking perpendicular transects of the forest on a 10-m grid. Positions of identified groves and trees were determined by at least four circumfirential readings from a

---

1. Telazol, Fort Dodge Animal Health, Overland Park, Kansas, USA; 25 mg.
2. Telinject, Agua Dulce, California, USA.
3. Micro-hematocrit tubes, Clay Adams, Parsippany, New Jersey, USA.
4. Miltex Surgical, York, Pennsylvania, USA.
5. School of Veterinary Medicine, University of California, Davis, California, USA.

hand-held global position system[6] and these positions were plotted on the general reserve map using a proprietary software program created by Williams.[7]

In September 2003, Jolly recorded fur condition in 357 individuals from 30 troops in the 4 different areas of the forest. In September 2004, Jolly again recorded fur condition in 431 individuals from 34 troops in the same four areas of the forest. General ringtailed lemur census methods are described in Jolly et al. (2002). Fur condition of individual lemurs was scored on a 6-point scale for body and for tail separately. Only the two worst conditions were classified as alopecia cases: lemurs with fur loss affecting more than 50% of either body or tail, or with extremely short (probably regrown) fur on more than 50% of either body or tail. All animals were scored by one observer (Jolly).

19.2.3. Statistical Analysis

Red cell mass and plasma proteins in affected and unaffected lemurs were compared using the Mann–Whitney U-test. The prevalence of alopecia in the high-density areas of Ankoba and tourist front was compared with the lower density areas of gallery and scrub by means of two-way contingency comparisons: chi-square analysis or the Fisher exact test was used to compare proportions. The prevalences of alopecia in lemurs with and without leucaena in their home ranges were compared by means of two-way contingency comparisons: chi-square analysis or the Fisher exact test was used to compare proportions. The proportions of nursing females were compared by means of two-way contingency comparison: chi-square analysis was used to compare the proportions. When possible, odds ratios (OR) and 95% confidence intervals (CI) were calculated as a measure of the association between parameters. All statistical analyses were performed using commercial software[8] with p values <0.05 considered significant.

19.3. Results

19.3.1. Veterinary Evaluation

In all evaluated lemurs, there was no external evidence of infectious or parasitic disease. Median red cell mass and plasma proteins were significantly higher in lemurs with normal fur relative to lemurs with abnormal fur (39 vs. 33%, $U = 222$, $p = 0.001$; and 6.9 vs. 6.0 mg/dL, respectively; $U = 257$, $p = 0.004$). Histopathologic evaluation of skin biopsies collected in March 2003 from lemurs with both normal and abnormal fur condition revealed no evidence of infectious, parasitic, or inflammatory disease and hair follicles were in the active anagen phase of growth.

---

6. Magellan SporTrak, Alexandria, Virginia, USA.
7. G.W. Williams, http://fontforge.sourceforge.net (1995).
8. SPSS, Chicago, Illinois, USA.

Histopathologic evaluation of skin biopsies collected in September 2004 showed no evidence of infectious, parasitic, or inflammatory disease. However, hair follicles were in the telogen and catagen phase, consistent with follicular arrest.

19.3.2. Mapping of Lemur Fur Condition

In 2003, 24% (58 of 240) of lemurs in the high-density zones of Ankoba and tourist front and 0% (0 of 118) in the lower density gallery and scrub zones had alopecia; the prevalances were significantly different ($p < 0.001$) (Table 19.1). Reserve wide, 38% (52 of 135) of lemurs with leucaena in their home range had alopecia, significantly different than 3% (6 of 223) prevalence of alopecia seen in lemurs without leucaena in their home range ($p < 0.001$). The crucial comparison is within the high-density areas. There, alopecia was seen in 42% (52 of 125) of lemurs with leucaena in their home range, significantly different than the 5% (6 of 115) prevalence of alopecia recorded in lemurs without exposure to leucaena ($p < 0.001$). The two high-density areas show similar proportions of leucaena-exposed and non-leucaena-exposed animals, even though they are different basic forest types: Ankoba as 50-year-old second-growth, adjacent to a vegetable garden; tourist front as ancient gallery forest next to a tourist complex (Table 19.2). The other forest types, gallery and scrub, are both spared, though offering widely different natural food resources.

Reserve-wide, 28% (22 of 78) of adult females exposed to leucaena were nursing infants, significantly different than the 54% (57 of 106) of unexposed adult females found to be nursing infants ($p < 0.001$). In the high-density zones, 28% (20 of 72) of adult females exposed to leucaena were nursing infants, significantly different than the 58% (30 of 52) of unexposed adult females found to be nursing infants ($p < 0.001$).

In 2004, 10% (31 of 325) of lemurs in the high-density zones and 2% (2 of 106) in the lower density zones had alopecia; the prevalences were significantly

TABLE 19.1. Alopecia (bald or short-furred) in *L. catta* by population density and access to *Leucaena leucocephala*, Berenty, Madagascar (September 2003 and 2004).

Density Year	Leucaena access			No leucaena access		
	Troops (no.)	Lemurs (no.)	Alopecia (no.)	Troops (no.)	Lemurs (no.)	Alopecia (no.)
2003						
High	9	125	52[a]	8	115	6[a]
Low	1	10	0	12	108	0
Total	10	135	52[b]	20	223	6[b]
2004						
High	11	150	31[c]	10	175	0[c]
Low	1	7	2[d]	12	99	0[d]
Total	12	157	33[e]	22	274	0[e]

[a-e] Values with superscipts within columns differ significantly ($p < 0.001$).
Note: Only high-density 2004 censuses are complete for troops in a region. Other counts are of those troops that could be found.

TABLE 19.2. Prevalence of alopecia (bald or short-furred *L. catta*) by habitat, year, and access to *Leucaena leucocephala* (September 2003 and 2004).

		Leucaena access			No leucaena access		
Habitar (yr)	Density (no./ha)	Troops (no.)	Lemurs (no.)	Alopecia (%)	Troops (no.)	Lemurs (no.)	Alopecia (%)
Ankoba (2003)	500	5	72	39	4	55	4
Ankoba (2004)	500	7	97	16	7	116	0
Tourist (2003)	500	4	53	45	4	60	7
Tourist (2004)	500	4	53	28	3	59	0
Gallery (2003)	250	1	10	0	6	56	0
Gallery (2004)	250	1	7	29	6	55	0
Scrub (2003)	100	0	0	0	6	52	0
Scrub (2004)	100	0	0	0	6	44	0

Note: Only high-density 2004 censuses are complete for troops in a region. Other counts are of those troops that could be found.

different ($p = 0.012$). Reserve wide, 21% (33 of 157) of lemurs with leucaena in their home range had alopecia, significantly different than the 0% (0 of 274) prevalence of alopecia seen in lemurs without leucaena in their home range ($p < 0.001$). Again the crucial comparison is within the high-density areas. Alopecia was seen in 21% (31 of 150) of lemurs with leucaena in their home range, significantly different than the 0% (0 of 175) prevalence of alopecia recorded in lemurs without exposure to leucaena ($p < 0.001$). (Tables 19.1 and 19.2).

Reserve-wide, 23% (17 of 74) of adult females exposed to leucaena were nursing infants, significantly different than the 39% (51 of 130) of unexposed adult females found to be nursing infants ($p = 0.027$). In the high-density zones, 23% (16 of 69) of adult females exposed to leucaena were nursing infants, significantly different than the 43% (35 of 82) of unexposed adult females found to be nursing infants ($p = 0.019$).

19.4. Discussion

Reserve-wide, the prevalence of severe alopecia was 16% in 2003 and 8% in 2004. The difference between years may reflect the usual year-to-year differences in diet in an erratic climate. In 2003, the reserve-wide prevalence of alopecia in lemurs with leucaena in their home range was 39%, significantly greater than the prevalence of 3% in lemurs without leucaena in their home range. In 2004, alopecia was seen only in lemurs with leucaena in their home range.

Adjacent troops could be very different in fur condition: T1A in 2004 would confront with T2, looking like a group of third-world famine posters ranged against an American football team. Differing fur conditions between adjacent troops is a clear confirmation of the importance of access to different resources on a troop-by-troop scale. These microdifferences also seem to confirm the importance of highly localized resources like leucaena groves in the diet, rather than more widespread factors such as malnutrition.

This raises the question whether leucaena is the likely sole cause of bald lemur syndrome. If the condition is seen in even a few lemurs adjacent to the troops with leucaena in their core ranges, how did these lemurs get it? One possibility is raiding across frontiers to the next troop's range. Feeding ranges in ringtailed lemurs overlap extensively, so the classification of being with or without leucaena cannot be absolute (Pride, this volume). Another possibility is that those particulular animals were suffering from general malnutrition. It is likely that there is a complex relationship between the proportion of natural foods in the diet, seasonal food stress, and recourse to introduced rather than forest foods. There may also be a vicious circle as leucaena affects gut function, thus in turn increasing malnutrition and stress.

Seasonal appearance and recovery of the syndrome follows the proportion of leucaena in the diet. Soma (this volume) reports that leucaena leaves, pods, and flowers were eaten by CX troop for 35–45% of the animals' feeding time in May–August, 30% in September, 10% in October, and minimally in December. For much of the year they were above the toxic limit of 10% of diet for non-ungulates. Thus leucaena was a staple throughout the dry season, dropping off with the flush of wet season growth of the natural forest in late October. Soma's troop was one of the less affected leucaena troops in 2003, which suggests the others may depend on it even more.

Many questions remain. First what are the physiological consequences of mimosine in a primate? Second, why do some individuals become severely bald and thin but others in the same troops seem unaffected? Preliminary results suggest that these condition differences do not reflect simple dominance rank (Tew, 2004) Third, what is the effect on infants?. Mimosine consumption peaks during females' gestation. Exposure of females to leucaena appears to negatively impact infant recruitment. In both 2003 and 2004, the percentage of adult females nursing infants was significantly lower in females with leucaena in their home range. Part of the reason could be simple lack of maternal fur. Infants usually cling to mothers' bellies during their first 2 weeks, suckling frequently. Two apparently normal newborns climbed up to the remaining fur tufts on their bald mother's backs from birth, but survived less than 2 days (Tew, Ichino, obs.). Finally, why does there seem to be less effect on the reserve's other folivorous primates, *Eulemur fulvus*, *Propithecus verreauxi*, and *Lepilemur leucopus*?

In 2004, the Berenty Reserve management removed stands of leucaena from the ranges of half of the affected troops, although an unusually rainy wet season has promoted regrowth. We will continue to monitor whether leucaena removal results in overall improvement of the animals' condition or whether the loss of a protein source in the dry season instead decreases their overall nutrition, along with changes in ranging and intertroop competition.

To the best of our knowledge, this is the first report of leucaena toxicity in a wildlife population. However, given the global distribution of leucaena, the problem may already be widespread among wildlife living in forests adjacent to leucaena trees introduced for agroforestry. Leucaena toxicosis is likely to become even more important as natural forests fragment, and the fragments come under pressure, so that browsing animals turn to feeding on planted trees. The authors hope that publication of this report will help scientists to recognize potential effects of leucaena toxicity in other animals worldwide.

Acknowledgments. We are grateful to the de Heaulme family for their continued support of Berenty Reserve and its visiting scientists. We appreciate the assistance of Drs. Linda Munson and Verena Affolter of the University of California at Davis, School of Veterinary Medicine.

References

Brewbaker, J. (1987). *Leucaena*: A multipurpose tree genus for tropical agroforestry. In: Steppler, H. A., and Nair, P. K. (eds.), *Agroforestry: A Decade of Development*. Nairobi, ICRAF.

Hughes, C. E. (1998). *Leucaena, A Genetic Resources Handbook*. Oxford, Oxford Forestry Institute.

Jolly, A., Dobson, A., et al. (2002). Demography of *Lemur catta* at Berenty Reserve, Madagascar: Effects of troop size, habitat and rainfall. *Int. J. Primatol.* 23:327–354.

Koyama, N., Nakamichi, M., et al. (2002). The population and social dynamics changes in ring-tailed lemur troops at Berenty, Madagascar between 1989–1999. *Primates* 43:291–314.

Pride, R. E. (2003). The socio-endocrinology of group size in *Lemur catta*. *Ecology and Evolutionary Biology*. PhD Dissertation, Princeton University. Princeton, N. J., Princeton University.

Rasamimanana, H., Ratovonirina, R., et al. (2000). Storm damage at Berenty Reserve. *Lemur News* 5:7–8.

Sethi, P., and Kulkarmi, P. R. (1995). *Leucaena leucocephala*: A nutrition profile. *Food Nutr. Bull.* 16(3).

Shelton, H. M. (1994). *Leucaena leucocephala:* The most widely used forage tree. In: Gutteridge, R. C., and Shelton, H. M. (eds.), *Forage Tree Legumes in Tropical Agriculture*. Wallingford, CAB International.

Simmen, B., Hladik, A., et al. (2003). Food intake and dieteary overlap in *Lemur catta* and *Propithecus verreauxi* and introduced *Eulemur fulvus* at Berenty, Madagascar. *Int. J. Primatol.* 24(5):949–968.

Tew, A. (2004). Social relationships and social correlates of hair loss in a troop of ring-tailed lemurs at Berenty Reserve, Madagascar. *Anthropology*. MSc Dissertation, Oxford, UK, Oxford Brookes.

20
Temporal Change in Tooth Size Among Ringtailed Lemurs (*Lemur catta*) at the Beza Mahafaly Special Reserve, Madagascar: Effects of an Environmental Fluctuation

FRANK P. CUOZZO AND MICHELLE L. SAUTHER

20.1. Introduction

Ringtailed lemurs (*Lemur catta*) are among the best-known Malagasy primates (Jolly et al., 2004), with nearly four decades of continuous field research, beginning with Jolly's (1966) seminal work (see Sauther et al., 1999 for a review of ringtailed lemur research). In this way, ringtailed lemurs are comparable to several anthropoid primates, including baboons and chimpanzees, both of which have been intensively studied in the wild since the 1960s (e.g., Altmann, 1980; Goodall, 1986). Despite some notable differences (e.g., Wright, 1999), ringtailed lemurs, in addition to being diurnal, semiterrestrial, and omnivorous, share a number of social attributes with many Old World Monkeys (i.e., living in large, multimale female resident groups; having more than one matriline in each group) (e.g., Hladik, 1975; Sussman, 1992; Sauther et al., 1999). With their long history of study, and their similarities to anthropoid primates, information on ringtailed lemurs is especially important for broad comparisons of primate biology, including hominid paleobiology, as recently seen in discussions of fossil hominid conspecific care (e.g., Lebel and Trinkhaus, 2002; DeGusta, 2003; Cuozzo and Sauther, 2004a, in press).

Among ringtailed lemurs, several populations have been the focus of long-term study (see Sauther et al., 1999 for a review). The ringtailed lemur population at the Beza Mahafaly Special Reserve (BMSR) in southern Madagascar (23°30′S latitude, 44°40′E longitude) is one such group (e.g., Ratsirarson, 1985; Sauther, 1989, 1991, 1992, 1993, 1994, 1998; Sussman, 1991, 1992; Gould, 1996, 1997; Yamashita, 1998, 2000, 2003; Gould et al., 1999, 2003; Sauther et al., 1999, 2001a, 2001b, 2002, this volume; Cuozzo and Sauther, 2004a, 2004b, 2005, in press; see Sussman and Rakotozafy, 1994; Sauther et al., 1999; Gould et al., 1999, 2003; and Ratsirarson, 2003 for detailed descriptions of Beza Mahafaly). Among primates, the population of ringtailed lemurs at BMSR is rare in that detailed dental data (including sets of dental casts) are available from two different points in time (1987/1988 and 2003/2004). This type of information exists for few other primate populations (see Dennis et al., 2004; King et al., 2005; and Lawler et al., 2005 for additional examples). Of special interest, the time interval between the 1987/1988 and 2003/2004 data sets spans a severe drought that occurred in 1991/1992, which resulted in a significant population decline and eventual population rebound among ringtailed lemurs (Gould et al., 1999, 2003). This time interval also spans a nearly complete replacement of this population (Gould et al., 2003; Sauther et al., in preparation).

Recent work on ringtailed lemurs living within the reserve at Beza Mahafaly has produced detailed information on patterns of intraspecific dental variation (Sauther et al., 2001a; Cuozzo et al., 2004), dental health (Sauther et al., 2002; Cuozzo and Sauther, 2004a, 2004b, 2005), and tooth use (Yamashita, 1998, 2003; Cuozzo and Sauther, 2004a, 2004b, 2005). Because of the detailed ecology, life history, and habitat information available for this group of ringtailed lemurs (Sauther et al., 1999), this population provides a rare opportunity to fully explore questions relating to dental variation, dental health, life history, ecology, and evolution. Subsequently, this population also provides an opportunity to investigate examples of microevolution ("allochronic" studies [e.g., Hendry and Kinnison, 1999]) as it relates to short-term, environmental changes such as the drought that severely affected southern Madagascar in 1991 and 1992 (e.g., Sauther, 1998; Gould et al., 1999, 2003; Jolly, 2004).

20.1.1. *Research Background*

20.1.1.1. Microevolution and Short-term Environmental Perturbation

The term "microevolution" refers to changes within populations or species (Hendry and Kinnison, 1999). Recent studies of micro- or "contemporary" evolution (i.e., observable evolution in heritable traits across a limited number of generations [e.g., Stockwell et al., 2003]) suggest that short-term environmental perturbations (e.g., droughts) can have rapid and measurable effects on living vertebrate populations (see reviews in Hendry and Kinnison, 1999, and Stockwell et al., 2003). The most famous example is from longitudinal studies of the ground finches of the Galápagos Islands (genus *Geospiza*), where research indicates a strong relationship between changes in beak size and drought, with larger beak size being selected as a response to the dominance of harder seeds that remained following the drought (e.g., Grant, 1985; Grant and Grant, 1995). There are numerous other case studies of contemporary evolution, for example those of introduced populations of *Anolis* lizards in the Caribbean (e.g., Losos et al., 2001). Recent reviews of contemporary evolution in response to environmental perturbations among fish, birds, and some mammals, including those of an anthropogenic nature, provide a number of additional examples (Kinnison and Hendry, 2001; Stockwell et al., 2003). Although contemporary evolution resulting from natural selection has been documented among modern humans (see review in Endler, 1986), to our knowledge this has been addressed for few extant nonhuman primates (DeGusta et al., 2003; see review in Endler, 1986).

20.1.1.2. Dental Change Over Time

Studies of dental change over time are important for a number of questions in evolutionary biology. Because mammalian tooth size is highly heritable (e.g., Gingerich, 1974b; Hillson, 1986; Hlusko et al., 2002) and crown size does not change after tooth formation (e.g., Swindler, 2002; DeGusta et al., 2003) except by attrition or pathology (e.g., Perzigian, 1975; DeGusta et al., 2003), changes in tooth size in the fossil record of primates and other mammals have successfully been used to address questions of phylogeny, adaptation, and climate change (e.g., Gingerich, 1974a, 1979a, 1979b, 1985, 1994; Gingerich and Schoeninger, 1977; Bown et al., 1994; Cuozzo, 2002). For example, among early Eocene mammals (e.g., the condylarthran genus *Hyopsodus*), changes in tooth size show a strong correlation with temperature fluctuations and their corresponding biostratigraphic units (e.g., Gingerich, 1974a; Bown et al., 1994; Cuozzo, 2002). This suggests that mammalian tooth size can reflect biological responses (e.g., an increase in body size) to a changing environment over time. Also, several studies (e.g., Kurten, 1957; Van Valen, 1963; Marcus, 1969) have used changes in tooth size across age cohorts in assemblages of fossil mammals (including primates) to document examples of natural selection. However, there are few empirical studies that demonstrate microevolution for dental characteristics in extant mammals (see review in Endler, 1986). One of the few examples comes from work on several modern human populations, where selection for larger tooth size occurred in response to intense tooth wear and severe crown attrition (Greene et al., 1967; Perzigian, 1975). In addition, a recent study of tooth size in an extant howler monkey population (*Alouatta palliata*) demonstrated that individuals with smaller molars had significantly decreased fitness, thereby suggesting selection for larger teeth (DeGusta et al., 2003). Given its high heritability, investigating possible changes in tooth size in a single population of primates—especially when the temporal interval spans a severe environmental change (i.e., drought)—allows us to explore the impact, and possible selective pressure (i.e., directional selection) of ecological changes and subsequent behavioral modifications (i.e., increased competition and/or aggression) on contemporary evolution.

20.1.1.3. Drought and Patterns of Mortality at Beza Mahafaly Special Reserve

Southern Madagascar underwent a severe drought in the early 1990s (e.g., Sauther, 1998; Gould et al., 1999, 2003) that, among other results, led to a widespread human famine (Jolly, 2004). During and following this drought, the ringtailed lemurs at Beza Mahafaly experienced a significant population decline, with the adult population dropping from 85 individuals in early 1991 to 51 in 1994 (Gould et al., 1999). By 2001, the adult population (n = 61) had recovered to near that of 1987 (n = 65), although still below that of the pre-drought peak in 1991 (Gould et al., 2003). In addition, as of 2001, this population had undergone nearly a complete replacement since 1987 (Gould et al., 2003; Sauther et al., in preparation). This drought had a particularly severe impact on adult females, infants, and juveniles (when compared with predrought years), with 21% of all adult females, 80% of all infants, and 57% of juveniles in three focal groups, having died during the 6 months from September 1992 through March 1993 (Gould et al., 1999). In addition, female mortality increased to 29% in 1993/1994, the year following the drought (Gould et al., 1999). Also of note, all females that died in 1992/93 had infants and were lactating (Gould et al., 1999).

20.1.1.4. Enamel Thickness, the Drought, and Food Availability

Despite this species possessing among the most thin enamel of all extant primates (e.g., Shellis, 1998; Martin et al., 2003; Godfrey et al., 2005), ringtailed lemurs living in and around areas of gallery forest across southern Madagascar have a diet dominated by tough, hard fruit of the tamarind tree, *Tamarindus indica* (e.g., Jolly, 1966; Sauther, 1998; Yamashita, 2000, 2003, in preparation; Simmen et al., this volume). Although not a perfect relationship (e.g., Martin et al., 2003), primate enamel thickness generally exhibits a strong correspondence with diet (e.g., Shellis et al., 1998). *T. indica* reproduces asynchronously (Sauther, 1998). Tamarind fruit is therefore available year round, hence being a ringtailed lemur keystone food source (Jolly, 1966; Sauther, 1998; Sauther et al., 2002; Cuozzo and Sauther 2004a; Simmen et al., this volume). In addition, during the dry season, tamarind is the primary food used at Beza Mahafaly (Sauther, 1998; Simmen et al., this volume). Tamarind fruit is also larger than all other foods used by the Beza Mahafaly ringtailed lemurs (e.g., Sauther, 1992), and thereby presents a very challenging food source (Figures 20.1a, 20.1b, and 20.2; see also Figure 8.1 in Mertl-Millhollen et al., this volume).

Ringtailed lemurs primarily process tamarind pods with their postcanine teeth (e.g., Sauther et al., 2002; Yamashita, 2003; Cuozzo and Sauther, 2004a, 2004b, 2005, in press), and it is this region of the mouth where severe attrition and tooth loss most often occur (e.g., Cuozzo and Sauther, 2004a, 2004b, 2005, in press). Although hard tamarind seeds are passed through the digestive system primarily unscathed (e.g., Yamashita, 2000; Simmen et al., this volume), accessing these seeds takes a severe toll on ringtailed lemur teeth. The outer casing of ripe tamarind pods is both hard and tough, in fact the hardest and toughest of all foods

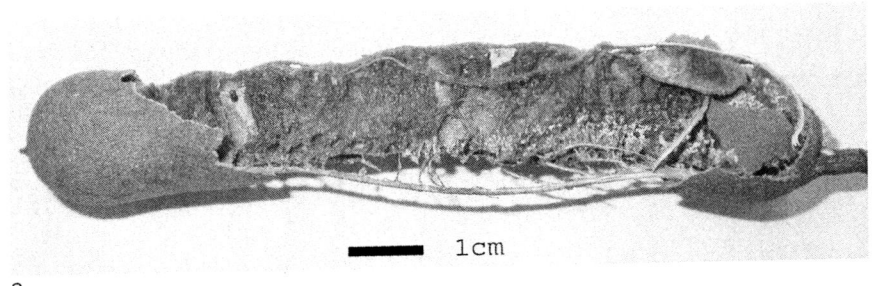

a

FIGURE 20.1. (a) A tamarind pod (*Tamarindus indica*) from Beza Mahafaly, with outer casing intentionally removed to show enclosed fruit (scale bar = 1 cm).

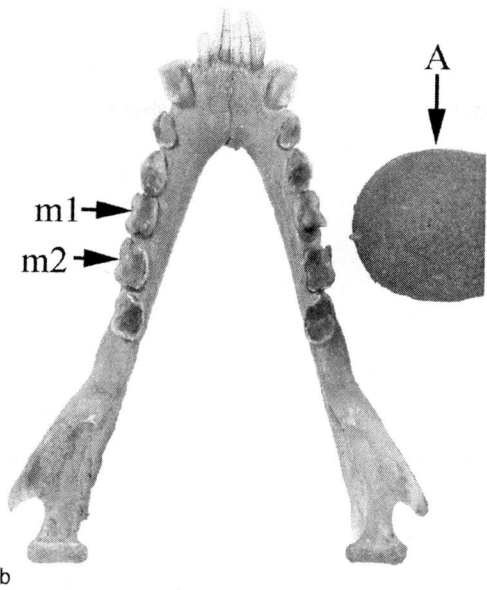

FIGURE 20.1. (*Continued*) (b) Illustration of the size of a tamarind pod (A) relative to tooth and mandible size in a ringtailed lemur skeletal specimen (BMOC 67).

consumed by ringtailed lemurs (Yamashita, 2000, in preparation; Cuozzo and Sauther, in press). When processing these pods, the outer casing is initially broken open in the region of the first and second molars, as well as the adjacent premolars, with the pod often being bitten down upon several times in order to initiate crack formation (Cuozzo and Sauther, 2005, in press). In addition, extraction of the seeds from the pod requires additional tooth use, with the hard, tough outer casing of the pod and the tough internal fibers (Figures 20.1a and 20.2) continually making contact with the surface of the teeth (Cuozzo and Sauther, 2004b, 2005, in press). Although tooth wear is a complex process resulting from the interaction of numerous variables (e.g., Maas and Dumont, 1999), the excessive amount of tooth wear and subsequent tooth loss seen among the ringtailed lemurs at Beza Mahafaly (Figures 20.3, 20.4, 20.5a, and 20.5b) is largely caused by processing the hard, tough pods of the tamarind tree (Cuozzo and Sauther, 2004a, 2004b, 2005, in press).

Despite the large size of tamarind pods (Figures 20.1a, 20.1b, and 20.2), ringtailed lemurs have very small maxillary first molars (relative to skull and palate length) when compared to other living and extinct lemurs (Godfrey et al., 2002). Therefore, ringtailed lemurs have molars with a small food processing area relative to the size of their keystone food (Figure 20.1b). When combined with thin enamel, early relative first molar eruption among lemurids (e.g., Eaglen, 1985; Godfrey et al., 2001, 2004), and a diet dominated by a hard, tough keystone food

FIGURE 20.2. A partially processed tamarind pod (white arrow) being held by a ringtailed lemur. Note the relative size of this partial pod, as well as the tough internal fibers that remain after initial processing. Photo, Michelle Sauther.

(tamarind fruit), the small size of the first molars (with a small food processing area and limited enamel surface) likely contributes to their high frequency of severe wear (see Figures 20.3, 20.4, 20.5a, and 20.5b) and eventual antemortem loss (Cuozzo and Sauther, 2004a, 2004b, 2005, in press). Given the large size of tamarind pods relative to the size of ringtailed lemur mouths and teeth (see Figures 20.1b and 20.2), larger molars—with a larger surface area and increased processing platform—would be beneficial during mastication (e.g., Perzigian, 1975). As discussed by Janis and Fortelius (1988) and Lucas (2004), increased tooth size is one way (along with increased enamel thickness) to increase the functional longevity of teeth. With the intense nutritional stress due to resource scarcity during the drought, the ability to effectively process this limited food resource could be a selective factor. Hence, we tested for changes in dental size.

20. Temporal Change in Tooth Size Among *Lemur catta* 349

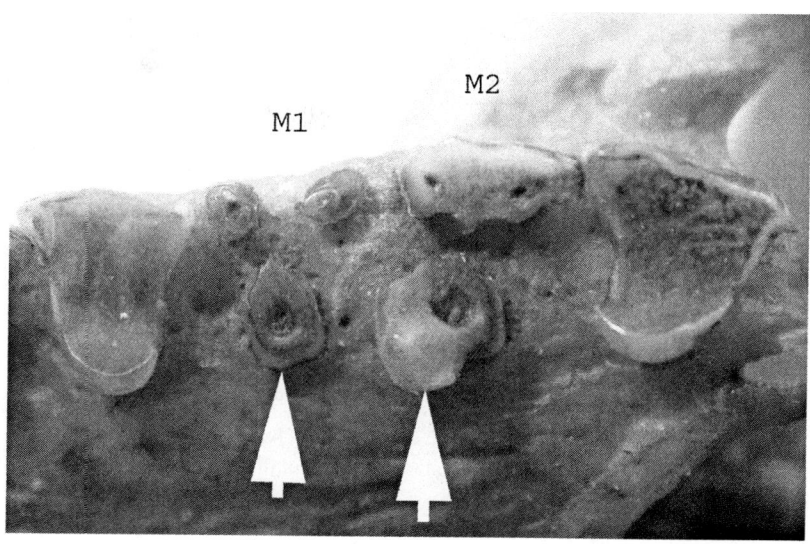

FIGURE 20.3. Tooth wear in a ringtailed lemur skeletal specimen from the Beza Mahafaly Osteological Collection (BMOC 67). Note the extensive wear and damage to M1 and M2, with white arrows marking the remaining lingual areas of the tooth crowns.

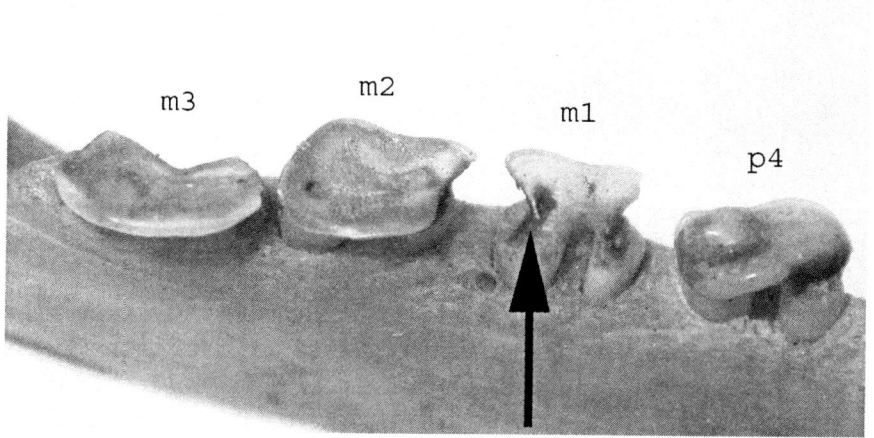

FIGURE 20.4. Tooth wear in a ringtailed lemur skeletal specimen from the Beza Mahafaly Osteological Collection (BMOC 70). Note the extensive wear and damage to m1, with black arrow marking the damaged lingual portion of the crown.

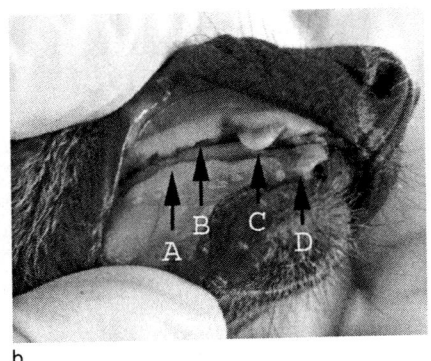

FIGURE 20.5. (a) Unworn teeth in a living 3-year-old ringtailed lemur (Yellow 187). A, Unworn right maxillary canine. B, Unworn right mandibular canine (toothcomb). (b) Severe wear and tooth loss in a living ringtailed lemur (Blue 132). A, Right mandibular gumline with no teeth present. B, Right maxillary P2 worn to the gumline, with only worn roots remaining. C, Heavily worn right maxillary canine. D, Worn right mandibular canine (toothcomb).

20.1.2. Research Questions

As seen in Hendry's (2005) recent discussion of the power of natural selection, determining the strength of natural selection can be elusive (see Lawler et al., 2005 for a study of the strength of selection in an extant primate population). Therefore, our primary goal in this study is to investigate the possible role of a severe drought as a selective pressure within a living population of ringtailed lemurs. Here we compare tooth size between the 1987/1988 and 2003/2004 Beza Mahafaly ringtailed lemur samples in order to address the following questions:

1. Did particular tooth positions exhibit a size increase in the reserve population between 1987/1988 and 2003/2004, following the drought? If so, what ecological, behavioral, and/or mechanical factors would lead to changes in tooth size following the drought?
2. Did specific tooth positions experience a size increase in either males or females selectively between 1987/1988 and 2003/2004, following the drought?

Answers to these questions provide a baseline for future work, as the ringtailed lemur population at Beza Mahafaly is currently the focus of a longitudinal study of ecology and dental life history (e.g., Cuozzo and Sauther, 2004a, 2004b, 2005, in press; Sauther et al., in preparation). In addition, this project provides a direct investigation of the role of ecological change as a selective force in mammalian evolution, which is less common in studies of contemporary evolution, as seen in Endler's (1986) compilation of studies of natural selection in wild populations (see summaries of more recent work in Hendry and Kinnison, 1999, and Stockwell et al., 2003).

20.2. Materials and Methods

In 1987/1988 and 2003/2004, dental data, including complete sets of dental impressions, were collected from sedated lemurs at Beza Mahafaly. Methods of impression collection and cast production for the 1987/1988 data have previously been described in detail (Sauther et al., 2001a). For the 2003/2004 sample, impressions were made using custom-built impression trays and Presidents Jet Regular Body polyvinalsiloxane impression material. Casts were made from Coecal™ Type III dental stone, similar to the dental stone used in 1987/1988 (Sauther et al., 2001a). The sample size from 1987/1988 includes 45 individuals, with 39 adults; the 2003/2004 sample consists of 83 adults among the 92 individuals captured and studied. No lemurs from 2003/2004 had dental data collected in 1987/1988 as only one individual included among the 2003/2004 data was alive (as a subadult) in 1987/1988 (Sauther and Cuozzo unpublished data).

Metric data were collected from the casts of both data sets using Fowler digital needle-point calipers measured to the nearest 0.01 mm. The one exception is toothcomb breadth from 2003/2004, which was collected from sedated lemurs in the field, using dial calipers measured to the nearest 0.05 mm. The same individual (F.C.) collected all measurements, thereby eliminating the potential for interobserver error. Due to variations in cast quality and individual dental pathology (e.g., dental abscesses, tooth wear, tooth loss) sample sizes vary for each variable. Measurements collected include (1) maxillary toothrow length (measured from the anterior margin of the canine to the distal border of M^3), (2) mandibular toothrow length (measured from the anterior margin of P_2 [the mandibular canine in *Lemur catta* is part of the toothcomb, as in most strepsirrhine primates] to the distal border of M_3), (3) palate breadth (measured from the lateral borders of M^3), (4) toothcomb breadth (measured from the lateral borders of the mandibular canines), (5) P_2 length (measured mesiodistally at the base of the tooth), (6) lengths of M^1 and M^2 (the maximum mesiodistal length measured across the lingual cusps), and (7) lengths of M_1 and M_2 (measured mesiodistally from the anterior margin of the trigonid to the distal border of the talonid). Mesiodistal tooth lengths were selected as indicators of overall tooth size due to their limited metric variability when compared to buccolingual width in this population of lemurs (Sauther et al., 2001), as well as in mammals in general (e.g., Gingerich, 1974b).

As crown size does not change after tooth formation (e.g., Swindler, 2002), except as a result of attrition or pathology (e.g., Perzigian, 1975; DeGusta et al., 2003), measurements from the permanent teeth present in subadults (maxillary and mandibular first and second molar lengths, and toothcomb breadth) are included in the data set. The presence of either deciduous or partially erupted adult maxillary canines (as well as adult P_2) affects toothrow length. Hence, this measure was not collected for subadults. Metric data for the two overall temporal samples were compared in order to test whether any measures increased following the drought. Sex-specific metric data were also compared between 1987/1988 and 2003/2004, in order to test whether males or females exhibited a change in tooth size across the temporal interval. We investigated directional change in lengths of the first and second maxillary and mandibular molars (which are central in processing tamarind pods), toothrow length (which is in part a product of tooth size), toothcomb breadth (as toothcombs are used in food acquisition [e.g., Sauther et al., 2002; Yamashita, 2003]), and caniniform P_2 (which is involved in food acquisition [e.g., Sauther et al., 2002] and sometimes food processing). In order to evaluate whether changes in tooth size were a function of change in overall cranial size, we also examined size change in palate breadth (measured at M^3), as this measure provides a strong indicator of skull width (and therefore skull size). All comparisons were tested for significant differences using unpaired student's t-tests ($p = 0.05$), and were conducted using Statview statistical and data analysis software (Haycock et al., 1992).

Following standards outlined by the U.S. CITES Management Authority (a unit of the U.S. Fish and Wildlife Service), as well as the Institutional Animal Care and Use Committee (IACUC) of the University of Colorado, each member of the research team wore protective covering such as surgical masks and gloves during initial data collection, in order to preclude disease transfer while handling lemurs. Furthermore, all methods and materials received approval by and followed standard animal handling guidelines (University of Colorado IACUC).

20.3. Results

Metric data for the 1987/1988 and 2003/2004 overall samples are compared in Table 20.1, and sex-specific temporal comparisons are presented in Tables 20.2 and 20.3. Of the nine variables studied, significant ($p < 0.01$) increases occurred in P_2, M_1, and M_2 length in the overall population. Somewhat unexpectedly, M^2 showed a significant ($p = 0.0325$) decrease in length between 1987/1988 and 2003/2004 in the population. Among males, as in the overall sample, M_1 and M_2 length showed significant ($p < 0.05$) increases. However, in contrast to the overall population, neither P_2 nor M^2 length ($p > 0.10$) displayed a significant change in size. Females exhibited a pattern similar to the overall and male samples, with a significant ($p = 0.0158$) increase in the length of M_2. Females, in congruence with the overall sample, experienced a significant ($p = 0.0248$) increase in P_2 length. In contrast, neither M_1 length (although larger in 2003/04 than 1987/88 [$p = 0.2009$] as with both the overall and male samples) nor M^2 length (shorter in 2003/2004 [$p = 0.0673$] as in the overall sample) exhibited a significant change. Palate breadth did not exhibit a significant temporal size change in either the overall population ($p = 0.6864$), or in the sex-specific samples (males, $p = 0.2178$;

TABLE 20.1. Ringtailed lemur tooth size compared between 1987/1988 and 2003/2004.

Variable	1987/1988[a]				2003/2004[d]				p value[e]
	n	Mean[b]	STD	CV[c]	n	Mean[b]	STD	CV[c]	
Maxillary toothrow length	23	35.13	0.94	2.70	42	35.20	0.96	2.70	0.7763
Mandibular toothrow length	21	31.07	0.62	2.00	47	30.74	0.79	2.60	0.0935
Palate breadth at M^3	21	26.28	0.77	2.90	32	26.20	0.65	2.50	0.6864
P_2 length	16	4.64	0.20	4.20	42	**4.86**	0.24	4.90	**0.0081**
Toothcomb breadth	20	7.26	0.25	3.40	81	7.15	0.28	4.00	0.1130
M^1 length	27	4.87	0.23	4.80	37	4.88	0.19	3.90	0.9393
M_1 length	22	5.04	0.18	3.60	34	**5.19**	0.20	3.90	**0.0071**
M^2 length	26	**5.26**	0.23	4.30	39	5.16	0.13	2.50	**0.0325**
M_2 length	26	5.39	0.22	4.00	45	**5.57**	0.16	2.90	**0.0002**

[a] Values presented for 1987/1988 differ slightly from those published in Table 3 of Sauther et al. (2001a) for this population due to the addition of data from (1) several adults and (2) the permanent teeth of non-adults, not previously included.
[b] All means in mm.
[c] CV = standard deviation ÷ mean × 100.
[d] Boldfaced values indicate a significantly different mean (p < 0.05).
[e] p value for t-tests of means between 1987/1988 and 2003/2004.

TABLE 20.2. Ringtailed lemur tooth size compared for males between 1987/1988 and 2003/2004.

Variable	1987/1988[a]				2003/2004[d]				p value[e]
	n	Mean[b]	STD	CV[c]	n	Mean[b]	STD	CV[c]	
Maxillary toothrow length	13	35.03	1.11	3.20	22	35.36	0.99	2.80	0.3690
Mandibular toothrow length	13	31.14	0.66	2.10	23	30.92	0.73	2.40	0.3955
Palate breadth at M^3	12	26.50	0.71	2.70	18	26.16	0.72	2.80	0.2178
P_2 length	10	4.72	0.21	4.50	21	4.87	0.25	5.20	0.1025
Toothcomb breadth	10	7.26	0.17	2.40	39	7.12	0.28	3.90	0.1459
M^1 length	11	4.83	0.21	4.20	16	4.86	0.21	4.40	0.6329
M_1 length	11	5.00	0.20	4.00	13	**5.21**	0.23	4.30	**0.0211**
M^2 length	14	5.22	0.25	4.70	19	5.14	0.12	2.30	0.2029
M_2 length	14	5.37	0.20	3.70	21	**5.56**	0.18	3.30	**0.0082**

[a] Values presented for 1987/1988 differ slightly from those published in Table 3 of Sauther et al. (2001a) for this population due to the addition of data from (1) several adults and (2) the permanent teeth of non-adults, not previously included.
[b] All means in mm.
[c] CV = standard deviation ÷ mean × 100.
[d] Boldfaced values indicate a significantly different (p < 0.05).
[e] p value for t-tests for means between 1987/1988 and 2003/2004.

TABLE 20.3. Ringtailed lemur tooth size compared for females between 1987/1988 and 2003/2004.

Variable	1987/1988[a]				2003/2004[d]				p value[e]
	n	Mean[b]	STD	CV[c]	n	Mean[b]	STD	CV[c]	
Maxillary toothrow length	10	35.26	0.72	2.00	20	35.02	0.91	2.60	0.4851
Mandibular toothrow length	8	30.98	0.59	1.90	24	30.57	0.82	2.70	0.2061
Palate breadth at M^3	9	25.99	0.79	3.00	14	26.25	0.58	2.20	0.3665
P_2 length	6	4.61	0.16	3.60	21	**4.84**	0.23	4.70	**0.0248**
Toothcomb breadth	10	7.24	0.31	4.30	42	7.19	0.29	4.00	0.4046
M^1 length	16	4.91	0.25	5.10	21	4.89	0.17	3.50	0.7800
M_1 length	11	5.09	0.16	3.20	21	5.18	0.19	3.70	0.2009
M^2 length	12	5.29	0.20	3.80	20	5.18	0.14	2.70	0.0673
M_2 length	12	5.42	0.24	4.40	24	**5.58**	0.15	2.60	**0.0158**

[a] Values presented for 1987/1988 differ slightly from those published in Table 3 of Sauther et al. (2001a) for this population due to the addition of data from (1) several adults and (2) the permanent teeth of non-adults, not previously included.
[b] All means in mm.
[c] CV = standard deviation ÷ mean × 100.
[d] Boldfaced values indicate a significantly different mean ($p < 0.05$).
[e] p value for t-tests for means between 1987/1988 and 2003/2004.

females $p = 0.2178$). This indicates that changes in tooth size were not a function of overall skull size change.

20.4. Discussion

20.4.1. Ontogeny, Weaning, and Increased Tooth Size

What factors would lead to increased tooth size in this population after a drought? Eaglen (1985) and Godfrey et al. (2001) discussed the role of natural selection as it relates to dental development, weaning, and ecology, specifically in terms of lemur biology and evolution. Ringtailed lemurs exhibit a pattern of rapid growth and development, with dramatic increases in body mass occurring during the first seven months of life (e.g., Pereira, 1993). This pattern of growth accelerates in the fourth month (Pereira, 1993), which roughly corresponds with both eruption of the adult first molars and weaning (e.g., Eaglen, 1985; Godfrey et al., 2001). Therefore, M_1 is important for the transition to an adult diet (as is true for primates in general [e.g., Godfrey et al., 2001]). As M_2 erupts in month seven (Eaglen, 1985), each of the two molar positions (M_1 and M_2) that experienced a size increase in the Beza Mahafaly ringtailed lemur population are present during the period of rapid development and body mass increase described by Pereira (1993). Of importance for our discussion, this period of rapid growth and development corresponds to the time during 1992/1993 when infants experienced 80% mortality (Gould et al., 1999). Given the high mortality of infants at this time (only six of 30 infants survived [Gould et al., 1999]), any trait that produced an advantage when processing keystone foods would have aided survival during this period. We argue that larger molars would be such a trait. Considering that notable wear of adult M_1 is present in subadults, and that marked wear is observable on deciduous teeth (e.g., Cuozzo and Sauther, 2004a, 2005, in press), the importance of larger permanent first and second mandibular molars becomes apparent (see Janis and Fortelius, 1988; Lucas, 2004). The amount of wear on subadult and deciduous teeth also indicates the importance of processing adult foods (dominated by tamarind fruit) by juveniles. As juveniles also experienced high mortality (57%) during late 1992 and early 1993 (Gould et al., 1999), larger teeth would have contributed to their likelihood of survival during this period of intense resource stress. Because infants and juveniles would not only have been competing for resources with members of their cohorts, but also with adults, any slight advantage in processing fallback foods (such as larger teeth) would have been especially beneficial at a time when younger individuals are under tremendous nutritional pressure.

20.4.2. Socioecology and Increased Tooth Size

As outlined earlier, larger teeth would provide a food processing advantage for ringtailed lemurs, given the dominance of a relatively large, tough, and hard keystone food. However, the mechanical advantage of larger teeth during the drought years would also have been enhanced by ringtailed lemur socioecology. Sauther (1993) noted that ringtailed lemurs are under continual feeding stress throughout the year. As conditions during the drought severely affected this population, as seen in increased mortality (with at least one entire troop disappearing [Gould et al., 1999]) and increased exploitation of poor quality foods (e.g., Sauther, 1998; Gould et al., 1999, 2003), resource competition was likely exaggerated, thereby leading to an increase in interindividual competition for resources. As Sauther (1993) reported, agonism among ringtailed lemurs often consists of aggressive agonism surrounding resource competition. Wright (1999), in a review of the effects of drought on lemurs, noted that increased aggression and resource competition occurred among lemurs in the dry forests of southern Madagascar. This was also noted by Jolly et al. (1993) at Berenty Reserve in southeastern Madagascar, where within-group competition and displacement of lower ranking female ringtailed lemurs increased during the 1991/1992 drought. During a period of resource reduction and scarcity, products of the tamarind tree, as the primary food source, would likely have become emphasized (see Lambert et al., 2004 for a recent discussion of the role of fallback foods in primate dental evolution). As females have first choice of food (male displacement of females is rare [Sauther, 1993]), males are often left to feed on poorer quality foods, such as leaves, during the dry season when food resources are limited (Sauther, 1994; Sauther et al., 2002). This suggests that larger food processing teeth would be advantageous for males, when viewed in the socioecological context of female dominance during a time of resource limitations. Although male mortality rates are not available due to continual male migration (Gould et al., 1999), it is likely that adult males, similar to adult females, infants, and juveniles, experienced increased mortality resulting from the drought (Gould et al., 1999).

However, the advantage that larger teeth would provide is not limited to males. The reproductive pattern of ringtailed lemurs, like many other Malagasy primates (e.g., Wright, 1999), is linked to resource availability, with different stages of the reproductive cycle (i.e., mating, gestation, birth, lactation) showing a strong correspondence to availability of specific food sources (Sauther, 1998). Ringtailed lemurs, similar to other Malagasy primates (e.g., Wright, 1999), exhibit reproductive synchrony, with females at Beza Mahafaly usually giving birth in October/November (e.g., Sauther, 1998; Sauther et al., 1999). Female ringtailed lemurs are pregnant during the dry months of the austral winter, when food resources are limited (e.g., Sauther, 1998), and hence give birth and begin lactation with little or no nutrient reserves (Sauther, 1998). Although several foods are available periodically, the primary food consumed during these times of nutritional stress is tamarind fruit (e.g., Sauther, 1998; Simmen et al., this volume). This situation was likely exacerbated during the drought of 1991/1992 (Sauther, 1998). In a time of resource scarcity, larger teeth, and an improved ability to process keystone and/or fallback foods, would benefit females who, despite having feeding priority, were under tremendous nutritional stress during gestation and lactation. As females experienced very high mortality during and following the drought (21% in 1992/1993; 29% in 1993/1994) at Beza Mahafaly (e.g., Gould et al., 1999), especially among lactating females with infants, the nutritional pressure on females would have been exaggerated. Although a number of foods are available during the various stages of the reproductive cycle, only tamarind is available for long periods and throughout the year (Sauther, 1998; Simmen et al., this volume). In this context, even slight advantages, whether physical (e.g., larger teeth for processing tamarind fruit) or behavioral (e.g., dominance rank), likely played a key role in determining which individuals survived the drought, and which would successfully reproduce (Sauther et al., in preparation).

Although our primary goal in this study was to investigate the possible selective pressure of a severe drought, rather than address the power of selection, we have computed values for selection intensity in order to place our data in a broader context. Selection intensity (see discussions in Endler, 1986 and Futuyma, 1998), also known as the directional selection differential (e.g., Grant, 1985), is a value that compares the intensity of quantitative change in terms of standard deviations. In our sample, the values for selection intensity (i) for M_1 length ($i = 0.71$) and M_2 length ($i = 0.90$) indicate that these traits increased by close to one full standard deviation. In his review of estimates of selection intensity, Endler (1986) noted that studies indicating intense selection exhibited values for i ranging from one half to sometimes two full standard deviations (see review in Futuyma, 1998). These data indicate that selection for increased M_1 and M_2 length in our study is comparable to a number of other previous studies. Thus, it appears that directional selection for larger teeth has occurred at a number of tooth positions in this population of lemurs, affecting both males and females (albeit caused by different yet compounding selective pressures) with an ecological perturbation (i.e., drought) being a primary catalyst.

20.4.3. *Increased Tooth Size in a High-Attrition Environment*

Mammalian teeth provide a faithful record of an individual's growth and development, evolutionary relationships, and life story (e.g., Morbeck, 1997; Schwartz and Dean, 2000). Previous data on the patterns of tooth use, wear, and eventual loss in this population illustrate that *L. catta* teeth directly reflect their interaction with the environment of Beza Mahafaly (e.g., Sauther et al., 2002; Cuozzo and Sauther, 2004a, 2004b, 2005). This population of ringtailed lemurs is notable for their high frequency of severe wear and tooth loss (Sauther et al., 2002; Cuozzo and Sauther, 2004a, 2004b, 2005, in press; see Figures 20.3, 20.4, and 20.5b). This pattern contrasts with sympatric *Propithecus verreauxi*, which exhibits far less wear and few missing teeth (Cuozzo and Sauther, in press). Excessive tooth wear has also been observed among the ringtailed lemurs at Berenty Reserve in southeastern Madagascar (Soma, pers. comm.; Crawford, pers. comm.) where, similar to Beza Mahafaly, tamarind provides a keystone food source (e.g., Jolly, 1966; Simmen et al., this volume). Perzigian (1975), in a study of natural selection in a historic population of modern humans, argued that larger teeth would be advantageous, and therefore "of some survival value especially where attrition is very pronounced." Greene et al. (1967) came to a similar conclusion for a Mesolithic human population. Lucas (2004) has also discussed increased tooth size as one possible response to continued wear. In contrast to other primate populations, as well as sympatric Verreaux's sifaka (Cuozzo and Sauther, in press), in which tooth loss is often a product of tooth damage and disease (e.g., Schultz, 1935; Smith et al., 1977; Lovell, 1990), tooth loss among the Beza Mahafaly ringtailed lemurs is primarily a product of excessive wear (e.g., Cuozzo and Sauther, 2004a, 2004b, 2005, in press). Among these ringtailed lemurs, M_1 begins to wear shortly after eruption, is usually the first tooth lost, and is the most frequently missing tooth in the population (Cuozzo and Sauther, 2004a, 2005, in press). In a population where M_1 is often severely worn and frequently absent, M_2 (even when worn) becomes especially important for mastication, as it often remains functioning long after M_1 (as well as P_3 and P_4) is lost (Cuozzo and Sauther, 2005, in press). In this context, larger first and second molars would be quite advantageous.

A similar argument can also be made for increased size in P_2. Ringtailed lemurs primarily process tamarind pods with their postcanine teeth (e.g., Sauther et al., 2002; Yamashita, 2003; Cuozzo and Sauther, 2004a, 2005, in press). However, the anterior teeth (e.g., P_2) are often used in the initial acquisition of this food (e.g., Sauther et al., 2002). In individuals with severe tooth wear and antemortem tooth loss, anterior teeth become important for food processing. This is seen in the individual lemur shown in Figure 20.5b, in which the maxillary canine is clearly worn (the end of the tooth is short, and quite rounded, indicating wear rather than breakage), and is not uncommon in this population. Among tooth positions, P_2 is among the least frequently missing teeth in ringtailed lemurs, although sometimes being severely worn (Cuozzo and Sauther, 2004a, 2005, in press). Therefore, a larger P_2 (even if damaged) would provide an extended surface for food processing in individuals whose postcanine teeth have been severely impaired, as often seen at Beza Mahafaly (Sauther et al., 2002, Cuozzo and Sauther, 2004a, 2004b, 2005, in press).

Selection for larger (and possibly longer lasting) teeth in a high attrition environment primarily relates to long-term survival, rather than to the effects of a severe but short-lived event such as the drought of 1991/1992. However, given the high frequency of excessive tooth wear in this population (e.g., Sauther et al., 2001a, 2002; Cuozzo and Sauther, 2004a, 2004b, 2005, in press)—which indicates the importance of tamarind fruit—the scenario described by Greene et al. (1967) and Perzigian (1975) among modern human populations provides an important context with which to understand the possible selective pressure of diet on dental evolution. Given this scenario, we might expect that rapid and excessive tooth wear in ringtailed lemurs could lead to selection for larger food processing teeth, as discussed by Lucas (2004). This becomes more likely when viewed in context of the intense nutritional and reproductive stress experienced by ringtailed lemurs during and following the drought of 1991/1992. This high attrition environment provides a constant pressure in this population of ringtailed lemurs, and the added stress resulting from the drought likely exacerbated this pressure. As our long-term research plans at Beza Mahafaly include the continued collection of longitudinal dental data (at both the individual and populations levels), we will be able to further explore the relationship between tooth wear and possible changes in tooth size, including studies of tooth size across a temporal span that is not affected by severe drought.

20.4.4. M^2 Size Reduction: Selection for Improved Occlusion?

Having provided explanatory scenarios for increased tooth size in this population, we must now address the unexpected decrease in M^2 length. Contrary to our expectations, M^2 experienced a significant ($p = 0.0325$) length decrease in the overall sample since 1987/1988. The pattern of simultaneous size increases in some tooth positions with size decreases in others is not without precedent. Both Kurten (1957), in a study of tooth size change in the European cave bear (*Ursus spelaeus*), and Van Valen (1963) in the Miocene horse *Mercyhippus primus*, documented this type of apparent conflict. Van Valen (1963), in discussing the simultaneous trends of smaller maxillary teeth yet larger mandibular teeth in *M. primus*, suggested that the decrease in maxillary tooth size might have represented a local or temporary reversal, as this species was increasing in overall size. Kurten (1957), when analyzing size changes in molar cusps, noted that cave bear

M^2 paracones became smaller, while the corresponding "valley" areas of M_2 became larger. This apparent paradox likely resulted from selection on the masticatory functional complex resulting from occlusion between the two teeth, with the two size trends leading to more efficient occlusion. The trend seen in our current data may reflect a similar pattern of selection, as the morphology of ringtailed lemur molars reflects their functional occlusion (e.g., Yamashita, 1998). *L. catta* M_2 basins are quite deep relative to other lemurids and may function as food retainers in which the breakdown of food is enhanced (Yamashita, 1998). Yamashita (1998) also noted that ringtailed lemur second molars have long crests with acute cusps, and that there is a "loose fit" between the M_2 hypoconid and M^2 trigon. It is possible that the increase in M_2 length accompanied by a decrease in M^2 length documented in our data reflects the type of functional selection postulated by Kurten (1957), given the function of ringtailed lemur second molar morphology discussed by Yamashita (1998). It is also interesting that the amount of variation in M^2 length (compare standard deviations for 1987/1988 and 2003/2004 in Table 20.1) significantly decreased (F ratio [variance], p = 0.0043), suggesting directional selection not only for size, but also for a reduction in variability (i.e., possibly targeting a "tighter" metric distribution and greater occlusal efficiency). Because the types of morphometric data collected by Kurten (1957) and Yamashita (1998) are accessible in the dental casts used in our study from both 1987/1988 and 2003/2004, we plan to further investigate this hypothesis as part of our continued research.

One other possible explanation for the reduction in M^2 length in this population was discussed by Brace et al. (1987), in which dental reduction in modern humans, albeit over a much longer period of time, resulted from the "Probable Mutation Effect" (Brace, 1963). In this scenario, reductions in tooth size may be caused by mutation alone, given an absence of natural selection (i.e., relaxed selection) (Brace et al., 1987). It is therefore possible that directional selection did not impact the maxillary teeth.

20.5. Conclusions

In a broad discussion and review of lemur ecology and evolution, Wright (1999) posed the question "What effect does drought have on a tropical fauna?" Wright (1999) argued that many of the characteristics unique to the Malagasy strepsirrhines (e.g., female dominance, reproductive synchrony) evolved in response to the challenges of Madagascar's unpredictable environment, for example seasonal fluctuations, cyclones, and droughts. Given the strong link between lemur biology, ecology, behavior, and the environment, investigating the response of lemurs to environmental changes in the wild, especially among populations for which longitudinal data are available, provides the opportunity to document examples of contemporary evolution. As noted earlier, the drought of 1991/1992 had a major impact on southern Madagascar (e.g., Sauther, 1998; Gould et al., 1999, 2003; Jolly, 2004). The ringtailed lemurs at Beza Mahafaly are no exception, as witnessed by a significant decrease in the adult population, and dramatic increases in adult female, infant and juvenile mortality in the years during and following the drought (Gould et al., 1999, 2003). This population also experienced a change in its dental characteristics, as P_2, M_1, and M_2 lengths each significantly increased ($p < 0.01$) in the overall population between 1987/1988 and 2003/2004.

Studies of tooth size change over time provide an opportunity to explore a number of questions in primate evolution and evolutionary biology (see earlier references). Our data, from a living population of ringtailed lemurs, have allowed us to conduct an allochronic study of primate dental size, and to investigate the impact of a severe drought on the contemporary evolution of this population. Because mammalian tooth size is highly heritable (e.g., Gingerich, 1974b; Hillson, 1986; Hlusko et al., 2002), these data indicate that environmental fluctuations (e.g., drought), combined with the use of a challenging keystone food, can provide important selective pressures on the evolution of primate teeth (as recently suggested by Lambert et al., 2004 in the evolution of primate enamel thickness), and more broadly, can lead to observable changes in a population in contemporary time. These data correspond to other studies of contemporary evolution in vertebrate populations (e.g., Galápagos finches), and illustrate the effect that rapid ecological changes can have on living populations (e.g., Grant and Grant, 1995; see reviews in Hendry and Kinnison, 1999, and Stockwell et al., 2003). In addition, our results indicate that socioecology (e.g., resource competition, interindividual aggression) can be an important variable when investigating natural selection, environmental change, and contemporary evolution. Although it is not possible to completely rule out genetic drift (e.g., immigration of larger-toothed individuals) as a cause of the increased tooth size seen in this population, it is unlikely. New data on male migration (females do not usually migrate [e.g., Sussman, 1992]) indicate that individuals at Beza Mahafaly tend to migrate within a limited area, often only migrating to adjacent troops (Sauther and Cuozzo, unpublished data). Our data also reflect the pattern described for several modern human populations (e.g., Greene et al., 1967; Perzigian, 1975), in which an increase in tooth size corresponds to high levels of attrition and tooth wear, a condition common to the ringtailed lemurs at Beza Mahafaly (e.g., Sauther et al., 2001a, 2002; Cuozzo and Sauther, 2004a, 2004b, 2005, in press).

We recognize that these questions require additional research, and one of our long-term goals is to continue our longitudinal study of ringtailed lemur dental variation, health, and feeding ecology at Beza Mahafaly. This work will include an emphasis on individual dental life stories, in the broader context of understanding *L. catta* ecology, evolution, and life history. In addition, we plan to expand our research on temporal change and focus on the impact of human populations on the environment surrounding the Beza Mahafaly Reserve (see Whitelaw et al., 2005). As human activity has had a dramatic impact on the environment and fauna of Madagascar over the past two thousand years (e.g., Godfrey et al., 1997; Godfrey and Jungers, 2003), and has likely influenced contemporary evolution among lemurs, illustrating the effects that rapid ecological change can have on a living species has a number of direct conservation implications for Malagasy primates.

Acknowledgments. We thank Enafa Efitroaromy, Ehandidy Ellis, Razanajafy Olivier, Emady Rigobert, and Elahavelo of the Beza Mahafaly Ecological Monitoring Team and Krista Fish, Mandala Hunter, Kerry Sondgeroth, James Loudon, Heather Culbertson, Rachel Mills, and David Miller for their assistance with data collection at Beza Mahafaly during the 2003 and 2004 field seasons. We thank Robert Sussman, Jeff Kaufman, Behaligno, and Manjagasy for their help with collection of data in 1987 and 1988. We especially thank Robert Sussman and Jeff Kaufmann (1987/1988) and Krista Fish (2003/2004) for their assistance with preparing dental casts. We thank Robert Sussman, Ingrid Porton, Randy Junge, Joel Ratsirarson, Jo Ajimy, Randrianarisoa Jeannicq, and Ibrahim Jacky Youssouf and Rafidisoa Tsiory (ANGAP) for their strong support and facilitation of our ongoing project. Our appreciation also goes to the Département des Eaux et Forêts, Ecole Superieur des Sciences Agronomiques, Université d'Antananarivo, and ANGAP for allowing us to continue our research at Beza Mahafaly. We thank Rich Lawler for reviewing our chapter and for his important suggestions, which have greatly improved our paper. Alison Jolly and Debbie Guatelli-Steinberg also provided helpful comments on this manuscript. We thank the anonymous reviewer of this volume for his/her effort. Funding for this study came from Primate Conservation Inc., the Lindbergh Fund, the Saint Louis Zoo, the John Ball Zoo Society, the National Science Foundation, the National Geographic Society, the Leakey Foundation, Washington University, and the University of Colorado, Boulder.

References

Altmann, J. (1980). *Baboon Mothers and Infants.* Harvard University Press, Cambridge and London.
Bown, T. M., Holroyd, P. A., and Rose, K. D. (1994). Mammal extinctions, body size, and paleotemperature. *Proc. Nat. Acad. Sci. USA.* 91:10403–10406.
Brace, C. L. (1963). Structural reduction in evolution. *Am. Nat.* 97:39–49.
Brace, C. L., Rosenber, K. R., and Hunt, K. D. (1987). Gradual change in human tooth size in the late Pleistocene and post-Pleistocene. *Evol.* 41:705–720.
Cuozzo, F. P. (2002). Dental variation and temporal change in early Eocene *Hyopsodus* (Mammalia, Condylarthra) from the Powder River Basin, Wyoming. *PaleoBios* 22(2):1–9.
Cuozzo, F. P., and Sauther, M. L. (2004a). Tooth loss, survival, and resource use in wild ring-tailed lemurs (*Lemur catta*): implications form inferring conspecific care in fossil hominids. *J. Hum. Evol.* 46:623–631.
Cuozzo, F. P., and Sauther, M. L. (2004b). Patterns of tooth wear and their relation to specific feeding behaviors in extant *Lemur catta* (Mammalia, Primates): implications for primate paleobiology. *J. Vert. Paleo.* 24(3):49A.

Cuozzo, F. P., and Sauther, M. L. (2005). Tooth loss in wild ring-tailed lemurs (*Lemur catta*): a function of life history, behavior, and feeding ecology. *Am. J. Phys. Anthropol. Supplement* 40:90.

Cuozzo, F. P., and Sauther, M. L. Severe wear and tooth loss in wild ring–tailed lemurs (*Lemur catta*): a function of feeding ecology, dental structure, and individual life history. *J. Hum. Evol.* In Press.

Cuozzo, F. P., Sauther, M. L., and Fish, K. D. (2004). Dental variation and dental health in a wild population of ring-tailed lemurs (*Lemur catta*) from Beza Mahafaly Special Reserve, Madagascar. *Am. J. Phys. Anthropol. Suppl.* 38:81.

Dennis, J. C., Ungar, P. S., Teaford, M. F., and Glander, K. (2004). Dental topography and molar wear in *Alouatta palliata* from Costa Rica. *Am. J. Phys. Anthropol.* 125:152–161.

DeGusta, D., Everett, M. A., and Milton, K. (2003). Natural selection on molar size in a wild population of howler monkeys (*Alouatta palliata*). *Proc. R. Soc. Lond B Suppl.* 270:S15–S17.

Dumont, E. R. (1995). Enamel thickness and dietary adaptation among extant primates and chiropterans. *J. Mammal.* 76:1127–1136.

Eaglen, R. H. (1985). Behavioral correlates of tooth eruption in Madagascar lemurs. *Am. J. Phys. Anthropol.* 66:307–315.

Endler, J. A. (1986). *Natural Selection in the Wild*. Princeton University Press, Princeton.

Futuyma, D. J. (1998). *Evolutionary Biology*. Sinauer Associates, Inc., Publishers, Sunderland, Mass.

Gingerich, P. D. (1974a). Stratigraphic record of early Eocene Hyopsodus and the geometry of mammalian phylogeny. *Nature* 248:107–109.

Ginigerich, P. D. (1974b). Size variability of the teeth of living mammals and the diagnosis of closely related sympatric fossil species. *J. Paleontol.* 48:895–903.

Gingerich, P. D. (1979a). Phylogeny of middle Eocene Adapidae (Mammalia, Primates) in North America: *Smilodectes* and *Notharctus*. *J. Paleontol.* 53:153–163.

Gingerich, P. D. (1979b). Paleontology, phylogeny, and classification: an example from the mammalian fossil record. *Syst. Zool.* 28:451–464.

Gingerich, P. D. (1985). Species in the fossil record: concepts, trends, and transitions. *Paleobiol.* 11:27–41.

Gingerich, P. D. (1994). New species of *Apheliscus*, *Haplomylus*, and *Hyopsodus* (Mammalia, Conylarthra) from the late Paleocene of southern Montana and early Eocene of northwestern Wyoming. *Contrib. Mus. Paleontol. Univ. Mich.* 29:119–134.

Gingerich, P. D., and Schoeninger, M. (1977). The fossil record and primate phylogeny. *J. Hum. Evol.* 6:483–505.

Godfrey, L. R., and Jungers, W. L. (2003). Subfossil lemurs. In: Goodman, S. M., Benstead, J. P. (eds.), *The Natural History of Madagascar*. The University of Chicago Press, Chicago and London, pp. 1247–1252.

Godfrey, L. R., Jungers, W. L., Reed, K., Simons, E. L., and Chatrath, P. S. (1997). Subfossil lemurs: inferences about past and present primate communities in Madagascar. In: Goodman, S. M., Paterson, B. D. (eds.), *Natural Change and Human Impact in Madagascar*. Smithsonian Institution Press, Washington, D.C., pp. 218–256.

Godfrey, L. R., Samonds, K. E., Jungers, W. L., and Sutherland, M. R. (2001). Teeth, brains, and primate life histories. *Am. J. Phys. Anthropol.* 114:192–214.

Godfrey, L. R., Petto, A. J., and Sutherland, M. R. (2002). Dental ontogeny and life history strategies: the case of the giant extinct indroids of Madagascar. In: Plavcan, J. M., Kay, R. F., Jungers, W. L., and van Schaik, C. P. (eds.), *Reconstructing Behavior in the Primate Fossil Record*, Kluwer Academic/Plenum Publishers, New York, pp. 113–157.

Godfrey, L. R., Samonds, K. E., Jungers, W. L., Sutherland, M. R., and Irwin, M. T. (2004). Ontogenetic correlates of diet in Malagasy lemurs. *Am. J. Phys. Anthropol.* 123:250–276.

Godfrey, L. R., Semprebon, G. M., Schwartz, G. T., Burney, D. A., Jungers, W. L., Flanagan, E. K., Cuozzo, F. P., King, S. J. (2005). New insights into old lemurs: the trophic adaptations of the Archaeolemuridae. *Int J Primatol.* 26:812–854.

Goodall, J. (1986). *The Chimpanzees of Gombe: Patterns of Behavior.* Belknap Press of Harvard University Press, Cambridge and London.

Gould, L. (1996). Male affiliative relationships in natural occurring ringtailed lemurs (*Lemur catta*) at Beza Mahafaly Reserve, Madagscar. *Am. J. Primatol.* 39:63–78.

Gould, L. (1997). Intermale affiliative relationships in ring-tailed lemurs (*Lemur catta*) at the Beza Mahafaly Reserve, Madagascar. *Primates* 38:15–30.

Gould, L., Sussman, R. W., and Sauther, M. L. (1999). Natural disasters and primate populations: the effects of a 2-year drought on a naturally occurring population of ring-tailed lemurs (*Lemur catta*) in southwestern Madagascar. *Int. J. Primatol.* 20:69–84.

Gould, L., Sussman, R. W., and Sauther, M. L. (2003). Demographic and life-history patterns in a population of ring-tailed lemurs (*Lemur catta*) at Beza Mahafaly, Madagascar: a 15-year perspective. *Am. J. Phys. Anthropol.* 120:182–194.

Grant, B. R. (1985). Selection on bill characters in a population of Darwin's finches: *Goespiza conirostris* on Isla Genovesa, Galápagos. *Evol.* 39:523–532.

Grant, P. R., and Grant, B. R. (1995). Predicting microevolutionary responses to directional selection on heritable variation. *Evol.* 49:241–251.

Greene, D. L., Ewing, G. H., and Armelagos, G. J. (1967). Dentition of a Mesolithic population from Wadi Halfa, Sudan. *Am. J. Phys. Anthropol.* 27:41–56.

Haycock, K. A., Roth, J., Gagon, J., Finzee, W. F., and Soper, C. (1992). Statview. Abacus. Concepts, Berkeley, Calif.

Hendry, A. P. (2005). The power of natural selection. *Nature* 433:694–695.

Hendry, A. P., and Kinnison, M. T. (1999). The pace of modern life: measuring rates of contemporary microevolution. *Evolution* 53:1637–1653.

Hillson, S. (1986). *Teeth.* Cambridge University Press, Cambridge.

Hladik, C. M. (1975). Ecology, diet, and social patterning in old and new world primates. In: Tuttle, R. H. (ed.), *Socioecology and Psychology of Primates.* Mouton Publishers, The Hague, pp. 3–35.

Hlusko, L. J., Weiss, K. M., and Mahaney, M. C. (2002). Statistical genetic comparison of two techniques for assessing molar crown size in pedigreed baboons. *Am. J. Phys. Anthropol.* 117:182–189.

Janis, C. M., and Fortelius, M. (1988). On the means whereby mammals achieve increased functional durability of their dentitions, with special reference to limiting factors. *Biol. Rev.* 63:197–230.

Jolly, A. (1966). *Lemur Behavior.* University of Chicago Press, Chicago.

Jolly, A. (2003). *Lemur catta*, Ring-tailed lemur, *Maky.* In: Goodman, S. M., and Benstead, J. P. (eds.), *The Natural History of Madagascar.* University of Chicago Press, Chicago, pp. 1329–1331.

Jolly, A. (2004). *Lords and Lemurs.* Houghton Mifflin Company, Boston.

Jolly, A., Rasamimanana, H. R., Kinnaird, M. F., O'Brien, T. G., Crowley, H. M., Harcourt, C. S., Gardner, S., and Davidson, J. M. (1993). Territoriality in *Lemur catta* groups during the birth season at Berenty, Madagascar. In: Kappeler, P. M., and Ganzhorn, J. U. (eds.), *Lemur Social Systems and Their Ecological Basis.* Plenum Press, New York and London, pp. 85–109.

Jolly, A., Gould, L., Koyama, N., Rasamimanana, H., and Sussman, R. W. (2004). Interpreting *Lemur catta. Folia. Primatol.* 75(suppl 1):156.

Kay, R. F. (1981). The nut-crackers—a new theory of the adaptations of the Ramapithecinae. *Am. J. Phys. Anthropol.* 55:141–151.

Kay, R. F. (1985). Dental evidence for the diet of australopithecines. *Ann. Rev. Anthropol.* 14:315–341.

King, S. J., Arrigo-Nelson, S. J., Pochron, S. T., Semprebon, G. M., Godfrey, L. R., Wright, P. C., and Jernvall, J. (2005). Dental senescence in a long-lived primate links infant survival to rainfall. *Proc. Nat. Acad. Sci. USA*. 102:16579–16583.

Kinnison, M. T., and Hendry, A. P. (2001). The pace of modern life II: from rates of contemporary evolution to pattern and process. *Genetica* 112–113:145–164.

Kurten, B. (1957). A case of Darwinian selection in bears. *Evolution* 11:412–416.

Lawler, R. R., Richard, A. F., and Riley, M. A. (2005). Intrasexual selection in Verraux's sifaka (*Propithecus verreauxi verreauxi*). *J. Hum. Evol.* 48:259–277

Lambert, J. E., Chapman, C. A., Wrangham, R. W., and Conklin-Brittain, N. L. (2004). Hardness of cercopithecine foods: implications for the critical function of enamel thickness in exploiting fallback foods. *Am. J. Phys. Anthropol.* 125:363–368.

Lebel, S., and Trinkaus, E. (2002). Middle Pleistocene human remains from the Bau de l'Aubesier. *J. Hum. Evol.* 43:659–685.

Liu, W., and Zheng, L. (2005). Tooth wear difference between the Yuanmou hominoid and *Lufengpithecus*. *Int. J. Primatol.* 26:491–506.

Losos, J. B., Schoener, T. W., Warheit, K. I., and Creer, D. 2001. Experimental studies of adaptive differentiation in Bahamian Anolis lizards. *Genetica* 112–113:399–514.

Lovell, N. C. (1990). *Patterns of Injury and Illness in Great Apes: A Skeletal Analysis.* Smithsonian Institution Press, Washington, D.C.

Lucas, P. (2004). *Dental Functional Morphology.* Cambridge University Press, Cambridge.

Marcus, L. F. (1969). Measurements of selection using distance statistics in the prehistoric orangutan *Pongo pygmaeus paleosumatrensis*. *Evolution* 23:301–307.

Martin, L. B., Olejniczak, A. J., and Maas, M. C. (2003). Enamel thickness and microstructure in pithecin primates, with comments on dietary adaptations of the middle Miocene hominoid Kenyapithecus. *J. Hum. Evol.* 45:351–367.

Morbeck, M. E. (1997). Life history in teeth, bones, and fossils. In: Morbeck, M. E., Galloway, A., and Zihlman, A. L. (eds.), *The Evolving Female: A Life History Perspective.* Princeton University Press, Princeton, N.J., pp. 117–131.

Pereira, M. E. (1993). Seasonal adjustment of growth rate and adult body weight in ringtailed lemurs. In: Kappeler, P. M., and Ganzhorn, J. U. (eds.), *Lemur Social Systems and Their Ecological Basis.* Plenum Press, New York and London, pp. 205–221.

Perzigian, A. J. (1975). Natural selection on the dentition of an Arikara population. *Am. J. Phys. Anthropol.* 42:63–70.

Ratsirarson, J. (1985). Contribution a l'etude comparative de l'eco-ethologie de *Lemur catta* dans deux habitats differents de la Réserve Spéciale de Beza-Mahafaly. memoire de Find'Etudes. Universite de Madagascar.

Ratsirarson, J. (2003). Réserve Spéciale de Beza Mahafaly. In: Goodman, S. M., and Benstead, J. P. (eds.), *The Natural History of Madagascar.* University of Chicago Press, Chicago, pp. 1520–1525.

Sauther, M. L. (1989). Anitpredator behavior in troops of free-ranging *Lemur catta* at Beza Mahafaly Special Reserve, Madagascar. *Int. J. Primatol.* 10:595–606.

Sauther, M. L. (1991). Reproductive behavior of free-ranging *Lemur catta* at Beza Mahafaly Special Reserve, Madagascar. *Am. J. Phys. Anthropol.* 84:463–477.

Sauther, M. L. (1992). *Effect of Reproductive State, Social Rank and Group Size on Resource Use Among Free-Ranging ring-tailed Lemurs (Lemur catta) of Madagascar.* Unpublished PhD thesis, Department of Anthropology, Washington University, St. Louis.

Sauther, M. L. (1993). The dynamics of feeding competition in a wild population of ring-tailed lemurs (*Lemur catta*). In: Kappeler, P. M., and Ganzhorn, J. U. (eds.), *Lemur Social Systems and Their Ecological Basis*. Plenum Press, New York and London, pp. 135–152.

Sauther, M. L. (1994). Changes in the use of wild plant foods in free-ranging lemurs during lactation and pregnancy: Some implications for hominid foraging strategies. In: Etkin, N. L. (ed.), *Eating on the Wild Side: The Pharmocologic, Ecologic, and Social Implications of Using Noncultigens*. University of Arizona Press, Tuscon, pp. 240–246.

Sauther, M. L. (1998). The interplay of phenology and reproduction in ring-tailed lemurs: implications for ring-tailed lemur conservation. *Folia. Primatol.* (Supplement) 69:309–320.

Sauther, M. L., Sussman, R. W., and Gould, L. (1999). The socioecology of the ring-tailed lemur: Thirty-five years of research. *Evol. Anthropol.* 8:120–132.

Sauther, M. L., Cuozzo, F. P., Gould, L., Sussman, R. W., Ratsirarson, J., and Bauer, R. Surviving a drought: individual life stories and ecological change among ring-tailed lemurs (*Lemur catta*) at Beza Mahafaly Special Reserve, Madagascar. In preparation.

Sauther, M. L., Cuozzo, F. P., and Sussman, R. W. (2001a). Analysis of dentition of a living, wild population of ring-tailed lemurs (*Lemur catta*) from Beza Mahafaly, Madagascar. *Am. J. Phys. Anthropol.* 114:215–223.

Sauther, M. L., Steckler, J. A., and Sussman, R. W. (2001b). A biometric analysis of sexual dimorphism in wild ring-tailed lemurs (*Lemur catta*). *Am. J. Phys. Anthropol. Suppl.* 32:130.

Sauther, M. L., Sussman, R. W., and Cuozzo, F. (2002). Dental and general health in a population of wild ring–tailed lemurs: a life history approach. *Am. J. Phys. Anthropol.* 117:122–132.

Sauther, M. L., Fish, K. D., and Cuozzo, F. (2004). Biological variability of wild ring-tailed lemurs, *Lemur catta*: effects of habitat and sex. *Folia Primatol.* 75(suppl 1):159–160.

Schultz, A. H. (1935). Eruption and decay of the permanent teeth in primate. *Am. J. Phys. Anthropol.* 19:489–588.

Schwartz, G. T. (2000). Taxonomic and functional aspects of the patterning of enamel thickness distribution in extant large-bodied hominids. *Am. J. Phys. Anthropol.* 111:221–244.

Schwartz, G. T., and Dean, C. (2000). Interpreting the hominid dentition: ontogenetic and phylogenetic aspects. In: O'Higgin, P., and Cohn, M. (eds.), *Development, Growth, and Evolution*. Academic Press, San Diego, pp. 207–233.

Shellis, P., Beynon, A. D., Reid, D. J., and Hiimae, K. (1998). Variations in molar enamel thickness among primates. *J. Hum. Evol.* 35:507–522.

Smith, J. D., Genoways, H. H., and Jones, J. K. (1977). Cranial and dental anomalies in three species of platyrrhine monkeys from Nicaragua. *Folia Primatol.* 28:1–42.

Stockwell, C. A., Hendry, A. P., and Kinnison, M. T. (2003). Contemporary evolution meets conservation biology. *Trends Ecol. Evol.* 18:94–101.

Sussman, R. W. (1991). Demography and social organization of free–ranging *Lemur catta* in the Beza Mahafaly Reserve, Madagascar. *Am. J. Phys. Anthropol.* 84:43–58.

Sussman, R. W. (1992). Male life history and intergroup mobility among ringtailed lemurs (*Lemur catta*). *Int. J. Primatol.* 13:395–413.

Sussman, R. W., and Rakotozafy, A. (1994). Plant diversity and structural analysis of a tropical dry forest in southwestern Madagascar. *Biotropica* 26:241–254.

Swindler, D. R. (2002). *Primate Dentition: An Introduction to the Teeth of Non–Human Primates*. Cambridge University Press, Cambridge.

Teaford, M. F., and Ungar, P. S. (2000). Diet and the evolution of the earliest human ancestors. *Proc. Nat. Acad. Sci. USA.* 97:13506–13511.

van Valen, L. (1963). Selection in natural populations: *Mercyhippus primus*, a fossil horse. *Nature* 197:1181–1183.

Whitelaw, D., Sauther, M. L., Loudon, J. E., and Cuozzo, F. (2005). Anthropogenic change in and around Beza-Mahafaly Reserve: methodology and results. *Am. J. Phys. Anthropol. Suppl.* 40:222.

Wright, P. C. (1999). Lemur traits and Madagascar ecology: Coping with an island environment. *Yrbk. Phys. Anthropol.* 42:31–72.

Yamashita, N. (1998). Functional dental correlates of food properties in five Malagasy lemur species. *Am. J. Phys. Anthropol.* 106:169–188.

Yamashita, N. (2000). Mechanical thresholds as a criterion for food selection in two prosimian primate species. In: *Proceedings of the 3rd Plant Biomechanics Conference, Freiburg-Badenweiler*. Thieme Verlag, Stuttgart, pp. 590–595.

Yamashita, N. (2003). Food procurement and tooth use in two sympatric lemur species. *Am. J. Phys. Anthropol.* 121:125–133.

Yamashita, N. Food physical properties and their relationship to morphology: The curious case of kily. In: Vinyard, C. J., Ravosa, M. J., and Wall, C. E. (eds.), *Primate Craniofacial Function and Biology*. Kluwer Academic Press, New York. In preparation.

Index

Page numbers followed by f and t indicates figures and tables, respectively.

A

Abdominal paunch, 324, 328
Acacia rovumae, 91
Advance vegetation index (AVI), 19
Agave rigida, 87
Aggression, 256–260
　spring, 262
　study in ringtailed lemurs, 240–241, 277
　unprovoked, occurred only before, or during and not even soon after reversals, 255t
Agonistic behavior, in lemurs, 300–301
　changes in, predicted after adolescent reversals, 248f
　intensities of, 251t
　proximate causes for, 252t
　of ringtailed lemurs, 283t
Allelochemicals, 161, 162
Allometric equation, to calculate energy expenditure in *Lemur catta*, 276, 277
Alluaudia procera (Didieriaceae), 58t
Alopecia syndrome in ringtailed lemur, 332
Alouatta palliata, 345
Andringitra Massif, high mountain population of, ringtailed lemur in, 7–9
　dietary elements of, 7
　elevational gradients, 7
　external measurements of two, 9t
"Animal minute," 123
Ankoba, 34
　lobe, population density of *L. catta* in, 17, 156, 225
Anthropogenic savanna, ringtailed lemur, 4
Antserananomby forest, *L. catta* population in, 26t
Aotus trivirgatus, 177, 289t
Aphloia theiformis (Aphloiaceae), 7, 8t

Argemone mexicana (Papaveraceae), food source of *L. catta*, 61
AVI: *see* Advance vegetation index
Azadirachta indica (Meliaceae)
　food source of *L. catta*, 59t
　foraging on, 221, 222f, 227
Azima tetracantha, 46, 60t, 70, 124, 127t

B

"Bald lemur syndrome," 332, 341
Bare soil index (BI)
　definition of, 19
　in FCD computation, 19
Bealoka Forest, *L. catta* densities in, 32
Bealoka parcel, distribution of *L. catta* in, 40
Behavioral dominance of brown lemurs over ringtailed lemurs, 134
Bemananteza Forest, troops of *L. catta* in, 10
benono, 37, 91, 100
Berenty area, *L. catta* troops in, 10
Berenty forest, *L. catta* population in, 40
Berenty Reserve, Madagascar
　area classification based on vegetation, 87
　climate of southern Madagascar, 32, 172, 188
　fauna in, 43, 47
　habitat zones of, 33–34, 36, 39
　impact of introduced tree species on *L. catta* feeding strategy at, 141, 156
　kily trees in, 91, 97
　lemur species in, 141–142
　ringtailed lemur studies at, 40
　studies of home ranges, 88
　studies of large trees, 88
　use of kily in, 98
　vegetation in, 135

Berenty Reserve, Southern Madagascar, resource competition between ringtailed lemurs and brown lemurs at, 121, 136
Bet-hedging strategy, 195
Beza Mahafaly forest, *L. catta* population in, 26t
Beza Mahafaly Osteological Collection (BMOC 67), tooth wear in a ringtailed lemur from, 349f
Beza Mahafaly Reserve, Madagascar
   climates of, 44
   fauna at, 46–47
   gallery forest, 46
   lemur studies at, 48
   xerophytic habitat in, 47
Beza Mahafaly Special Reserve (BMSR), ringtailed lemur population at
   drought and patterns of mortality of, 345
   enamel thickness, the drought, and food availability, 346
   tooth size change among *Lemur catta*, temporal, 358
BI: *see* Bare soil index
Biogeography, ringtailed lemur, 3–7
Birth/lactation season, food availability in, 209
Bougainvilea (Nyctaginaceae), food source of *L. catta*, 63
Brown lemurs (*Eulemur fulvus rufus*), 119, 122, 141
   behavioral dominance of, over ringtailed lemurs, 134
   feeding habits of, 142
Brown lemurs (*Eulemur fulvus rufus*), feeding competition between ringtailed lemurs and, at Berenty, 119, 122
   behavioral dominance of brown lemurs over ringtailed lemurs, 134
   comparison of diet overlap and activity patterns, 122
   diet overlap across troop-pairs and habitats, 125
   diurnal activity budgets, 130
   feeding height, 128
   interspecific diet overlap, 132
   interspecific differences in competitive ability, 133
   interspecific differences in dietary flexibility, 132
   overall diet and diet overlap, 122
   resource types fed on by, 127t
   seasonal food scarcity and *Tamarindus indica* as a keystone species, 134
   study site, 121, 162–163
Buddleja madagascariensis (Loganiaceae), 7, 8t

C

Callithrix argentata, 177
Canopy cover in closed canopy forest, 75f, 76f
Carnivora predators, 8
Cattle forage, use of, by ringtailed lemurs within the degraded habitat, 317f
Cedrelopsis grevei, 8, 46, 60t
Celtis bifida (Ulmaceae)
   food source of *L. catta*, 61
   pulps, 166
Celtis philippensis (Ulmaceae), food source of *L. catta*, 61
Characterization methods of forest condition
   approximation of deforestation methodology, 17, 22
   forest canopy density, computation of, 19, 21f
   image pre-processing, 19
   lemur density and forest canopy density, relationship between, 24
   problems in, 24
Chimpanzees (*Pan troglodytes*), 154, 160, 257
Chi-square tests in ringtailed lemurs, 277
Chorda tympani proper nerve, 160, 177
Coecal™ Type III dental stone, 351
Cognitive ethology, in primates, 259, 261
Cohesive sociality, in primates, 258
Commicarpus commersonii (Nyctaginaceae), food source of *L. catta*, 60t
Condensed tannins (CT), seasonal variations in, 166
Contemporary evolution, 344
Cordia caffra (Boraginaceae), 57t, 61, 146t, 147t
   food source of *L. catta*, 156
Cordia sinensis, 63, 141, 146
   feeding by *L. catta*, 63
   as food source of *L. catta*, 57
Crateva sp. (Capparaceae), 63
   food source of *L. catta*, 57
Cryptoprocta ferox, 8, 40, 47

D

Didiereaceae plant family, distribution of ringtailed lemur and, in Madagascar, 4, 5f, 6
Dietary elements in high elevational zones, ringtailed lemur
   Aphloia theiformis, 7, 8t
   Asteropeia micraster, 7, 8t
   Buddleja madagascariensis, 7, 8t
   Ficus pyrifolia, 7, 8t
   Ficus spp., 7, 8t, 127
   Locusta migratoria, 7

Dietary elements in high elevational zones, ringtailed lemur (*Cont'd*)
  Maesa lanceolata, 7, 8t
  Solanum auriculatum, 7, 8t
  V. secondiflorum, 7
  Vaccinium emirnense, 7, 8t
Dietary flexibility, interspecific differences in, 132
Diet breadth, 124–125, 126f
Diet overlap
  across lemur troop-pairs and habitats, 124, 125f
  and activity patterns in brown and ringtailed lemurs, comparison of, 122, 123
  interspecific, in brown and ringtailed lemurs, 128
Directional selection differential, 356
Distasteful stimuli, 164, 175
Diurnal activity budgets, in brown and ringtailed lemurs, 130
Dominance contest, in primates, 245, 246, 252
Dominance reversals, in lemurs, 228
Drought, effect on *L. catta* population, 187, 190

E

Ecological model, for *L. catta* behavior, 208
Ectoparasites, 317, 322
Edge habitat in ringtailed lemurs' resilience to food scarcity, role of, 135
Eliurus myoxinus (Madagascar tree-rat), 39, 47
Encounters, intertroop, in ringtailed lemurs, 226
Environmental perturbation, short-term, 344
Eulemur fulvus rufus: *see* Brown lemurs
Eulemur fulvus ssp., 276
European cave bear, tooth size change in, 358
Evolutionary theory, 264

F

Fauna, at Beza Mahafaly Reserve
  Acacia bellula, 46
  Acacia rovumae, 46
  Alluaudia procera, 46
  Amphiglossus splendidus, 48
  Azima tetracantha, 46
  Cedrelopsis grevei, 46
  Cheirogaleus medius, 47
  Commiphora spp., 46
  Crateva excelsa, 46
  Echinops telfairi, 47
  Euphorbia tirucalli, 46
  Felis spp., 47
  Gardenia spp., 46

Fauna, at Beza Mahafaly Reserve (*Cont'd*)
  Gelonium adenophorum, 46
  Geogale aurita, 47
  Grewia spp., 46, 47
  Gyrocarpus americanus, 46
  Hipposideros commersoni, 47
  Lemur catta, 47
  Lepilemur leucopus, 47
  Microcebus griseorufus, 47
  Microchiroptera, 47
  Mormopterus fugularis, 47
  Mus musculus, 47
  Propithecus verreauxi, 47
  Pteropus rufus, 47
  Quivisianthe papinae, 46
  Rattus rattus, 47
  Rhigozum madagascariensis, 46
  Rhopalocarpus lucidus, 46
  Salvadora angustifolia., 46
  Setifer setosus, 47
  Stereospermum variablile, 46
  Tadarida jugularis, 47
  Tamarindus indica, 44, 46–47
  Taphozous mauritianus, 47
  Tarenna pruinosum, 46
  Tenrec ecaudatus, 47
  Terminalia spp., 46
  Viverricula indica, 47
Fauna, in Berenty Reserve
  gray-and-red mouse lemur, 39
  gray mouse lemur, 39
  hybrid brown lemurs, 39
  Indian civet, 39
  large tenrec, 39
  lemur predators, 40
  Madagascar giant fruit bat, 39
  Madagascar radiated tortoise, 40
  Madagascar tree-rat, 39
  ringtailed lemur, 39
  shrew-like tenrec, 39
  spider tortoise, 40
  spiny tenrec, 39
  white-footed lepilemur, 39
  white sifaka, 39
FCD: *see* Forest canopy density
Fecal analysis, nonintrusive, 261, 262
Feeding height, comparison of, in ringtailed lemur and brown lemur, 128
Female eviction, in ringtailed lemurs, 233–235, 237
  case study of, 235
Ficus spp., 7
Fisher exact test of alopecia, 338
"Flighty mother," 286

Food availability, serial specialization in, 142
Food chemical composition, 163
Food patch, 123, 134
Food plants, food items, and seasonal variations in, *L. catta*
  Argemone mexicana (Papaveraceae), 60t
  Azadirachta indica (Meliaceae), 63–64
  Bougainvilea (Nyctaginaceae), 60t, 63
  Celtis bifida (Ulmaceae), 61
  Celtis philippensis (Ulmaceae), 61
  Commicarpus commersonii (Nyctaginaceae), 60t, 62f
  Cordia caffra (Boraginaceae), 57t, 61
  Cordia sinensis, 57t, 61, 63
  Crateva sp. (Capparaceae), 63
  Enterospermum pruinosum (Rubiaceae), 60t, 61
  Eucalyptus sp., 63
  Grewia spp. (Tiliaceae), 61
  Hildebrandtia spp. (Convolvulaceae), 61, 63
  Justicia glabra (Acanthaceae), 61, 62f
  kily ripe pods, 56
  Leucaena leucocephala (Mimosaceae), 59t, 63–64
  Neotina isoneura (Sapindaceae), 60t, 63
  Opuntia rackeets, 63
  Quivisianthe papinae (Meliaceae), 61, 62f
  Rinorea greveana (Violaceae), 61
  Rynchosia sp. (Fabaceae), 61
  Salvadora angustifolia (Salvadoraceae), 60t, 61
  Secamone sp. (Asclepiadaceae), 61
  Senna siamea, 63
  Talinella dauphinensis (Portulacaceae), 61
  Tamarindus indica (Caesalpiniaceae), 56
Food resource distribution, in primates, 222
"Food-scarce" weaning season, 209
  ranging patterns in, 220f
Food scarcity condition, atypical, 213
  ranging patterns in, 220f
Foraging time
  effect of lemur group size on, 221
  in large groups, 227
Forest canopy density (FCD), 18–26
  measurement, flowchart of, 20f
  population density of *L. catta*, comparing known density and, 17
  relationship between lemur density and, 24
  values of, from Landsat 5 TM images, 19, 21f
Frugivorous–folivorous species, 141, 172
Fur loss: *see Alopecia* syndrom in ringtailed lemur

G

Gallery forests, 113, 115, 119; *see also* Ringtailed lemurs, gallery troops, effect of eratic climate on population of
  diet of brown lemurs and ringtailed lemurs in, 119
Gallery habitat
  diet overlap between ringtailed and brown lemurs in, 125f, 126f
  feeding heights of ringtailed and brown lemurs in, 129f
Geochelone radiata. (Madagascar radiated tortoise), 40
Gorilla gorilla, 154, 174
Gray-and-red mouse lemur: *see Microcebus griseorufus*
Gray mouse lemur: *see Microcebus murinus*
Gregarious lemurs, 257
Grewia spp. (Tiliaceae), food source of *L. catta*, 61
Gusto-facial reflex, 175

H

Habitat structure in *L. catta*, 224
Habitat zones, in Berenty Reserve, 33
Habitat zones for lemur demography
  front, 37
  gallery, 37
  scrub, 37
  spiny forest, 37
Hematocrit values, and ringtailed lemurs, 319
Heritable territoriality, 205
Hildebrandtia spp. (Convolvulaceae), food source of *L. catta*, 63
Home range, definition of, 88
Horn's Index of Overlap, 124–127
Hybrid brown lemurs: *see* Fauna, Berenty Reserve, *Eulemur fulvus rufus x collaris*
Hyopsodus, 345

I

Image pre-processing, TM images, 19
Immigrant males, affiliative behavior of, 306
Indri indri, 155
Intergroup encounter, for food resources in lemurs
  effect of group size and, 208
  effect of location on rate of, 208, 210f
  participation in, 219t
  role of adult lemur females participation in, 218f, 222f, 223

Intergroup encounter, for food resources in lemurs (*Cont'd*)
  role of membership in large groups in reducing costs of, 217–218
  role of membership in large groups in winning, 213–217
Introduced tree species, 141
  efficiency of food items by, 152
  impact on *L. catta* feeding ecology: see Ringtailed lemur, impact of introduced tree on feeding ecology of
  spatial distribution of, 155
  territory, home range, and distribution of, 155

K

Kily fruits, near Mandrare River, harvest of, 92, 93f
Kily ripe pods, food source of *L. catta*, 56
Kily trees, 86
  grown in the ranges of ringtailed lemurs, 99
  population dynamics of, 97–98
Kily trees in the broader study area of 30.4 hectares, population of
  cohorts and DBH, 91
  distribution pattern, 92, 93f
  estimation of birth year based on the growth rate, 94
  population dynamics, 97
Kily trees within the main study area of 14.2 hectares, population dynamics of
  grown in the ranges of ringtailed lemur troops, 97
  harvest of kily fruits, 97
Kirindy, *L. catta* distribution at, 4
"K selection" in lemurs, 195, 196

L

L. catta
  in Ankoba and Malaza lobes of Berenty Reserve, spacing of, 38f
  at Berenty, diet of, 124
  feeding of *T. indica*, 144
  plant species consumed by, in Berenty, Antserananomby, and Beza-Mahafaly, 57t–60t
L. leucocephala leaves, 150
Lactation period, level of activity of *Lemur catta* during, 271
Lemur(s)
  density, 18
  and large trees, study populations of, 91
  metabolic mechanisms in, 262

Lemur(s) (*Cont'd*)
  prediction map using transition function and FCD, 22
Lemur biological data
  collection of, 318
  data analysis of, 318
Lemur catta, feeding trends of
  food plants, food items, and seasonal variations in, 56, 57t–61t
  in Madagascar, 55
Lemur catta, male sociality and integration during dispersal process in, methods of
  age-class and migration status of two males, 297
  data collection, 297
  study animals/study groups, 298
Lemur catta, tooth size change among temporal, 343
  increased tooth size in a high-attrition environment, 357
  lengths of M_1 and M_2, 351
  mandibular toothrow length, 351
  materials and methods, 351
  maxillary toothrow length, 351
  metric data for 2003/2004 sample, 351
  M_2 size reduction: selection for improved occlusion?, 358
  ontogeny, weaning, and increased tooth size, 354
  palate breadth, 351
  P_2 length, 351
  sex-specific temporal comparisons of, 354
  socioecology and increased tooth size, 355
  toothcomb breadth, 351
Lemur catta (*L. catta*); see also Ringtailed lemur
  male sociality and integration, 296
Lemur predators in, Berenty reserve
  Buteo madagascariensis, 40
  Milvus migrans, 40
  Polyboroides radiatus, 40
Lemur studies, at Beza Mahafaly Reserve
  ecology and behavior of ringtailed lemurs, 48
  socioeconomic studies of neighbors of, 48
Lepilemur leucopus: see White-footed sportive lemur
Lepilemur ruficaudatus, 276, 288, 289t
Leucaena and lemur fur condition, mapping of, 337
Leucaena leucocephala (Mimosaceae), 332
  food source of *L. catta*, 59t, 63
Leucaena toxicity, 341
Life-history modulations, 260

M

Macaca fuscata, 154
Macaques (*Macaca mulatta*), 160
Madagascar, 3–7
   anthropogenic savanna, 4
   Berenty Reserve, climates of, 32
   cognitive level in lemurs of, 286
   Didiereaceae plant family, 4
   freshwater sources, 3, 9
   humid forest zones, 6
   ringtailed lemur, occurrence of, 3, 16
   spiny bush elements, 4
   tree-rat, 39
Maesa lanceolata (Myrsinaceae), 7
Malagasy
   lemurs, 271
   primate female dominance, 271
Malaza forest at Berenty
   changes in, 70
   reduction of *L. catta* troops in, 69
   tamarind recruitment in, 69
   tamarind regeneration, 77
   Tamarindus indica, 69
   vegetation zones in, 71
Malaza forest at Berenty, changes in
   canopy cover and species composition, differences in, 74
   detection methods for, 71
   Mandrare River, closed canopy forest of, 81
   pattern of vegetation, 72
Mandrare River, 32, 37, 38f, 69, 73f, 81–82, 107, 145f, 332
   distribution of *L. catta* in, 39
   harvest of kily fruits near, 97, 99–100
Mangoky River basin, ringtailed lemur in, 4, 6, 11–12
Mann-Whitney U test, 89, 94
   and bald lemur syndrome, 332
Marmosets (*Callithrix jacchus*), 160
Menabe region, ringtailed lemur in, 4
Microcebus griseorufus, 39, 47, 141
Microcebus murinus, 39
Microevolution, 344
Microgale sp. (shrew-like tenrec), 39
Mimosine, 156
   detoxification, 337
   toxicity, 336
"Minimax" model, 246
"Miracle tree"; see also *Leucaena leucocephala*
   and ringtailed lemur, 335
Mortality rates, of lemur, 114

N

Neotina isoneura (Sapindaceae), 91t, 149
   food source of *L. catta*, 64
Nitrogen content, in tamarind leaves, 108
Nomadic group/troops, 236f
   in the 14.2-ha study area during the 12.5-year period, 234
Nomadic phase of social groups in, ringtailed lemurs, 242
Nutritional differences by habitat, ringtailed lemurs, 320t
Nylon collars, 299

O

Opuntia rackeets, food source of *L. catta*, 63

P

PAL: see Physical activity level
Pan paniscus, female dominance in, 271
Pan troglodytes, 154
***Pearson regression analysis, 89
Pelomedusa subrufa (terrapin), 40
Phromnia, 123
Physical activity level (PAL), 277
Physiologicogenetic systems, 260
Plant species, ingested by *L. catta*, 146t
Population density of *L. catta*, at Berenty, 208
Population dynamics of kily trees, 93–94
Postmigration in ringtailed lemurs, 296–297
Postmigration period, *Lemur catta*
   affiliative behavior during the postmigration period, 302–303
   agonistic behavior during the postmigration period, 303
"Premigration behavior" and ringtailed lemurs, 300
Propithecus diadema edwardsi, sexual differences in, 271
Propithecus verreauxi, 39, 49, 154, 357
Propithecus verreauxi verreauxi, 276; see also Verreaux' sifaka
Pyxis arachnoides (spider tortoise), 40

Q

Quinine, in wild *Lemur catta*, 176–177
   rejection threshold for quinine hydrochloride, 178f
Quivisianthe papinae (Meliaceae), 71, 124
   food source of *L. catta*, 59

R

Range takeover, in ringtailed lemurs, 233–234
  case study, 234, 235–237
Reaction-norm concept, 260–261
Resource defense by ringtailed lemur, group size role in, 208, 222–224
  benefits of, 218
  maladaptive behavior of, in current conditions, but not under those in which *L. catta* involved, 224–225
  resource defense is less costly than permitting rivals to deplete resources, 225–226
  spatial dominance relation maintenance is less preferable over establishment, 226
Ring-tailed activities, indexes of energy expenditure for some, 276t
Ringtailed lemurs *(Lemur catta); see also* Fauna, in Berenty Reserve, ringtailed lemur
  affiliative and agonistic behavior during premigration period, 300
  affiliative behavior during migration and integration into green group, 301–302
  affiliative behavior of immigrant males, 306
  aggression study in, 245–260
  agonism and number of males in new group, 304–305
  agonistic behavior during migration/integration, 302
  antinutrient content of mature leaves eaten by, 169f
  basal metabolism of, 276
  behavioral differences during dispersal between natal and older male, 307
  behavioral differences in, 320–322
  behavioral differences in play and feeding and social agonism, 324f
  behavioral dominance of brown lemurs over, 134
  behavior of, 327
  bequeathing territory by, 200–203
  of Berenty, 282
  in Berenty Private Reserve, 272
  between-troop aggression, 201t
  biogeography of, 11–12
  birth season intertroop encounters in 1995–1997, 197t
  body weights, skinfold measurements, and selected linear measurements in, 319
  characterization methods of forest condition, 18
  comparison of the males and females' PAL, 285
  conservation implications for, 204
  demographic patterns of male migration, 297
Ringtailed lemurs *(Lemur catta); see also* Fauna, in Berenty Reserve, ringtailed lemur *(Cont'd)*
  demography of, 188–196
  dental health of, 318–319
  diet, 162–163, 167–169
  dietary elements in high elevation zones, 7
  distribution of female activities according to reproductive periods, 280f
  distribution of male activities according to reproductive periods, 280f
  ecological aspect of high mountain population of, 7–9
  effect of hydration and nutrition on, 319
  effect of the erratic climate of southern Madagascar on population, fertility, and survival, 188–196
  energy expenditure in, allometric equation to calculate, 275–284
  evolution and energetic strategy, 288–290
  feeding on green leucaena pods, 333f
  female dominance, genes responsible for, 290
  female dominance in, 280–281
  females spent in inactivity in regard to their dominance over males, 281f
  follow-up, 307–308
  food chemistry of, 166–170
  food elements, comparison of, 8t
  food intake during the wet season by, 225–226
  front troops, effect of eratic climate on population of, 188–196
  fur condition, mapping of, 339–340
  gallery troops, effect of eratic climate on population of, 188–196
  geographical distribution of, 3, 5f, 6–7
  group defense, 305–306
  G3 troop females' PAL in respect to intrasexual hierarchy, 285f
  habitat occupied by, 27t
  health and growth and development of, 324–327
  immigrant males, rank, nontransitive dominance relationships, 306–307
  individual variation, 284–285
  information about, 3–4
  intergroup agonistic behavior, 225
  interindividual differences in physical activity level, 287–288
  intertroop encounter behavior of, at Berenty Reserve, 198–199
  kily trees grown in the ranges of, 97
  length of male tenure in, 307–308
  level of activity of, during lactation period, 279f

Ringtailed lemurs (*Lemur catta*); see also
 Fauna, in Berenty Reserve, ringtailed
 lemur (*Cont'd*)
 level of females' activity in relation to their
  dominance, 283–284
 litter size in, 187
 male and female activities and inactivity,
  277–280
 male and female energy expenditure,
  281–284
 male and female physical activity level,
  284
 male dermatitis in, 325f
 male migration pattern in, 297–298
 matrilineal society in, 240
 mean distance covered by ring-tailed females
  during each reproductive period, 283f
 migration partners and choice of new group,
  304
 mothers, 272
 night-sleeping sites of, 8
 occurrence of, in areas without freshwater,
  9–11
 population density, 16–18
 population of non-infant, for Malaza regions,
  189f
 postmigration period, 302
 proximate causes of social behavior in, 258
 research site and study groups, 314
 sampling protocol, 314–316
 scrub troops, effect of eratic climate on
  population of, 188–196
 seasonal variations of quality of diet, 167f
 sex differences in, 271, 322–324, 327–328
 skin biopsies analysis of, 328
 social characteristics of, 233
 species and female feeding priority, 271
 territoriality of, at Berenty, 196–200
 time spent in activities according to their
  sex, 279f
 time spent in inactivity by females in regard
  to their intrasexual hierarchy, 281f
 tooth wear in, from Beza Mahafaly
  Osteological Collection (BMOC 67), 347f
 troop fissions in, 201t, 202–203
 unworn teeth in, 350f
 use of satellite imagery for measuring forest
  canopy density, 20–24
 use of satellite imagery for population
  density of, 28
 veterinary evaluation of, 338–339
Ringtailed lemur, female
 mortality in, 345–346
 pregnancy in, 356

Ringtailed lemurs, feeding competition between
 brown lemurs and, at Berenty, 119–121
 comparison of diet overlap and activity
  patterns, 122–123
 diet overlap across troop-pairs and habitats,
  123–126, 128
 diurnal activity budgets, 130–131
 feeding height, 128
 interspecific diet overlap, 132
 interspecific differences in competitive
  ability, 133–134
 interspecific differences in dietary flexibility,
  132–133
 overall diet and diet overlap, 124–125
 resource types fed on by, 127t
 role of edge habitat in ringtailed lemurs'
  resilience to food scarcity, 135–136
 seasonal food scarcity and *Tamarindus indica*
  as a keystone species, 134–135
 study site, 121–122
Ringtailed lemurs, female dominance of,
  271–272
 behavioral observations of, 275
 energy expenditure, 275–277
 statistical analyses, 277
Ringtailed lemurs, group size role in resource
  defense, 208
 advantages of large group through resource
  defense, 218–219, 221
 animal stay in large groups, 228–229
 association of group size and intergroup
  encounter rate, 213
 food intake rates estimation for, 212–213
 foraging efficiency compensation by large
  group, 227–228
 intergroup conflict estimation for analysis,
  211–212
 ranging for, 211
 role of large groups in reducing costs of
  intergroup competition, 217–218
 role of large groups in winning intergroup
  competition, 213–214, 217
 seasonal characterization for analysis, 209–213
Ringtailed lemurs, impact of introduced tree on
  feeding ecology of, 141–142
 activity budget and diet, 143–144
 availability and consumption of introduced
  tree species, 149–150
 change of feeding tradition in, 151–152
 home range and territory, 144, 145t
 introduced species impact on population and
  health of, 156
 strategy by, to deal the harsh season,
  154–155

Ringtailed lemurs, ranging behavior, influence of tamarind tree quality and quantity on, 102–103, 113–115
  behavioral observations, 104–105
  hydrologic environment, 106–107
  site and subjects for analysis of, 104
  tamarind fruit abundance, 109
  tamarind tree abundance and behavioral observations, 109–112
Ringtailed lemurs, social changes
  consequences of social changes, 240
  female eviction, 233–237, 241
  process of social changes and phases of social groups in, 239f
  range takeover, 233–234, 237, 239–240
  troop fission, 233–234, 240–241
Ringtailed lemurs, taste perception in, 160–162
  adaptive value of taste sensitivity and feeding strategy, 176–179
  feeding behavior and food chemistry, 162–163
  food context effect on taste preference/aversion thresholds for various taste stimuli, 174–176
  secondary metabolites role, 161
  taste perception of captive lemurs, 170
  taste perception of wild lemurs, 170, 172
  taste thresholds for fructose, quinine hydrochloride, and tannic acid, 163–164
  taste thresholds of captive lemurs, 164–165
  taste thresholds of free-ranging lemurs, 165–166
Ringtailed lemur studies, Berenty Reserve, 40–41
Rinorea greveana fruit, 149–150
  foraging on, 221
Rinorea greveana (Violaceae), food source of *L. catta*, 61
Rynchosia sp. (Fabaceae)., food source of *L. catta*, 61

S

Sakamena River in Beza Mahafaly Reserve, distribution of *L.catta* in, 44
Salvadora angustifolia (Salvadoraceae), 70
  food source of *L. catta*, 61
Scaled shadow index (SSI)
  definition of, 20
  in FCD computation, 19–20
Scaled vegetation density (SVD)
  definition of, 20
  in FCD computation, 20
Scrub habitat, diet overlap between ringtailed and brown lemurs in, 126f

Seasonal food
  scarcity, adaptations to, 134–135
  use in *Lemur catta*, 62f
Seasonal variation of diet, in lemurs, 163
Secamone sp. (Asclepiadaceae), food source of *L. catta*, 61
Secondary metabolites, 161
Selection intensity of a severe drought, 356
Senna siamea, food source of *L. catta*, 63
Setifer setosus (spiny tenrec), 39
Sex differences in ringtailed lemur, 322–326
Sex ratios, in *L. catta* population, 190, 191f
Shadow index (SI)
  definition of, 20
  in FCD computation, 20
SI: *see* Shadow index
Sisal, 87
Skin biopsies, 337
Socio-bioenergetics, 275
Socioecology, 355
Solanum auriculatum (Solanaceae), 7
Spatial dominance relationships in *L. catta*, 226
Spiny bush elements, ringtailed lemur, 3
SSI: *see* Scaled shadow index
Standard venipuncture techniques, 317
Statsoft, 277
Status striving, 246–249
  testing for hypothesis of, 251–252
Student t-test in ringtailed lemurs, 277
Survivorship, in *L. catta* after erratic climate, 193–194
SVD: *see* Scaled vegetation density

T

Talinella dauphinensis (Portulacaceae), food source of *L. catta*, 61
Tamarind *(Tamarindus indica)*, 44, 46–47, 69–70, 76, 102
  correspondence with primate enamel thickness, 346
  fruit abundance effect on feeding behavior of lemurs, 109, 111–112, 111t, 114
  fruiting patterns, 114
  germination rate of control, by *L. catta*, 66f
  influence of quality and quantity of, on *Lemur catta* behavior: *see* Ringtailed lemur ranging behavior, influence of tamarind tree quality and quantity on
  L. catta, as food resource of, 56
  leaf, water content in, 113
  for reducing resource competition in lemurs, 128, 130

Tamarind regeneration, distribution and total lengths in, 77t
Tamarind regeneration, in Malaza forest at Berenty
  distance from mature tamarinds, 80, 80t
  distribution of young tamarinds, 78t
  edge effects, 79–80, 80t
  methods for, 77–78
  survival and growth rates for, 79
Tannic acid, in wild *Lemur catta*, 176–177
  rejection threshold for, 178f
Tannin analyses, 163
Taste thresholds
  of captive lemurs, 164–165
  of free-ranging lemurs, 165–166
  for various taste stimuli according to food context, 174–176
Tel-inject blow gun system, 315
Tenrec ecaudatus (large tenrec), 39, 47, 195
Territoriality of ringtailed lemurs at Berenty, 196–200
Themaic mapper (TM) images, for ringtailed lemurs, 19
TM images: *see* Themaic mapper images
Tolerance levels among female lemurs, 259–260
Tooth size change among *Lemur catta*, temporal, 355
  increased tooth size in a high-attrition environment, 357–358
  lengths of M_1 and M_2, 351
  mandibular toothrow length, 351
  materials and methods, 351–352
  M_2 size reduction: selection for improved occlusion?, 358–359
Total phenolics (TP), seasonal variations in, 166
Trachypithecus geei, 155
Troop fission, in lemurs, 233–234, 241
  case studies for ringtailed lemurs, 235–239, 241
  in cercopithecine species, 241
  nomadic phase of social groups in, 241–242
  role of male dominance in, 241

Troop fission, in lemurs (*Cont'd*)
  troop histories of ringtailed lemur population within the 14.2-ha study area for 12.5-years—from September 1989 to January 2002, 236f
Two-bottle test, 164, 165f, 171f
  on captive ringtailed lemurs using fructose solutions, 171f
  on captive ringtailed lemurs using tannin mixtures, 172f
  on free-ranging ringtailed lemurs, 173f

U
Ursus spelaeus, tooth size change in, 358

V
Van Valen, tooth size change in, 358
VDS: *see* Vegetation density
Vegetation density (VDS)
  definition of, 20
  in FCD computation, 20
Vegetation types in Malaza forest, 72t
  area of, 74t
Vegetation zones in Malaza forest, 73f
  brush and scrub, 71
  brush and scrub–spiny forest, 71
  closed canopy tamarind forest, 71
  open neotina–tamarind forest, 71
Verreaux' sifaka (*Propithecus verreauxi verreauxi*), 141, 357
Viverricula indica (Indian civet), 39, 47
voleli, 91, 100

W
Water conservation, for brown lemurs, 136
White-footed sportive lemur (*Lepilemur leucopus*), 39, 141
White sifaka: *see Propithecus verreauxi*

Printed in the USA

Job #: 100276

Author Name: Jolly

Title of Book: Ringtailed Lemur Biology

ISBN #: 0387326693